Accounting, Finance, Sustainability, Governance & Fraud: Theory and Application

Series Editor

Kıymet Tunca Çalıyurt, Iktisadi ve Idari Bilimler Fakultes, Trakya University
Balkan Yerleskesi, Edirne, Turkey

This Scopus indexed series acts as a forum for book publications on current research arising from debates about key topics that have emerged from global economic crises during the past several years. The importance of governance and the will to deal with corruption, fraud, and bad practice, are themes featured in volumes published in the series. These topics are not only of concern to businesses and their investors, but also to governments and supranational organizations, such as the United Nations and the European Union. Accounting, Finance, Sustainability, Governance & Fraud: Theory and Application takes on a distinctive perspective to explore crucial issues that currently have little or no coverage. Thus the series integrates both theoretical developments and practical experiences to feature themes that are topical, or are deemed to become topical within a short time. The series welcomes interdisciplinary research covering the topics of accounting, auditing, governance, and fraud.

More information about this series at https://link.springer.com/bookseries/13615

İsmail İyigün · Ömer Faruk Görçün
Editors

Logistics 4.0 and Future of Supply Chains

Springer

Editors
İsmail İyigün
Trakya University
Edirne, Turkey

Ömer Faruk Görçün
Kadir Has University
Istanbul, Turkey

ISSN 2509-7873 ISSN 2509-7881 (electronic)
Accounting, Finance, Sustainability, Governance & Fraud: Theory and Application
ISBN 978-981-16-5643-9 ISBN 978-981-16-5644-6 (eBook)
https://doi.org/10.1007/978-981-16-5644-6

This Springer imprint is published by the registered company Springer Nature Singapore Pte Ltd.
The registered company address is: 152 Beach Road, #21-01/04 Gateway East, Singapore 189721, Singapore

Acknowledgments

Initially, we must thank Prof. Kıymet Tunca Çalıyurt for her patience and dedication, as the book would not have been sufficiently well-structured without her perseverance and belief. Additionally, we have to thank all authors who made close and fruitful collaboration with editors of this book; indeed, this book has become efficient, thanks to their devoted contributions and they tried contributing as much as possible and in their forces. Their efforts were really incredible and admirable. In addition, we appreciate all scientists and practitioners who have contributed to the technological improvements in creating a more livable world. Finally, we have to thank for the tolerance of our family, they did show great tolerance uncomplainingly.

İsmail İyigün, Ph.D.
Ömer Faruk Görçün, Ph.D.

Contents

Editors and Contributors

About the Editors

Dr. İsmail İyigün was born in 1971 in Uzunköprü. He has been working as a lecturer in Edirne Vocational School of Social Sciences at the Trakya University since 2006. His current research areas are marketing, logistics, and supply chain management. He worked at the Uzunköprü Vocational School Marketing Program in the same university from 1995 to 2005. He received his Ph.D. degree with business administration from Gebze Technical University in 2016. The focus of his Ph.D. thesis is the effects of sales and logistics competencies on the relationship continuity in supply chain and corporate governance: an application in logistics enterprises. He received his master's degree for economics from the Trakya University and researched the commodities market of historical process and application of Uzunköprü. He graduated high school in Kabatas in 1989 and received an undergraduate degree from the Marmara University, Dept. of Economics in 1994. He is married and has two children. He speaks English fluently.

Assoc. Prof. Ömer Faruk Görçün teaches at Faculty of Business Administration in Kadir Has University. He is also a member of the transportation council of the Ministry of Transportation of Turkey. In addition to that, he has performed some duties such as president and member in the committees of the Research & Development and the railway regulation. He had duties as a member of the board in Rail Systems Association and was advisor for the Crane Operator Association and Heavy Transport Association. As an aspect of his academic studies and works, he works on optimization and decision-making techniques in the fields of logistics and supply chain management. In addition to many studies published in reputable international and national journals, he wrote many scientific books, which were published by international and national publishers, such as Industry 4.0, Integrated Logistics Management, Supply Chain Management, Warehouse and Inventory Management, Railway Transportation. Currently, Görçün continues his academic life as a scientific committee member, editor, technical committee member in addition to his managerial and academic duties at the university. He is married with two wonderful daughters.

Contributors

Haci Mehmet Alakaş Kirikkale University, Department of Industrial Engineering, Kirikkale, Turkey

Ezgi Uzel Aydınocak Faculty of Business and Administrative Sciences, Beykoz University, Istanbul, Turkey

Ayşegül N. Bayraktar Beykoz Üniversitesi, Istanbul, Turkey

Gülüm Burcu Dalkiran Department of Tourism and Hotel Management, Vocational School of Social Sciences, Trakya University, Edirne, Turkey

Hazar Dördüncü Department of International Trade and Management, Nişantaşı University, Istanbul, Turkey

Tamer Eren Kirikkale University, Department of Industrial Engineering, Kirikkale, Turkey;
Department of Industrial Engineering, Kırıkkale University, Kırıkkale, Turkey

Ömer Faruk Görçün Department of International Trade and Logistics, Kadir Has University, Istanbul, Turkey

Özhan Görçün Gelişim University, Istanbul, Turkey

İsmail İyigün Balkan Campus, Trakya University, Edirne, Turkey;
Social Vocational College, Trakya University Edirne, Edirne, Turkey

Lokman Kantar Gelişim University, Istanbul, Turkey

Hande Küçükönder Department of Numerical Methods, Faculty of Economics and Administration, Bartin University, Bartın, Turkey

Mustafa Özan Dr. Istanbul Gelisim University Faculty of Applied Sciences Logistics Department, Instanbul, Turkey

Evrencan Özcan Department of Industrial Engineering, Kırıkkale University, Kırıkkale, Turkey

Barış Öztuna Faculty of Economics and Administrative Sciences, Department of Labor Economics and Industrial Relations, Çankırı Karatekin University, Çankırı, Turkey

Mahmut Tutam Department of Industrial Engineering, College of Engineering, Erzurum Technical University, Erzurum, Turkey

Duygu Yücel Department of Accounting, and Taxation, Vocational School of Social Sciences, Trakya University, Edirne, Turkey

List of Figures

List of Tables

Part I
Introduction Chapter

Chapter 1
Introduction

İsmail İyigün and Ömer Faruk Görçün

Abstract Over the past 40 years, we have seen dramatic changes and transformations in the operations carried out in the logistics and supply chain management. First of all, operations carried out today are not the same as those that were carried out in the past. Furthermore, they are more complicated in comparison with those of the past, and organization of these logistics operations is not possible unless logistics companies use advanced technologies. More importantly, while technology has rapidly developed, customer demands have also shown dramatic changes. Also, advanced technology usage has been more advantageous to companies. Even small-scale companies have become very competitive to stay in the business environment, thanks to their own technological developments. This, of course, assumes the companies and supply chains can easily compete depending on their technology usage in the highly competitive environment, which is often not the case. It is not sufficient to be aware of the power of technology; small enterprises should internalize the fourth logistics revolution as an important component of the industry 4.0. It is not an instrument for reducing cost alone, it is a circular process that requires an evaluation from a wider perspective. In addition to the theoretical background of logistics 4.0., real-world examples and case studies are available in this book. It has successfully examined the techno-logical improvements concerning the logistics management. It has proposed a methodological frame to understand the technological revolutions happening in the present day from the perspective of logistics management.

Keyword Supply chain · Logistics · Logistics 4.0 · Smart technologies

Over the past 40 years, we have seen dramatic changes and transformations in the operations carried out in the logistics and supply chain management. First of all,

İ. İyigün (✉)
Balkan Campus, Trakya University, Edirne, Turkey
e-mail: ismailiyigun@trakya.edu.tr

Ö. F. Görçün
Kadir Has University, Istanbul, Turkey
e-mail: omer.gorcun@khas.edu.tr

İ. İyigün and Ö. F. Görçün (eds.), *Logistics 4.0 and Future of Supply Chains*,
Accounting, Finance, Sustainability, Governance & Fraud: Theory and Application,
https://doi.org/10.1007/978-981-16-5644-6_1

operations carried out today are not the same as the operations carried out in the past. Furthermore, they are more complicated in comparison with the past, and well-organized logistics operations are not possible unless logistics companies can use advanced technologies.

More importantly, while technology has rapidly developed, customer demands have also shown dramatic changes. Also, advanced technology usage has led to more advantages to companies. Even small-scale companies have become to stay in very competitive business environment thanks to their own technological developments.

This, of course, assumes the companies and supply chains can easily become competitive depending on their technology usage in the highly competitive environment, which is often not the case. It is not sufficient to be aware of the power of technology, small enterprises should internalize the fourth logistics revolution as an important component of the Industry 4.0. It is not an instrument of reducing cost only, and it is a circular process that requires to evaluate from a wider perspective.

In addition to the theoretical background of Logistics 4.0, real-world examples and case studies are available in this book. It has successfully examined the technological improvements concerned with the logistics management. It has proposed a methodological frame to understand the technological revolutions happening in the present day from the perspective of logistics management.

INTRODUCTION The chapters of the book were selected delicately to respond to the requirements of readers on the information of Industry and Logistics 4.0. The book consists of 3 parts and 14 chapters and all of them have a valuable contribution to both the related literature and general readers. In the first chapter, Barış Öztuna examined the technological applications used in the field of logistics. He evaluated the new technological improvements from the simple to advanced in detail. In chapter two, İsmail İyigün assessed the technological improvements and requirements in the fields of logistics and supply chains in the general viewpoint, and Özhan Görçün and Hande Küçükönder presented a comprehensive analysis on big data and machine learning, which are the most important components of the Logistics 4.0 in Chap. 3. While Ayşegül Nuriye Bayraktar evaluated the impacts of 3D printing technology on the fields of logistics in addition to its fields of usage in Chap. 4, in Chap. 5, Ömer Faruk Görçün examined the autonomous robotics and its impacts on the fields of supply chain and logistics. In Chap. 6, Mahmut Tutam evaluated technological improvements in the field of warehousing with a new perspective.

Also, Evrencan Özcan and Tamer Eren evaluated the cloud Information Systems in a general viewpoint in Chap. 7. In Chap. 8, smart factories, which are the most important components of both Industry 4.0 and Logistics 4.0 were deeply analyzed by Hazal Dördüncü. Ezgi Uzel Aydinocak examined the Internet of Things (IoT) in Marketing Logistics in a wide perspective in Chap. 9. Hacı Mehmet Alakaş and Tamer Eren evaluated the integrated systems and their utilization in the field of logistics in Chap. 10. While Mustafa Özan assessed the sustainability of approaches related to Logistics 4.0 in Chap. 11, Lokman Kantarcı examined the different sides of Logistics 4.0 with respect to finance and economy in Chap. 12. Gülsüm Dalkıran evaluated the implications of Logistics 4.0 in the field of the tourism industry in Chap. 13. Finally,

Duygu Yücel examined the negative effects of Logistics 4.0 such as unemployment, social problems, and so on in Chap. 14.

In addition to technological improvements occurring in the last three decades, the global pandemic has caused changes in almost everything in our daily lives. It has also affected the global supply chains at a large scale, and it has forced the supply chains for transformation, structurally, since it has led to unbalanced situations in the global material flows. Taking benefits from the technological advances as much as possible is the best and optimal way for all parties of the supply chains such as suppliers, manufacturers, retailers, logistics service providers, and customers.

Nowadays, decision-makers who are responsible to decide are aware that it is required to make the investment in the technological infrastructure to enable dealing with the ambiguities which will be occurring in the future. Hence, having a more advanced technological infrastructure may provide a safer material flow and balance between demand and supply. In addition, it can help to construct a well-operated logistics system from suppliers to customers as it can lead to increase in communications and information sharing among all parties in a supply chain. Furthermore, it can help to reduce reversal material flows in a logistic system since it can lead to reducing misunderstandings and mistakes among all actors. Therefore, reducing the operational cost is dependent on the well-designed logistics system supported with technological instruments.

Today, we have to understand the technological improvements required for todays' conditions. In addition, companies, industries, and commercial actors are not only responsible to meet these requirements, but also all individuals should understand the paradigms of this century and the future right now. Thus, it is not possible to survive in the highly competitive business environment without well-understanding the technological developments. In fact, we are living in the future world and designing a new world. It is a fact well-known to everyone that the most important paradigms of the present and future are hidden into this expression: "there is no impossible". Indeed, it is true, everything we evaluated as impossible are realized at the present. Nowadays, usage of surgery robotics in medical operations is very usual for all of us. Furthermore, robotics can do it with slight medical mistakes. Also, computers and software can organize logistics and industrial activities, and robotics systems can perform their duties with almost zero mistake.

As is seen, this book aims to respond to many requirements of the interested readers who are curious about new developments in the fields of logistics and supply chains. Initially, we are trying to provide an insight on what will be happen in the future and what will be changed in our lives together with our readers. We hope that this book provides a different viewpoint for readers and can help to understand the developments for today and the future.

Dr. Ismail İyigün has been working as a lecturer in Vocational School of Social Sciences at the Trakya University since 2006. His current research areas are marketing, logistics and supply chain management. He graduated from Ph.D. Programme on Business Administration at Gebze Technical University in 2016. His dissertation is on the "Effects of Sales and Logistics Competencies on the Relationship Continuity on Supply Chain and Corporate Governance". He received master degree from Economics, Social Graduate School at the Trakya University. He researched the commodities market of historical process and application of Uzunköprü. He graduated from Economics Department, Marmara University in 1994. Dr. İyigün teaches on supply chain management, physical distribution systems, logistics information systems, selling techniques, and basic marketing courses in Trakya University and Kadir Has University.

Ömer Faruk Görçün who is currently a Faculty Member of the Faculty of Business Administration at Kadir Has University is also a member of transportation council of Ministry of Transportation of Turkey. In addition to that, he has performed as president and member in the committees of the Research & Development and the railway regulation. He acted as the member of the board in Rail Systems Association and was advisor for the Crane Operator Association and Heavy Transport Association. In aspect of his academic studies and works, he works on optimisation and decision-making techniques in the fields of logistics, supply chain management. In addition to many studies published in international and national reputable journals, he wrote many scientific books, which published by international and national publishers, such as Industry 4.0, Integrated Logistics Management, Supply Chain Management, Warehouse and Inventory Management, Railway Transportation. Currently, Görçün continues his academic life as scientific committee member, editor, technical committee member in addition to his managerial and academic duties at the university. He is married and has two wonderful daughters.

Part II
Logistics 4.0 and Future of the Supply Chains

Chapter 2
Logistics 4.0 and Technologic Applications

Barış Öztuna

Abstract Along with Industry 4.0, which was first outlined at the Hannover Expo in 2011, work–life balance has gone under a major transformation. With the 4th industrial revolution, new technologies, such as artificial intelligence, cloud computing, 3D printers and the Internet of Things started to take greater prominence in the world of business. One of the sectors affected by the process that started with Industry 4.0 is logistics. For that reason, a concept called Logistics 4.0 has emerged. The concept of Logistics 4.0 denotes a process that causes an increase in efficiency and a decrease in the costs of automation. The objective of this study is to make predictions about the innovations that will be brought to logistics with the new process started by Industry 4.0. This study covers information and predictions regarding the technologic applications brought by Logistics 4.0.

Keywords Industry 4.0 · Logistics 4.0 · Automation

2.1 Introduction

The influence of automation and digitization has huge impacts on work–life balance, social life and industry worldwide. Together with Industry 4.0, which was first outlined at the Hannover Expo in 2011, competition has come to a peak in many sectors. If a company intends to compete in the market, it has to adapt itself to newly arising technologies. In the modern age, it has become vital for companies to minimize their costs and increase their efficiency and speed.

In today's world, digitization is used for operational improvements and keeping up with changing customer expectations. Large sales data analysis makes it easier to understand customer expectation. Developing technologies helps companies to build specialized products in bulk.

B. Öztuna (✉)
Faculty of Economics and Administrative Sciences, Department of Labor Economics and Industrial Relations, Çankırı Karatekin University, Çankırı, Turkey
e-mail: barisoztuna@karatekin.edu.tr

İ. İyigün and Ö. F. Görçün (eds.), *Logistics 4.0 and Future of Supply Chains*,
Accounting, Finance, Sustainability, Governance & Fraud: Theory and Application,
https://doi.org/10.1007/978-981-16-5644-6_2

Also, digitization helps to build production processes that are environmentally friendly and more efficient. Industry 4.0 shows that logistics companies should keep up with such technological advances. Since logistics is a highly engaged and central sector that is crucial for industry, it must adapt itself to these innovations. Within this context, the concept of Logistics 4.0 emerged. This concept is closely linked with augmented reality, cloud computing, big data, autonomous devices, 3D printers, drones and the Internet of Things.

This study consists of three phases. After the introduction, the first phase explains the concept of Industry 4.0, stating the attributes and changes that Industry 4.0 may bring about. In the second phase, the concept of Logistics 4.0 and its effects on the sector are explained, and information is given regarding the technologies it uses. In the third phase, information on the applications of Logistics 4.0 is provided. Current and future examples of use are included.

2.2 The Concept of Industry 4.0

One of the main distinguishing features of the 4th industrial revolution is that there are interconnected machines. But the fourth industrial revolution is not limited to smart and interconnected systemsc (Fig. 2.1). There is also a mutual interaction of physical, digital and biological domains where technologies from various fields are intertwined, including genetic engineering, nanotechnologies and renewable energy (Schwab 2016).

Industry 4.0 means the integration of informatics, the internet, automation and data collection with new means of production. It means the implementation of virtual systems over the internet to highly physical structures for supply chains. In this

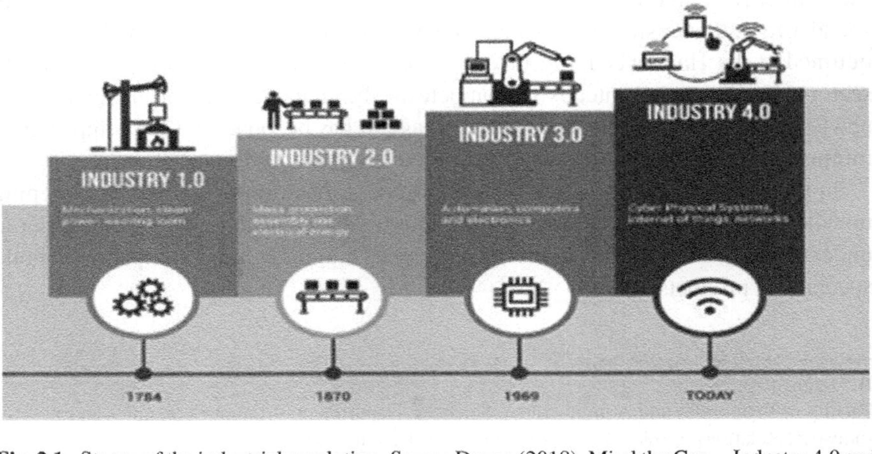

Fig. 2.1 Stages of the industrial revolution. *Source* Dyson (2018). Mind the Gap—Industry 4.0 and the Future of Manufacturing. https://www.ibtimes.co.uk/mind-gap-industry-4-0-future-manufacturing-1665243. Accessed 25 June 2018

context, a modern production model called "light out factory" or in certain cases "Smart Factory" is emerging (Banger 2016) (Fig. 2.2).

Industry 4.0 should not be thought of as merely digitization, AI, 3D printers, augmented reality, big data and the Internet of Things (Fig. 2.3).

Industry 4.0 is a process in which the interaction between humans and machines is maximized, and data flow is synchronized and systematic in order to achieve top quality and automatic operation. Industry 4.0 will change the way supply chains and logistics operates. In this process, besides the structural expansion of supply chains, communication between the agents of the supply chains will increase. In return, the functionality of supply chains gets simpler (Görçün 2016). The main features of Industry 4.0 include horizontal integration formed with supply chains, engineering product digital integration along the supply chain, vertical integration and network production systems. All of them are connected to one another using physical systems and perfected through virtual reality to make them suitable for servicing the customer. All of this integration comes together to form the Internet of Things. In this context, it can be said that the Internet of Things emerges from the connection of smart devices

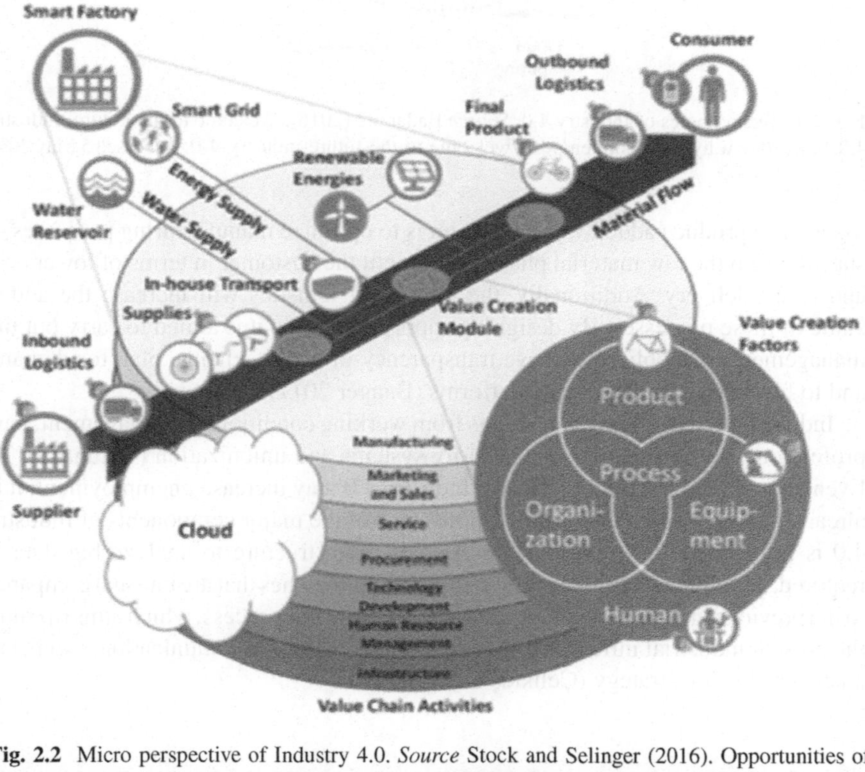

Fig. 2.2 Micro perspective of Industry 4.0. *Source* Stock and Selinger (2016). Opportunities of Sustainable Manufacturing in In-dustry 4.0. 13th Global Conference on Sustainable Manufacturing—Decoupling Growth from Resource Use. https://core.ac.uk/download/pdf/82023214.pdf Accessed 12 May 2018

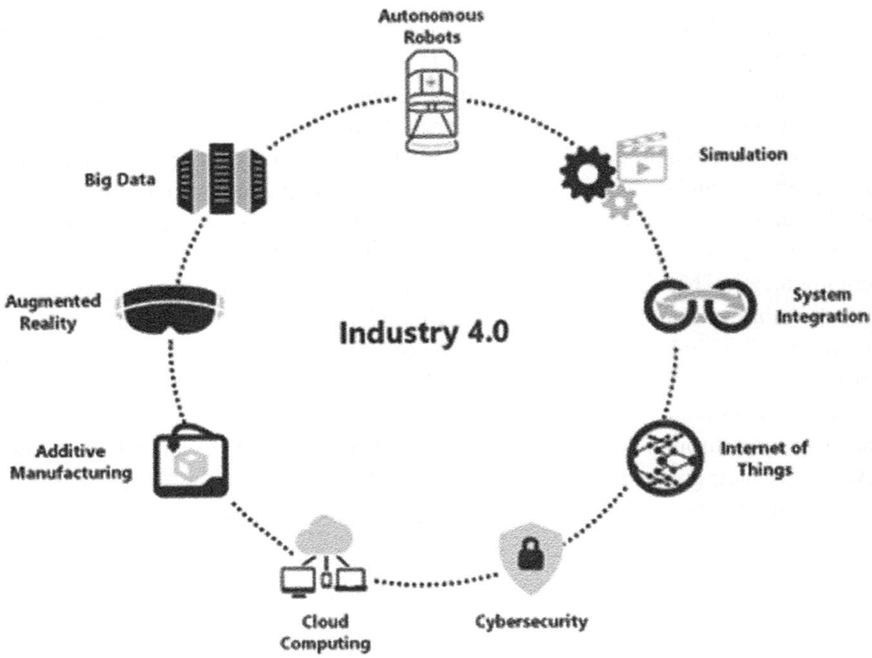

Fig. 2.3 Technologies of Industry 4.0. *Source* Badaraite (2016). Welcome to the Future: Industry 4.0. https://www.hylasoft.com/en/posts/welcome-to-the-future-industry-4-0. Accessed 5 May 2018

together to produce added value. The aim is to optimize manufacturing processes—starting from the raw material phase—to benefit the customer in terms of lower cost and faster delivery. Additionally, the improving logistics will increase the added value of these processes. By designing supply networks, it is aimed to carry out the management of suppliers, achieve transparency of supply chains, plan for demand and to form product-improving platforms (Banger 2017).

Industry 4.0 will affect many things from working conditions to management, new professions, employment, social security systems and unionization (Öztuna 2017). Even though at first it might look like Industry 4.0 may increase unemployment, it is already starting to create new professions. One of the major components of Industry 4.0 is information technology staff. Also an infrastructure to analyze big data is required. The 4th industrial revolution involves approaches that are unusual compared to the previous industrial revolutions. This revolutionary process, which aims to equip the present industrial infrastructure with AI and machine communication, aspires to a new production strategy (Çeliktaş et al. 2015).

2.3 The Concept of Logistics 4.0

Logistics businesses can create a difference by offering logistic services that goes beyond the expectation of customers. For example, delivering small-scale products to multiple points and the ability to differ the process with respect to other suppliers is essential to succeed in the globally competitive environment. In this context, using information technology makes it easier for businesses to keep payment, order and inventory records and tracking their customers. Information technology is especially to be used in individualization, cross docking, consolidation, labeling and packaging, which all offer solutions with significant benefits (Çağlar 2014). This transformation, called Logistics 4.0, includes independently working automation systems, just as Industry 4.0. It is possible to define Logistics 4.0 as a new logistic system that is flexible, adaptable to market fluctuations, that lowers costs and supplies customer needs in the fastest and plentiful way (Şekkeli and Zümrüt 2018). Research and development operations of companies, digitization and visualization, process integration via the internet and mobile apps, creating a network between items and studies regarding cloud computing, communication technologies, simulation and robotics systems causes companies to focus on Logistics 4.0 (Göçmen and Erol 2018). Below is a table that compares the logistics point of view of the previous industrial revolution with the Industry 4.0 version (Table 2.1).

All Industry 4.0 applications in the logistics industry are included in Logistics 4.0. In logistics, process efficiency is an important factor. For this reason, all Logistics 4.0 applications are intended to increase efficiency. This efficiency is obtained through

Table 2.1 Comparison of the logistics point of view of the previous industrial revolution with the Industry 4.0 version

Industry 3.0	Logistics process	Industry 4.0
Lower cost and risks	Supply	Sustainable strategic supplies
Logistics total cost management	Distribution	Web-based source optimization
Excel-based calculations	Inventory management	Web-based and communicated databases and real-time data management
RFID (Radio Frequency Identification) and similar technologies	Storage	Smart devices and storage facilities
Route and time optimization	Transportation	Smart devices and routes
ERP systems next to each other, link is just the order itself	Order management	Communicable, automated information and material flow
Focus on automation	ICT background	Focus on networking

Source Pesti and Nick (2016). Industry 4.0 From the Aspect of Logistics Innovations. So-ciety for Regional Science and Policy. 6th Winter Seminar of Regional Science, 9–12 March 2016, Slovakia. http://ersa.sk/Zbornik/files/Pesti_Nick.pdf. Accessed 15 May 2018

applying the principles of Industry 4.0 to the main logistics services in efficiency, storage, transportation, packaging, handling and information services. Innovations in this area are by technological developments (Horenberg 2017).

Supply chain integration is one of the factors that support the application of new technologies in the logistics industry. Lowering the cost, increasing the response and level of service are examples of Industry 4.0's applications in this sector. Digitizing information will help logistics firms to respond rapidly and be fast and ready in the global market. Supply chain integration has higher requirements for flexibility in real-time data sharing, response time and speedy management, and new technologies can meet these requirements (Ermolaeva 2017). Industry 4.0 technologies and concepts may be expected to have their greatest impact on supply, production and distribution in the supply chain. However, structural changes in the organization may be expected in manufacturing processes. Industry 4.0 in supply chain management includes the development and integration of innovative information and communication technologies in this sector. The main objective is to motivate a smart chain of product and processes through the value chain. Thanks to progress made in analytics and machine learning (ML), companies can analyze their operational data almost simultaneously and can use the results to reduce gaps and revenue losses—this impact can be as high as 10% in some sectors. From driverless transportation to smart containers, smart warehouses and smart ports, all of these things can be obtained from the 4th industrial revolution (Sharma 2018).

The rules for Distribution Centers (DC) to enable Industry 4.0 applications are as follows (Taliaferro et al. 2016):

- **Meet the current requirements in supply** Automation with Industry 4.0 can be helpful. This technology can be used to significantly reduce costs and risks. Technological advancements can have important results beyond current skills.
- **Develop strategies for finding skills** DCs that adopt the technology of Industry 4.0 have to evaluate what they can do from today in order to find the skills that can be supported for potentially higher levels of automation.
- **Follow the trends** Examine the new opportunities to improve distribution operations and gather competitive intelligence to monitor potential marketing threats. This will provide efficient planning and decision making over the next 10 years.
- **Plan for data management, credibility and security** Plan for data management, credibility and security in systems that get more and more complex, including data management systems and the application of Industry 4.0 technologies. Data management systems are important for the application of Industry 4.0 technologies, as they are very important in gathering, analyzing, applying data and preserving skills.

Smart storage and intercorporate logistics solutions will increase the competitive strength of manufacturers. Flexible architectural designs optimized for logistics operations can be tested by simulation applications. Augmented reality goggles, on the other hand, will help staff to correctly choose the next step by bringing logistical data and manufacture into sight. This twin development, by enabling less time to be spent on intercorporate logistics operations and shortening the storage cycle and delivery time, will increase business capital. Alongside this, the horizontal data and system management that the manufacturers will produce for suppliers will create a lot of common work areas. Errors will be minimized, thanks to these standardized processes and cooperating together in time. Suppliers will be able to regulate their operations in accordance with the orders that come from customers with the help of horizontal integration. This situation will help them to maximize their "on time" logistical potential and minimize logistical and operational costs (TÜSİAD 2016).

Traditional supply chain planning missions will be replaced by smart collaborative systems that are able to analyze databases. In this way, it will be possible to keep all processes from the supplier to the end customer—regarding raw materials and/or products—under control. At this point, there will be a transformation for employers from the management of business to the management of controlling the business. Items that make up the Internet of Things can check whether or not a product is left by detecting products that are running short in warehouses or in sales points by means of sales channels or by silently interrogating the warehouses. If the product is in surplus, the raw materials can be controlled. This control can be achieved by machines except for a few exceptions that need to be entered onto the system. If RFID is being used on market shelves, it can help to stock the shelves by checking the storage condition of the product without waiting for the cashier's report or sales report (MÜSİAD 2017).

It is important to locate the places where the loading is defined. Despite new technologies being used to determine the location, the prevalent technology for closed position systems uses RFID. Among other topics in Logistics 4.0 are Cyber-Physical Systems (CPR), the Internet of Things (IoT), Data Mining (DM) and Internet of Services (IoS). CPS is related to the integration of the digital and physical world using sensors. IoT enables communication with other systems and users. Due to large volumes of data and computing speed, DM must be managed with delicacy. IoS is provided by different/multiple providers and includes models, users and services. Smart services and products are important components of Logistics 4.0. Smart products are able to make connections, control processes and perform real-time capture. Smart services provide evaluation, pricing and information services. The most important thing is to define the company (Göçmen and Erol 2018).

The impacts of accelerating digital and automation technologies in the logistics sector with Industry 4.0, and its uncertainties today, are explained in the table below (Table 2.2).

The benefits of the application of Industry 4.0 technologies to the logistics sector are specified below (Çilekli 2018):

- Increase in process and labor efficiency in logistics, warehouse operations, order choosing/gathering, ability to give a secure, sustained service without accidents and crashes and reduction in product loss.
- Fuel-efficiency, resource productivity, decrease in delivery time, carbon emissions, transportation and storage costs in spare part manufacturing and some logistical services.

Table 2.2 Impacts and uncertainties of technologies in the logistics sector

Technology	Impact	Uncertainties
Physical Internet (According to IoT)	Advanced supply chain transparency, security, and efficiency Advanced environmental sustainability (more efficient source management)	Changes in social expectations regarding data confidentiality and security Regulation regarding data confidentiality and security may come or might become more strict Sector's desire and ability to collaborate and invest Whether or not international firms will go with standardization
IT standards	Activation of horizontal collaboration More efficiency and transparency	It is uncertain whether companies will adapt due to concerns regarding data security
Data Analysis	Better customer experience and an increase in operational efficiency More inventory management and monitoring Advanced "estimated maintenance"	Uncertainty in capacity to process data Questions regarding data security Changes in social expectations regarding data confidentiality and security Regulation regarding data confidentiality and security may come or might become more strict
Cloud	Enabling new platform-based business models and increase in efficiency	Uncertainty in costs Uncertainty in data security
Robotics and Automation	Advanced supply chain management (reduces forgery) Decrease in bottlenecks (3rd party documentation) Reduction in errors Increase in efficiency	Uncertainty in adaptation rate Unclear whether one or two dominant solutions will emerge or whether there will be multiple competing solutions
Autonomous Devices	Decrease in human labor force More efficiency in delivery processes	Regulatory environments not currently in place in most countries Liability issues not yet clear Ethical questions remain especially in relation to emergency situations
UAVs/Drones	Increase in cost efficiency (using condition, inventory, monitoring, and delivery) Decrease in labor force	Regulation in most countries not sufficient for commercial use in public areas like delivery Safety and privacy concerns may hamper market acceptance

(continued)

Table 2.2 (continued)

Technology	Impact	Uncertainties
3D printing	Lower transportation demand Transported goods would mostly be raw materials	Extent of speed, scale, purchase, and customer sectors are not yet clear

Source Tipping and Kauschke (2015). Shifting Patterns: The Future of the Logistics Industry. https://www.pwc.com/gx/en/industries/transportation-logistics/publications/the-future-of-the-logistics-industry.html. Accessed 8 May 2018

- Capacity planning, optimized use of assets, loading order and vehicle route optimization, full time scheduling and instantaneous time of arrival estimates.
- Delivery location clustering, time-preference delivery planning for each order, estimated delay warnings, optimized roundtrip planning, lowering of transport distances and costs, decreased waiting times and empty trips.
- Transparency of logistics operations, increasing traceability and reliability, protecting the goods from theft and damage.

In the Transportation and Logistics Sector Report of 2018, answers were received and evaluated from over 100 participants between March 01 and April 01 2018 to a Transportation and Logistics Sector Questionnaire online data form—composed of 26 questions from 109 distinct logistics businesses in Turkey. The questionnaire measured the present situation along with future trends and specified the ideas/opinions/suggestions of transportation and logistics sector service providers. Industry 4.0 and digitization, which are among the technological advancements that concern the transportation and logistics sector as closely as every other sector, were highlighted as the most important factors for service providers. Moreover, autonomous cars, drones and automation-based systems were raised among the critical developments that will affect the size of the logistics sector in the future. Securing the data at hand and generating the required databases necessitates the use of cloud and data analytics technologies by logistics businesses. A major proportion of the businesses that participated in the questionnaire surmise that the logistics sector will continue to progress using technologies providing data security (ULUK 2018).

The increasing accessibility of digital transformation technologies in the industry enables companies to keep up with trends like the circular economy, personalization and omni-channel strategies that revolutionize customer habits. The supply chain, which has become more transparent with the digital transformation, facilitates diverse value suggestions to customers. For example, Adidas can execute the omni-channel strategy which provides the opportunity to their customers to buy products either online or from a physical store. They are able to achieve this by using RFID (radio-frequency identification), which allows the products to be traced throughout the value chain and the "infinite corridor" that allows shopping even in situations where the product is out of stock in the store (TÜSİAD & BCG 2017). The greatest effects of Industry 4.0 technologies and concepts will especially apply to the supply, production and distribution in the supply chain. From a technological standpoint, the organization of supply will change with the execution of business intelligence (BI) technologies, smartphone applications, automatic identification, data capture (AIDC) and RFID technologies and the miniaturization of electronic devices. However, structural changes in organizations are expected mainly in production processes. The technologies affecting this are man to machine (M2M) communication and Smart Plant, including Smart Logistics. With the combined execution of smartphone applications and smart data tools, human interaction within the supply chain will have a great effect in customer integrated and organizational, boundary-free companies' sales departments (Pfohl and et al. 2015).

Advanced artificial intelligence solutions have numerous applications in the supply chain, particularly in the storage segment. This includes vehicles without the need for human input (autonomous cars). The concepts of robotics and automation are extensively used in the supply chain. Programming the most recent generation of robots is more flexible and less costly. Their mission is to assist workers in repetitive and physically challenging tasks. The Internet of Things (IoT) is a rather revolutionary technological solution in the field of logistics. IoT conducts interrelated computing devices via a network without human input. It assists companies in keeping inventory, managing storage stock and optimizing fleet routes (A&A 2017).

2.4 Logistics 4.0 and Technological Application

Autonomous vehicles are one of the greatest innovations of the 4th Industrial Revolution. Driverless trucks are no longer only seen in science-fiction movies. They are being tested on the roads in Europe and the USA. Combined with shipping systems brought out by Uber, a clearer picture emerges of how the product will be carried problem-free without any need for human input. There are also examples of smaller companies which use small robots for delivery. Starship Technologies, which was started in 2016 by the co-founders of Skype, is a company that develops and uses canine robots to deliver products to the doorstep. They are already testing around the globe and are active in south London (Twydell 2017). Tesla CEO, Elon Musk, asserted his position in the commercial vehicle market with a vision of an electric and autonomous future. Although not completely autonomous, Tesla points us to the future of truck transportation with a truck that travels 500 miles on a single charge. Tesla is not the only company that is developing electric trucks, but it is rebuilding the hype around the sector. DHL has placed an order for Tesla Electric Class 8 Semi Trucks. These electric trucks will be tested for the first time in the US in 2019. Self-driving technologies will assist drivers who have to work long hours. There is a shortage of drivers in markets like the USA and Great Britain. According to commercial truck fleet news site *truckinginfo.com*, there was a need for 100,000 drivers in 2017 due to an aging workforce (Meahl 2017).

There is currently no consensus on the hardware and software that will support fully automated driverless trucks. It will be determined by the combination of supply,

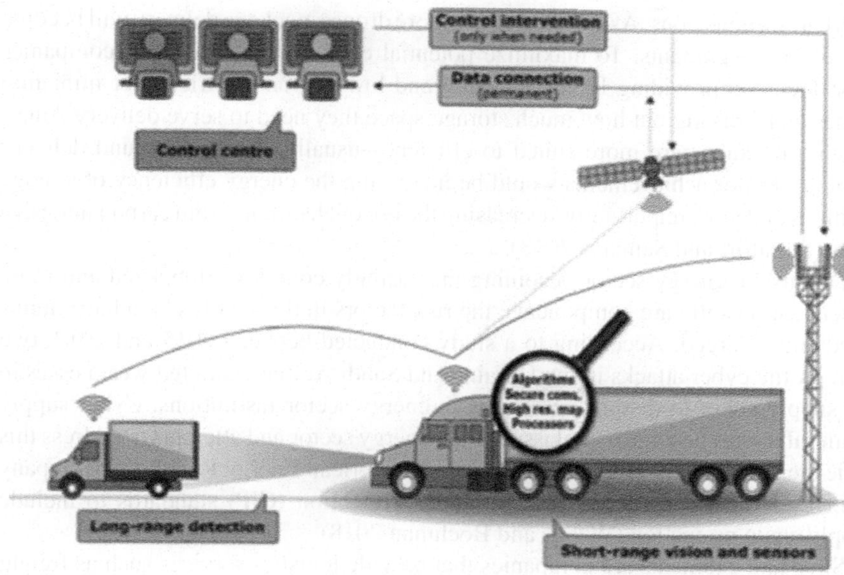

Fig. 2.4 Stylized driverless truck operating environment (with optional control center). *Source*: International Transport Forum (2017). Managing the Transition to Driverless Road Freight Transport. https://www.itf-oecd.org/managing-transition-driverless-road-freight-transport. Accessed 19 May 2018

demand and political factors mentioned below. Based on the present technological situation, the work environment of driverless trucks is presented below in Fig. 2.4 in a simplified manner. These remote drivers can either be a necessity (level 3 conditional automation) or for risk aversion at a higher level of automation (Level 4 or 5) (International Transport Forum 2017).

The three main components of a fully automated truck's potential work environment are integrated systems, support infrastructure and (remote) control center. As mentioned above, fully autonomous vehicles will require less supporting ICT infrastructure (International Transport Forum 2017).

Long distance intercity transportation is also affected by technology. Due to their inability to control factors such as another driver's error or unsuitable weather conditions, there is a risk of a traffic accident for all truck drivers. Accident scenarios usually demonstrate the difficulty of making sudden maneuvers with trucks. These vehicles are intrinsically heavy and carry heavy loads. Any collision will most likely result in harm done to other users of the road. Autonomous technology can help the driver react more swiftly by considering the driver's and the vehicle's conditions and calculating the quickest maneuver. This can decrease the intensity and number of accidents drastically, thus autonomous vehicles have the potential to be important and beneficial (DHL 2017).

Another new trend in the logistics sector is drones. Using electric drones to deliver some packages can reduce the need for transportation and fuel consumption along with carbon emissions. As companies use more drones, package delivery will become one of its assignments. To maximize potential environmental benefits, companies must focus on providing light packaging and low carbon electricity for miniature planes and working out how much storage space they need to serve delivery zones. Heavier packages are more suited to efficient—usually electric—ground delivery vehicles. Major achievements would be increasing the energy efficiency of storage facilities and most importantly, decreasing the use of electricity from carbon intensive fuels (Stolaroff and Samaras 2018).

Due to the energy sector becoming increasingly complex, globalized and more dependent on software components, the risk factors in the supply chain have multiplied and enlarged. According to a study conducted between 2015 and 2017, two high profile cyber attacks in the Ukraine and Saudi Arabia exploited weaknesses in the supply chain to disrupt operations in energy sector institutions. Cyber supply chain safety has become a vital issue in the energy sector and attempts to address this issue are increasing. For example, the North American Energy Reliability Company (NERC) is updating Critical Infrastructure Protection (CIP) standards to include supply chain protection (Woods and Bochman 2018).

Shyp and Zipments are companies that provide logistics services such as freight options or transportation capacity, but they do not own any assets, thus they have 50% less costs than the industry average, which allows them to be much more competitive because they have no costs regarding the protection of assets and economies of scale. This is an indication of an interesting future in which your typical logistics provider and supplier will adopt a more consultancy focused approach and use new business models that resemble project managers like e-brokerage rather than cargo shipping. Growth in e-retail, combined with connection technologies, will provide new solutions for transportation and logistics firms. The rise of digitization in truck transportation will force traditional freight brokers to align their business model to mobile-based freight commission agencies. Mobile applications are vitally important to problem-free and active brokerage systems, also known as the "Uberization" of truck transportation. In the future, it is expected that mobile-based freight brokers will form potentially synergistic partnerships and develop corporate software solutions with traditional freight brokers, OEMs and telematics providers who will facilitate this change. Imagine a scenario where a mobile application provides truck drivers with load ratios, routes and the senders' schedule requirements. This is expected to make a series of processes such as delivery, critical real-time information about the delivery, delivery status, shipments, load matching and paying the driver, virtually automatic. Empty miles and overcapacity problems cause $20 billion of lost revenue. The advantages of these types of business models will increase asset usage, fuel efficiency and cost minimization. In the future, we will witness online services providing more agile utilities and eliminate traditional freight brokerage firms (Singh 2016).

The DHL Global Trade Barometer application collects air and maritime transport market data from China, India, Japan, Germany, South Korea, the USA and the UK. The DHL Global Trade Barometer indicates a general solution beyond global trade. This application breaks down international supply chains into regions and countries, sectors and macroeconomic factors. It impressively demonstrates how the utilization

Table 2.3 DHL global trade barometer—world indexes

	Jan 2018	Feb 2018	Mar 2018
Global trade	64	66	66
Air trade	71	71	70
Ocean trade	60	62	63

Source Tike (2015). DHL (2018). DHL Global Trade Barometer—Global Index March 2018. https://www.logis-tics.dhl/global-en/home/insights-and-innovation/insights/global-trade-barome-ter.html?cid=Internal_DeliveringTomorrow_GTB2018EN_CTA-GTB_Text. Accessed 21 June 2018

of digitization in logistics, through the use of big data and projection analytics, can be employed to forecast the future and open up new opportunities (Scharwath 2018).

The DHL Global Trade Barometer index forecasts a strong positive growth trend due to maintaining a relatively high 66 during March (DHL 2018) (Table 2.3).

DHL Global Trade Barometer – World index are explained in the table below (Table 2.4).

The Deutsche Post DHL Group committed to a 2050 goal of zero-emission logistics in 2017. E-mobility is certainly not the only way to attain a zero-emission future, but it is a very promising start. After clearly defining goals, the next step, especially for those who shape civic mobility, is to re-evaluate today's business models and finding new ways to provide goods and services to the masses. Electric mobility, not only in the logistics industry but also in other industries, has the potential to become the new mobility standard. Deutsch Post DHL Group aims to eliminate emissions due to logistics by 2050 (Appel 2017) with the help of its environmental protection program called GoGreen. Sustainability has become one of the world's biggest problems. Worldwide contributions to a more sustainable future is through green logistics. Using bicycles and electric vehicles for the first and last miles of 70% of its deliveries will significantly reduce emissions (DHL 2017).

Table 2.4 DHL global trade barometer—world (Country Development)

	Jan 2018	Feb 2018	Mar 2018
China	59	59	61
Germany	69	67	64
USA	64	65	67
India	84	86	84
Great Britain	63	63	61
Japan	68	69	70
South Korea	65	61	69

Source Tike (2015). DHL (2018). DHL Global Trade Barometer—Global Index March 2018. https://www.logis-tics.dhl/global-en/home/insights-and-innovation/insights/global-trade-barome-ter.html?cid=Internal_DeliveringTomorrow_GTB2018EN_CTA-GTB_Text. Accessed 21 June 2018

Blockchain and DLT (Distributed Ledger Technology) can be utilized in shipping and the supply chain. DLT provides a reliable identification and verification method for obeying following distances, traffic rules and customs regulations, along with the goods and their status throughout the entire transport chain. This method introduces a way to handle systematic inefficiencies, fraud and theft, all of which increase the cost of managing the transport chain. Various players in the sector have either started their own DLT trials or have joined an alliance. This alliance accepts that blockchain technology not only simplifies the procedures in the transportation sector, but can also be used by the entire transportation sector. Its partners are global and include UPS, FedEx, Penske Logistic, GE Transport, SAP, Daimler, etc. (International Transport Forum 2018).

The logistics industry is a key player prepared to benefit from the IoT revolution. It is not very surprising that IoT and logistics are a perfect match since every day millions of consignments are being handled, carried, tracked and stacked by various machines, tools and humans. In logistics, IoT can connect different entities meaningfully throughout a supply chain and then, can analyze the data obtained from these new connections in order to spot potential improvements. Thus, IoT leads to higher operational efficiency for logistics providers by creating dynamic and automatic services that are specifically tuned for their customers. With this technology, the decline in costs will enable IoT to have an important impact on the logistics industry in the next 10 years. Most of the technologies behind IoT, including sensors, microprocessors and wireless connections, have been in use for several years in various logistics applications. In fact, the logistics industry was one of the first users of IoT technologies since the introduction of hand-held scanners—which have digitized the multi-sensors that are used to track the operational process, delivery process, cargo integrity and the performance of the delivery truck. However, even the early adopters are at the tip of the iceberg when it comes to the utilization potential of IoT in logistics (DHL&Cisco 2015). The trend report published by DHL and Cisco on the Internet of Things, predicts that there will be 50 billion interconnected devices in 2020, where today we have 15 billion. The report is focused on the meaning of this technological revolution for businesses. The increased income and reduced cost model will spread among companies due to the increased number of connections. Therefore, an important advantage will be gained via the use of the internet and network in supply chain points such as storage and cargo handling. IoT is anticipated to have a reforming impact on all the companies that have a supply chain or logistics operations, spanning specialized delivery options for clients, more efficient storage and transport (Tike 2015) (Fig. 2.5).

Although AR (Augmented Reality) is at its relatively early stages of adoption in logistics, it can provide some significant benefits. For example, AR can provide the logistics providers with fast access to expected information anytime and from anywhere. This is important in order to fully plan and execute intended duties such as delivery and loading optimization, and also in order to provide a better customer service. Most substantial AR solutions in logistics are the ones that optimize the collection process. The pick-by-paper approach is still in use in most of the warehouses in the developed world. However, a paper-based approach is slow and prone

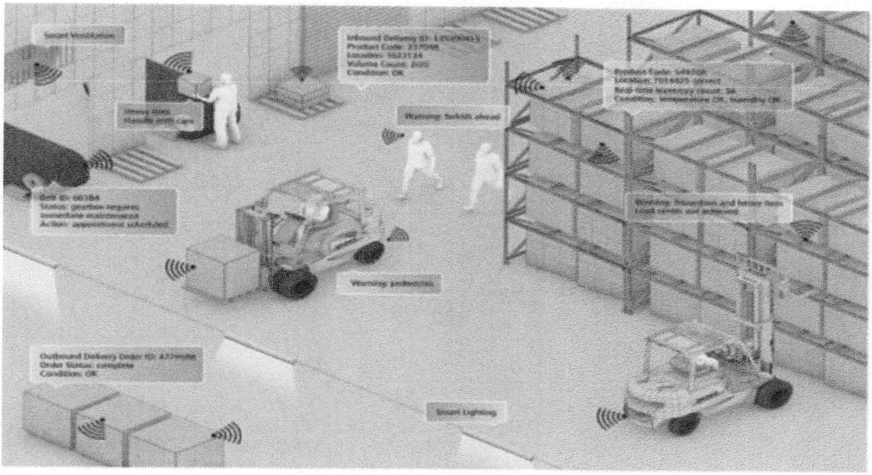

Fig. 2.5 Application of IoT technology in the logistics industry. *Source* DHL&Cisco (2015). Internet of Things Logistics. http://www.dhl.com/content/dam/Local_Images/g0/New_aboutus/innovation/DHLTrendReport_Internet_of_things.pdf Accessed 15 May 2020

to errors. In addition to that, to ensure the efficiency and accuracy of the collection process, it is handled by temporary workers who need little training. Systems designed by Knapp, SAP and Ubimax, which are composed of mobile AR technologies like head-mounted displays (HMD), cameras, wearable PCs and batteries that provide minimum energy, are still in the testing phase. This technology provides smooth integration of the information with the vision collection software, real-time object detection, barcode reading, closed navigation and Warehouse Management Systems (WMS) (DHL 2014a, b).

New trends have emerged on the Logistics Trend Radar. One identified and captured new trend is "batch size one" hyper customization. In the future, production and retail strategies are going to change in order to account for the increase in customer requests for product customization. That being said, some trends such as crypto-currencies and crypto-payment fell from the radar because new insights and developments in the logistics sector have reduced the interest in them. More extensive trends like the IoT are being followed on a larger scale. The Logistics Trend Radar 2016, as prepared by logistics experts, is presented below in Fig. 2.6 (DHL 2016):

2.5 Conclusions

Industry 4.0 is based on four basic technological advancements. There have been radical changes in production systems with the emergence of the first advance, namely 3D printers. With the second advance, the Internet of Things, the coordination of all

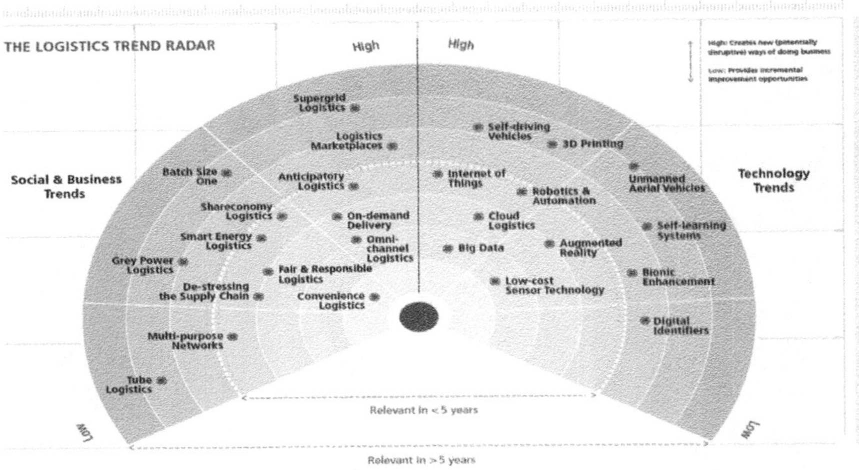

Fig. 2.6 The logistics trend radar 2016. *Source* DHL (2016). Logistics Trend Radar http://www.dhl.com/content/dam/downloads/g0/about_us/logistics_insights/dhl_logistics_trend_radar_2016.pdf Accessed 7 May 2018

the machines can be achieved. With another advance, augmented reality, the use of information in all of these processes was taken to another level via cameras and sensors, and the final advance is the use of autonomous robots in daily life as robot technology advances in every field.

Since logistics is in a key position within industry, it is of crucial importance that it adapts to the new technologies that come with Industry 4.0. Driverless trucks, which are used today and will be used in the future, are expected to lower driver employment while reducing the cost for logistics companies. They are also expected to reduce the number of collisions. For short distances, drone vehicles will be used by the logistics industry. Thus by reducing the transportation costs, carbon emissions can be reduced and fuel can be saved, and by using more company drones, traffic can be avoided and faster deliveries can be achieved.

It is accepted that the blockchain technology not only simplifies the procedures in the freight transport industry but it can be used in the entire transportation industry. The Internet of Things provides an increase in efficiency while creating customized automatic services for the customers of logistics providers. This technology reduces costs. Industry 4.0 ensures error-free operation of the process in the logistics industry by letting the users and transportation vehicles communicate using the Internet of Things. With Logistics 4.0, sustainability has become an important topic. With the increasing use of electric vehicles, fuel saving is achieved without damaging the environment.

Logistics 4.0 reduces carbon emissions, reduces costs, reduces stock costs, increases labor productivity, reduces risks, reduces delivery times and provides estimations of instantaneous arrival times and possible delay warnings.

References

A&A (2017) Top 10 future trends in supply chain and logistics. https://www.aacb.com/trends-in-supply-chain-and-logistics/.Accessed 17 June 2018

Appel F (2017) E-mobility has arrived. https://delivering-tomorrow.com/e-mobility-has-arrived/ Accessed 18 May 2020

Badaraite Z (2016) Welcome to the future: industry 4.0. https://www.hylasoft.com/en/posts/welcome-to-the-future-industry-4-0 Accessed 5 May 2020

Banger G (2016) Endüstri 4.0 ve Akıllı İşletme. Ankara: Dorlion Yayınevi

Banger G (2017) Endüstri 4.0 Ekstra. Ankara: Dorlion Yayınevi

Çağlar BM (2014) Lojistik İşletmelerinde Bilişim Teknolojilerinin Kullanımı, Müşteri Memnuniyeti ve İşletme Performansı İlişkisi: Bir Araştırma. Selcuk University J Inst Soc Sci 32:41–55

Çeliktaş MS, Sonlu G, Özgel S, Atalay Y (2015) "Endüstriyel Devrimin Son Sürümünde Mühendisliğin Yol Haritası. TMMOB Makina Mühendisleri Odası Mühendis Ve Makine 56(662):24–34

Çiçekli S (2018) Sanayi 4.0'ın Lojistik Sektörüne Etkileri. Anahtar Dergisi, (352). http://anahtar.sanayi.gov.tr/tr/news/sanayi-4-0in-lojistik-sektorune-etkileri/9635.Accessed 8 July 2020

DHL (2014). Augmented reality in logistics. http://www.dhl.com/content/dam/downloads/g0/about_us/logistics_insights/csi_augmented_reality_report_290414.pdf. Accessed 7 May 2020

DHL (2014) Self-driving vehicles in logistics. https://delivering-tomorrow.com/wpcontent/uploads/2015/08/dhl_self_driving_vehicles.pdf.Accessed 20 May 2020

DHL (2016) Logistics trend radar. http://www.dhl.com/content/dam/downloads/g0/about_us/logistics_insights/dhl_logistics_trend_radar_2016.pdf. Accessed 7 May 2020

DHL (2017) Environment & solutions. https://www.dpdhl.com/en/responsibility/environment-and-solutions.html. Accessed 3 May 2020

DHL (2018) DHL global trade barometer – global index March 2018. https://www.logistics.dhl/global-en/home/insights-and-innovation/insights/global-trade-barometer.html?cid=Internal_DeliveringTomorrow_GTB2018EN_CTA-GTB_Text. Accessed 21 June 2020

DHL&Cisco (2015) Internet of things logistics. http://www.dhl.com/content/dam/Local_Images/g0/New_aboutus/innovation/DHLTrendReport_Internet_of_things.pdf.Accessed 15 May 2018

Dyson S (2018) Mind the gap – industry 4.0 and the future of manufacturing. https://www.ibtimes.co.uk/mind-gap-industry-4-0-future-manufacturing-1665243. Accessed 25 June 2020

Ermolaeva A (2017) Industry 4.0 and HR in logistics. University of economics in prague international Bbusiness – central European business realities (Thesis). https://pdfs.semanticscholar.org/9b12/99bf48c2491ca0e8d66bf1e63e809cee9ccc.pdf. Accessed 1 June 2020

Göçmen E, Erol R (2018) The transition to industry 4.0 in one of the Turkısh logistics company. Int J 3D Print Technol Digit Ind 2(1):76–85

Görçün FÖ (2016) Dördüncü Sanayi Devrimi Endüstri 4.0. İstanbul: Beta Basım

Horenberg D (2017) Applications within logistics 4.0 A research conducted on the visions of 3PL service providers.In: 9th IBA Bachelor Thesis Conference, July 5th, 2017, Enschede, The Netherlands

International Transport Forum (2017) Managing the transition to driverless road freight transport. https://www.itf-oecd.org/managing-transition-driverless-road-freight-transport Accessed 19 May 2020

International Transport Forum (2018) Blockchain and beyond: encoding 21ST century transport. https://www.itf-oecd.org/sites/default/files/docs/blockchain-and-beyond-encoding-21st-century-transport.pdf Accessed 20 June 2020

Meahl B (2017) Are we driving into an autonomous future? https://delivering-tomorrow.com/are-we-driving-into-an-autonomous-future/ Accessed 18 May 2020

MÜSİAD (2017) Endüstri 4.0 ve Geleceğin Lojistiği 2017 Lojistik Sektör Raporu. http://www.musiad.org.tr/F/Root/Pdf/lojistik_raporlari_2017_12_25.PDF Accessed 5 May 2020

Öztuna B (2017). Endüstri 4.0 (Dördüncü Sanayi Devrimi ile Çalışma Yaşamının Geleceği. Ankara: Gece Kitaplığı

Pesti I, Nick AG (2016) Industry 4.0 from the aspect of logistics innovations. society for regional science and policy. In: 6th winter seminar of regional science, 9–12 March 2016, Slovakia. http://ersa.sk/Zbornik/files/Pesti_Nick.pdf.Accessed 15 May 2020
Pfohl CH, Yahsi B, Kurnaz T (2015) The impact of industry 4.0 on the supply chain.In: International conferance of logistics (HICL)-20. Hamburg. https://www.researchgate.net/publication/288466876_The_Impact_of_Industry_40_on_the_Supply_Chain. Accessed 12 May 2020
Scharwath T (2018) Leveraging logistics to forecast trade trends. https://delivering-tomorrow.com/leveraging-logistics-to-forecast-trade-trends/. Accessed 20 June 2018
Schwab K (2016) Dördüncü Sanayi Devrimi. Zülfü Dicleli (translated). İstanbul: Optimist Yayım ve Dağıtım
Şekkeli H, Zümrüt Bİ (2018) Endüstri 4.0'ın Etkisiyle: Lojistik 4.0 (By The effect of the industry 4.0 on logistics 4.0). J Life Econ 5(2):17–36
Sharma J (2018) Industry 4.0 effects logistic 4.0. https://www.ioscm.com/blog/industry-4-0-effects-logistics-4-0/. Accessed 5 July 2020
Singh S (2016) Future of logistics: five technologies that will self-orchestrate the supply chain. https://www.forbes.com/sites/sarwantsingh/2016/09/22/future-of-logistics-5-technologies-that-will-self-orchestrate-the-supply-chain/2/#780afd057f92. Accessed 17 May 2020
Stock T, Selinger G (2016) Opportunities of sustainable manufacturing in industry 4.0. 13th Global Conference on Sustainable Manufacturing - Decoupling Growth from Resource Use. https://core.ac.uk/download/pdf/82023214.pdf. Accessed 12 May 2020
Stolaroff J, Samaras K (2018) Delivering packages with drones may benefit the environment. https://www.weforum.org/agenda/2018/02/delivering-packages-with-drones-might-be-good-for-the-environment. Accessed 20 June 2020
Taliaferro A, Guenette AC, Agarwal A, Pochon M (2016) Industry 4.0 and Distribution Centers. https://www2.deloitte.com/content/dam/Deloitte/pe/Documents/technology/DUP_Industry-4-0-distribution-centers.pdf. Accessed 12 May 2020
Tike Z (2015) Nesnelerin İnterneti Tedarik Zinciri ve Lojistik Operasyonlarına 1.9 Trilyon Dolar Kazandıracak! http://www.turkishtimedergi.com/dijital-ekonomi/nesnelerin-interneti-1-9-trilyon-kazandiracak/. Accessed 8 May 2020
Tipping A, Kauschke P (2015) Shifting patterns: the future of the logistics industry. https://www.pwc.com/gx/en/industries/transportation-logistics/publications/the-future-of-the-logistics-industry.html. Accessed 8 May 2020
TÜSİAD & BCG (2017) Türkiye'nin Sanayide Dijital Dönüşüm Yetkinliği. https://tusiad.org/tr/yayinlar/raporlar/item/9864-tusiad-bcg-turkiye-nin-sanayide-dijital-donusum-yetkinligi. Accessed 25 May 2020
TÜSİAD (2016) Türkiye'nin Küresel Rekabetçiliği İçin Bir Gereklilik Olarak Sanayi 4.0. http://www.tusiad.org/indir/2016/sanayi-40.pdf. Accessed 5 May 2018
Twydell S (2017) What does industry 4.0 mean for the future of logistics? https://www.itproportal.com/features/what-does-industry-40-mean-for-the-future-of-logistics/. Accessed 8 June 2020
ULUK (II. Ulaştırma ve Lojistik Ulusal Kongresi) (2018) Ulaştırma ve Lojistik Sektör Raporu 2018. http://www.ulk.sakarya.edu.tr/wp-content/uploads/2018/05/Ula%C5%9Ft%C4%B1rma-ve-Lojistik-Sekt%C3%B6r-Raporu-2018.pdf. Accessed 15 June 2020
Woods B, Bochman A (2018) Managing risk in the energy sector's cyber supply chain. https://www.weforum.org/agenda/2018/06/managing-risk-in-the-energy-sector-s-cyber-supply-chain. Accessed 20 June 2020

Barış Öztuna is an Associate Professor at the Faculty of Economics and Administrative Sciences, Department of Labour Economics and Industrial Relations at the Çankırı Karatekin University, Turkey. He finished his B.A. at the Department of Business Administration, Celal Bayar University, the Faculty of Economics and Administrative Sciences in Manisa, Turkey in 2005. He worked his thesis "The Contribution of ISO 9001:2000 Quality Management System to Quality of Business Life: An Investigation" at the Total Quality Management Master Programme Dokuz Eylul

University Institute of Social Sciences in Izmir, Turkey in 2007. He had his Ph.D about Restructuring in Social Insurance and Effect to Institutional Performance at Dokuz Eylul University Institute of Social Sciences, Department of Labour Economics and Industrial Relations in Izmir, Turkey in 2013. His research interests on social policy, social security, logistics, industry 4.0. He authored two books publication "Industry 4.0 (the future of the working life with the fourth industrial revolution), Society 5.0 (Super Smart Society) and a number of articles in peer-reviewed journals.

Chapter 3
Technology, Supply Chain, and Logistics Management

İsmail İyigün

Abstract It can be stated that developments in information communication technologies and micro-electromechanical systems have opened the door to the last industrial revolution (Industry 4.0). The effects of Industry 4.0 show its effects in all areas of life in addition to the area of supply chain management. In this context, this study intends to go over the effects of Industry 4.0 on supply chain management. For this purpose, a meta-analysis of academic publications in the literature related to Industry 4.0 has been carried out. Academic publications published since 2013, when the concept of Industry 4.0 emerged, have been analyzed and their contents have been investigated. In particular, the key technologies that form the basis of Industry 4.0 have been put forward and the most studied technologies are listed under Industry 4.0. On the other hand, publications examining the concepts of logistics and supply chain management with Industry 4.0 have been also examined. As a result, it is aimed to reveal the most commonly used Industry 4.0 technologies in the Logistics 4.0 and SCM 4.0 applications. As a result of the meta-analysis, technologies including IoT, CPS, smart factory, big data and smart product have been at the forefront in the Industry 4.0. On the other hand, it is seen that many different technologies such as cloud computing, artificial intelligence and data mining have been used in the researches examined. It has been determined that the studies, especially about logistics and supply chain management, which were taken together with Industry 4.0, have been concentrated in 2017 and 2018.

Keywords Industry 4.0 · Supply chain management · Logistics management

3.1 Introduction

It can be stated that three different industrial revolutions have taken place from the past to the present. In this context, the first industrial revolution (Industry 1.0) is the industrialization process that we all know as the "industrial revolution" in the real

---

İ. İyigün (✉)
Social Vocational College, Trakya University Edirne, Edirne, Turkey
e-mail: ismailiyigun@trakya.edu.tr

İ. İyigün and Ö. F. Görçün (eds.), *Logistics 4.0 and Future of Supply Chains*, Accounting, Finance, Sustainability, Governance & Fraud: Theory and Application,
https://doi.org/10.1007/978-981-16-5644-6_3

sense and started at the end of the eighteenth century and showed a great impact in the nineteenth century. The symbol of this first industrial revolution was steam powered machines. The second industrial revolution (Industry 2.0), on the other hand, is the revolution that is characterized by mass production and electricity that took effect in the twentieth century. After that, the third industrial revolution (Industry 3.0) refers to the process of transition from analog systems to digital systems in production. The use of computers in this process has played a significant role in the digitalization of production (Witkowski 2017; Lu 2017). It is seen that these three industrial revolutions are characterized by mechanization, electricity use, and digitalization in production, respectively. These industrial revolutions have enabled enormous capacities to be obtained in the field of production and caused an increase in productivity. However, in addition to this productivity and capacity increase, this production capacity has brought some environmental, social, and technological problems with it. Productivity increase alone was not enough for companies to maintain sustainability in the global competitive environment. For this reason, companies have started to search for various ways in order to solve the problems and new production methods. In particular, they tried to develop collaborative, innovation-oriented production and distribution figures which take into consideration the entire life cycle of the products and virtual and physical structures accordingly. The trade boundaries between the post-Cold War countries have disappeared and trade have started to increase among these countries. In the 1960s, while customers only bought the existing products, on the other hand in the 2000s the wishes and expectations of customers changed the production process of the companies and made production process more complex. Thus, companies have now felt the necessity for interdisciplinary work and the Fourth Industrial Revolution (Industry 4.0), in which all objects are communicated and interacted over the Internet, emerged (Lu 2017).

In this part of this book, the concept of Industry 4.0 will be examined and the innovations, tools, and methods used in the scope of Industry 4.0 will be discussed. In this context, logistics technologies, supply chain management, Logistics 4.0, and SCM (Supply Chain Management) 4.0 applications will be discussed in detail, which technologies used in supply chain and logistics, and communication possibilities between these technologies will be discussed from Logistics 4.0 perspective and a meta-analysis will be done to analyze the studies in the literature.

3.2 Industry 4.0

Industry 4.0 that is to say the fourth industrial revolution is planned that all units that are directly or indirectly relevant to production are cooperating with each other and that digital data should be integrated with software and information technologies (Schuh et al. 2014). Industry 4.0 centers on the creation of clever goods, procedures, and processes. The first applications of Industry 4.0, which was first suggested at the Hannover Fair in 2011, were started in April 2013 (MacDougall 2014). Industry 4.0 aims to transform design patterns from mass to individuality by bringing together design innovations, new products to market faster and combining intelligent production and industrial components that attempt to satisfy individual customer demands (Pisching et al. 2015). The fourth industrial revolution includes a comprehensive digital transformation of many production activities. This revolution brings production to higher levels to create new formation, intelligent, interconnected, and interoperable production systems that can trace system performance in real time to manage costs, prevent malfunctions and reduce downtime (Giannetti and Ransing 2016). New production systems have been constructed with cyber-physical systems that can work with network connections and interact with people in complex intelligent factory environments. These systems have been benefitted from big data and predictive analyzes in a comprehensive way in an effort to manage manufacturing processes more effectively and to ensure the production of customized products with increased profitability and energy efficiency (Schlaepfer et al. 2015). New generation production systems maximize existing continuous improvement capabilities with new information creation and real-time decision-making from process data (Giannetti and Ransing 2016). The basis of Industry 4.0 composes of three elements: digitalization and combination of vertical and horizontal value chains, digitalization of goods and services, and the presentation of avant-garde business figures. (Gilchrist 2016).

For the realization of Industry 4.0, objects, sensors, and system-moving components must be in constant communication and exchange real-time data (Lasi et al. 2014). These tasks are carried out by the Cyber-Physical Systems (CPS) and Internet of Things (IoT). CPS is concerned with the combination of computational technologies into physical processes; it has relied on a structure consisting of fixed systems and physical surrounding, where information processing is always carried out (Pisching et al. 2015). The cyber-physical systems of Industry 4.0 are anticipated to produce a large amount of transaction data and lead to a revolution with the use of these data and knowledge in the manufacturing industry (Giannetti and Ransing 2016).

The smart product, one of the components of Industry 4.0, is the expansion of the character of the workpiece to an active part of the system. These smart brands can store the necessary resources and information in their own memories, and therefore they can direct the operations in order to complete their production processes thanks to the memory of their operational data and requirements (Loskyll et al. 2012). The concept of smart machine describes the process of machines becoming cyber-physical production systems (Zamfirescu et al. 2014). Smart machines are local control intelligent automatic components that can contact with other area tools, production modules, and products by means of open networks. When these machines are self-organized in the manufacturing network, production lines get very ductile and modular; even the smallest batch sizes can be generated economically. In addition, a modular production line based on cyber-physical system will allow for easy integration or replacement of new manufacturing units. The augmented operator targets the technological support of operators operating in the harsh environments of modular production systems. Operators working in Industry 4.0 are considered to be the most flexible components of Industry 4.0, which can adapt to the increasingly difficult working environment (Weyer et al. 2015).

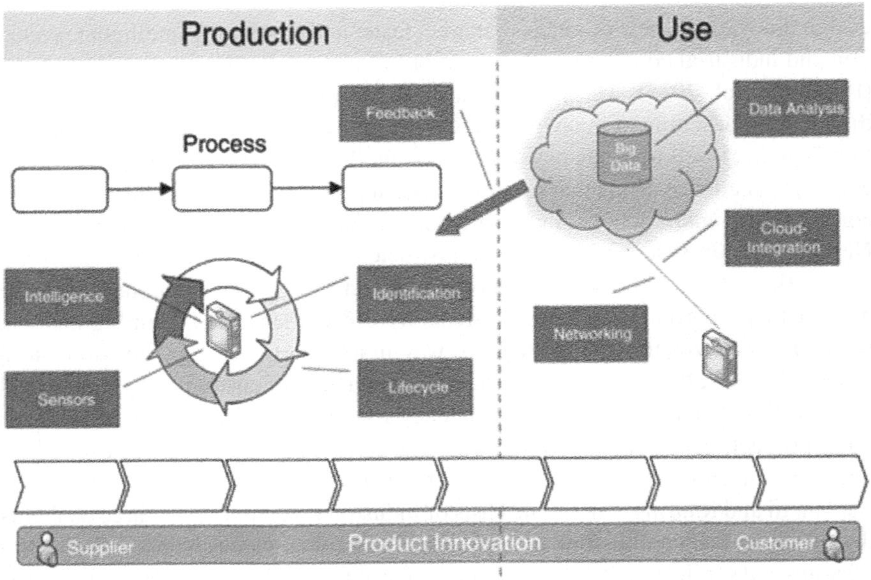

Fig. 3.1 Basic Structure of Industry 4.0 (Schmidt 2013)

3.2.1 Basic Components of Industry 4.0

Industry 4.0 is relied on several new emerging technologies. Mobile computing, cloud computing, and big data technologies are the most important ones of these Technologies (Schmidt and Möhring 2013). The emphasis of cloud-information and mobile computing for Industry 4.0 is much more than the large computing possibilities, so services are made globally via the Internet. As shown in Fig. 3.1, these services can be assimilated and used (Schmidt 2013). The easy combination of services makes it possible for all elements in the value chain to work in collaboration. This situation brings the relationship to the foreground rather than the transactions between the components. Cloud computing also enables new business models and processes to emerge. Components connected to each other through cloud computing enable foretelling maintenance and present information on the possibilities of optimization in production (Schmidt et al. 2015). Before the use of cloud computing and the Internet, the data gained during production first remained within the system and after a while the memory was deleted due to lack of memory. Today, thanks to integrated networks and systems, data collection and storage has become very easy. However, the flow and storage of large amount of data has made it important to continuously analyze big data and to optimize production (Abbott 2014). Therefore big data is another important component of Industry 4.0 (Fig. 3.1).

The machines used in the advanced industries are now equipped with cutting edge Internet technologies. Thus, the machines in the production process can be programmed, monitored and the output of the production can be controlled remotely.

Hermann et al. (2016) found that there are four main ingredients of Industry 4.0 by scanning academic and sectoral publications. These components are defined as Internet of things (IoT), cyber-physical systems (CPS), Internet of services (IoS), and smart factory. In the following part of this chapter these components will be examined in detail.

3.2.1.1 Internet of Things (IoT)

Internet of Things is a concept that has been used frequently in recent years. The term Internet of things attributes to various structures which can communicate with each other using wireless communication technologies (Atzori et al. 2010). The Internet of Things is a promising technology that offers many opportunities in terms of production and distribution channels today (Hofmann and Rüsch 2017). The notion Internet of Things was first introduced by British Entrepreneur Kevin Ashton in 1999. In this new approach, a system is envisaged where objects contact with one another with wireless sensors and data. It is seen that this system consists of very different components and these components change according to the application area. While objects can be a part of this system when necessary, animals, people, and many other elements can become a part of this system. At this point, all these elements should be able to exchange data with wireless networks (Witkowski 2017). This approach, which emerged with the spread of information and interaction technologies at the end of the twentieth century, has caught on since the beginning of the twenty-first century. In addition, this technology has had a key position in the transition from the 3rd Industrial Revolution to the 4th Industrial Revolution. Because the objects that are connected to each other and share the information they perceive have enabled simultaneous information to be integrated into production and supply chain processes (Hofmann and Rüsch 2017). From a technical point of view, the Internet of things can be summarized as sensing, information processing and communication systems embedded in objects that allow physical objects used in production and distribution processes to be interconnected by wireless networks and to enable information exchange.

IoT stands out as an abstract concept rather than a concrete concept. The idea of IoT emerged in an abstract way as a consequence of the idea of integrating information and communication technologies into "things" people use at work, at home, or in any other area of life. The need for labeling and tracking or monitoring "things", i.e., physical objects, was the basis for the emergence of this approach. Labeling and monitoring have been achieved with the introduction of low cost Radio frequency identification (RFID) devices. At first, thanks to these devices, systems operating autonomously within themselves have made an incredible progress with the development of Internet and information system infrastructures and the increase in smartphone technologies. After this point, a paradigm shift occurred in the field of IoT, and systems containing IoT have started to be used more and more, especially thanks to broadband Internet and advanced information processing systems. Simple IoT examples used in every field can be found today. However, IoT examples used in the industrial field are seen to be used in more complex devices. For example, IoT devices used in the industrial field can be used in actuators, robots, 3D printers, production assembly lines, chemical mixture tanks, milling machines, health devices, insulin and infusion pumps, cars, planes, and trains (Thames and Schaefer 2016).

3.2.1.2 Cyber-Physical Systems (CPS)

The integration of the physical world and the cyber world is one of the main features of the Industry 4.0 approach (Kagermann 2015). The main element of this integration is the Cyber-Physical Systems (CPS). CPS provides integration between the physical world and the digital world via information transfer from cyber to physical domain. Algorithms planned and programmed in the digital world are implemented in the physical world with CPS. The computers and networks, embedded in the systems used, control the physical processes, while feedback from the physical world also influences the processes in cyberspace (Lee 2008). The improvement of CPS occurred in three stages. The first generation CPS includes recognition of objects using RFID tags. Thanks to this, storage and analysis operations can be done centrally. In the second generation CPS, a limited number of functions were gained by using many sensors and actuators. The third generation CPSs are supplied with sensors and actuators which can be controlled by computer networks (Mario et al. 2015).

3.2.1.3 Internet of Services (IoS)

Today, people live in a world where services are becoming globally widespread (Andersson and Mattsson 2015). As a result, it is seen that Internet of Services (IoS) have emerged relied on the notion that services can be made easily via web technologies as in Internet of Things (Wahlster et al. 2014). The notion "Internet of Services" refers to the infrastructure allowing universal services to be provided to users (Schroth and Janner 2007). Internet services (IoS) allow the service providers to offer the services through the Internet (Hermann et al. 2016; Atzori et al. 2010). For the realization of Internet of Services, a business model, an infrastructure, and the presence of a service are required (Hermann et al. 2016). In the future, IoS can be said to have a key role in Internet-based market places for the industries (Barros and Oberle 2012).

3.2.1.4 Smart Factories

Today, most of the industrialized countries try to invest in advanced and modern production methods, innovation, and designs. The majority of these investments are made to establish smart factories based on Industry 4.0 and to follow smart production processes (Lu 2017). Industry 4.0 is based on smart manufacturing processes where objects interact with each other over the Internet and use modern technologies such as data mining, artificial intelligence, and three-dimensional printers. Smart factories where objects can communicate with each other have become an essential element in Industry 4.0. These factories are also called "dark factories" because there are systems that communicate with each other and no people are working. For instance, in a smart factory in China manufacturing mobile phones, it is observed that the human resource has decreased by approximately 90%, while the error rate in products has decreased from 25 to 5% compared to the past (Yıldız 2018).

In the previous parts of this study, IoT and CPS as Industry 4.0 components were explained. These components play a fundamental role in running a smart factory. CPSs that communicate over IoT can work in harmony within the factory without interfering with each other. For this reason, there is a distributed production system in smart factories where resources and production systems communicate with each other over a network. This system enables resources, machines, and transfer systems to work in a coordinated manner and destroys the logic of today's known production approaches. For this reason, the existence of smart factories is considered as a sign of a new industrial revolution (Industry 4.0). In smart factories, it is seen that the resources are automatically moved along the lines and turned into final products. This process is carried out with the CPS, with easily definable steps and communication between objects. In this way, very complex production processes become processes that people can manage, and the production process is carried out more efficiently and sustainably (Hofmann and Rüsch 2017).

Production lines developed depending on the production requirements are used in the smart factories of Industry 4.0. The machines that communicate with each other in these production lines take the production information they need from the big data in the cloud computing system. Communication and interaction between units in smart factories are provided by a wireless Internet network. In this context, IoT technology is used to detect and exchange information in machines and other parts (Kang et al. 2016). All production resources (sensors, machines, actuators, robots, etc.) not only do automatic information exchange, but also have the consciousness and intelligence to continue the production process. Thanks to this ability, the factory gains the feature of autonomous operation and can be called a smart factory. On the other hand, many production process components such as product design, production engineering, and production planning can be simulated modularly. Moreover, these processes can be integrated between end-to-end systems. This shows that inter-system connection is also provided and systems can work distributed (Qin et al. 2016).

Industry 4.0's "Smart Factories" include intelligent machines and systems determining business needs with sensors, contact with other production tools through the Internet, and pull the production information they need from "Big Data" in the cloud systems. In smart factories, communication and interaction between the way of production are provided by the Internet (Kang et al. 2016). All production sources (sensors, actuators, machines, robots, conveyors, etc.) not only exchange information automatically, but also they are conscious and intelligent enough to control the production process and anticipate and maintain the machines to manage the factory system. In addition, many production processes, such as product design, production planning, manufacturing engineering, production, and services, can be simulated modularly. Furthermore, these processes can be interconnected by an end-to-end system, which means that they are not only commanded by a centralized system, but are also interconnected (Qin et al. 2016).

3.2.1.5 Big Data

Performing systematic analyzes is a significant issue in many aspects, from energy efficiency to product optimization, and from service development to efficient use of equipment and machinery. In Industry 4.0, data gathered from different sources have been evaluated comprehensively for decision-making and real-time interaction (Rüßmann et al. 2015). The significance of big data is increasing with the increasing use of electronic devices that produce a very big amount of data every single second (Zhong et al. 2015). Big data refers to large datasets that are very comprehensive and complex, which are difficult to compute using traditional computing applications (Zhong et al. 2015). Data mining techniques are widely used in production processes prior to the big data period. In order to increase the efficiency of data mining, statistical, neural networks, decision trees, and genetic algorithm techniques are used for analyzing big data (Shahbaz et al. 2019).

The big data is an important business system or platform that offers more features for the acquisition, storage, and analysis of large amounts of data from various sources to obtain added value. Most of the data that generates big data is obtained from various devices around industrial enterprise and supply chain networks, including computer systems, embedded sensors, computerized devices, and smartphones. Besides, there are some limitations on the performance of big data (Addo-Tenkorang and Helo 2016). When comparing big data with traditional datasets, unconfigured data stacks are required which require more real-time analysis (Chen et al. 2014). Big data reveals the data that are not yet meaningful in capturing new opportunities and making strategic decisions (Addo-Tenkorang and Helo 2016). The service and manufacturing sectors are increasingly stacked data while they produce a service or a product. These data are collected in real time and continuously with information technologies such as mobile devices, sensing techniques from the air, microphones, cameras, Internet technologies of the objects, and wireless sensor networks (Zhong et al. 2016).

3.3 Logistics 4.0

The changes in the consumer needs and expectations, globalization, the spread of technology-based communication, the diversification of activities that need to be carried out, and the impact of technological innovations in recent years have led to various changes in logistics activities over time. In view of the chronological development of industrial revolutions, logistics processes have been affected in parallel with every revolution in the production industry (Timm and Lorig 2015). In other words, the logistic development process is closely related with industrial development process. The logistic development process (Fig. 3.2) consists of four stages as in the industrial development process (Galindo 2016: 25–30) (Fig. 3.2).

In the first industrial revolution (Industry 1.0), the industrialization of the steam engine began with the industrialization of rural life. Manufacture and transportation activities with manual vehicles or animal power have been started with machines. During this period, mostly road transport was carried out. With the introduction of the steam engine in transportation, railway and air transportation were carried out; a little density of road transport has decreased. During this period, the warehouses were designed as rooms for finished products or raw materials. Products going to or from warehouses are handled and transported manually by people. Transportation for distribution is provided by road, sea, or rail (Wang et al. (2016).

It would be better to define the technological changes in the second period (Industry 2.0) rather than as a revolution. In this period, new materials such as steel, copper, or aluminum were used in machine construction. There have been unprecedented developments in the chemical industry, and power supplies such as electricity and oil have been used in production, transportation, and communication. Again in this period, road transport continued; the transport of rail and steam ships became more widespread. The use of container ships in transport is one of the most important innovations of this period. The presence of electricity in the logistics sector and the use of electrical logistic equipment in warehouses have been used. In this way, the products can be placed and removed automatically on the shelves. Hand-held forklifts have also been replaced by motorized handling and transport vehicles (Wang et al. (2016).

In the third industry (Industry 3.0) revolution, there have been two main technological developments, namely, the invention of numerically controlled machines (CNC, DNC machine tools such as turning, milling, etc.) and industrial robots that enable automatic production. In logistics sector, the concept of technology-based logistics

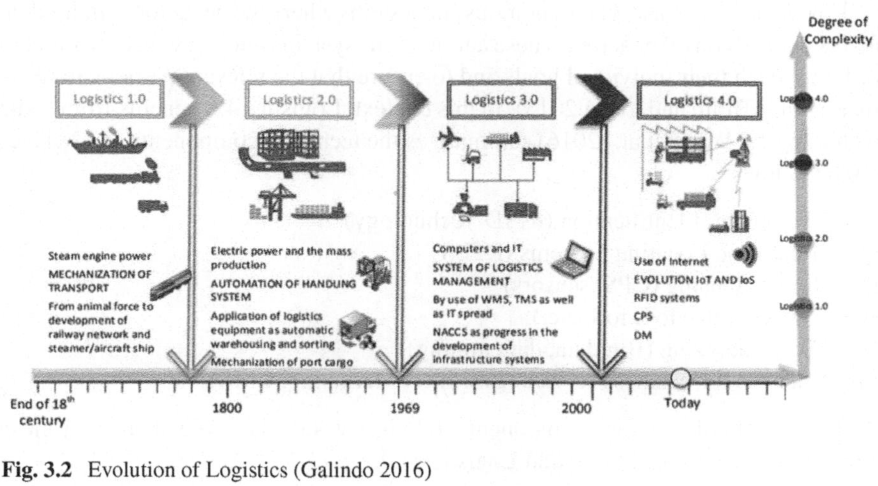

Fig. 3.2 Evolution of Logistics (Galindo 2016)

management systems has started to develop. With this effect, important software such as WMS (Warehouse Management System) and TMS (Transport Management System) have become widespread. These softwares are used in particular for planning the inbound logistics process so that orders to suppliers can be immediately accessed when needed. Again, thanks to the software, the plans, charts, and routes of fleet vehicles have already become available in the process of transporting the final product or raw material. In the process of production logistics, the products are mostly used for automatic moving belts or with the help of the forklift truck. In some enterprises, the routes are transported into the enterprise via the latest technology robots. Product delivery process is also managed according to these plans and schedules planned before production begins (Galindo 2016).

In technology-based physical logistics processes, with the effect of Industry 4.0, transformation from hardware-based applications to software-based applications has been initiated, and its computing and communication capabilities have been increased and all necessary information has been made easier to share with all stakeholders (Timm and Lorig 2015).

This transformation primarily begins with the integration of production and logistics processes with information technologies. This transformation process involves the exchange of information about the products owned by the company and the production, both within the enterprise and with customers and suppliers. Suppliers are expected to benefit from this information and exchange of information in the supply chain. This change and the expectations lead to the emergence of the concept of Logistics 4.0", in accordance with Industry 4.0, that brings smart innovations cutting edge Technologies in logistics. In the production process, people, machines, parts, and products will be communicated simultaneously. In this way, all networks will become standardized and the information will be accessible. The information will be stored in Internet environments called cloud to reduce errors and increase accessibility (Timm and Lorig 2015). As a result, production and distribution processes will be able to respond flexibly to all expected or unexpected changes.

Logistics 4.0 consists of autonomous subsystems where the behavior of individual actors depends on other actors. These autonomous systems interact with each other in order to reach their individual goals and to ensure that the relevant parties can reach their goals (Timm and Lorig 2015). In this context, Logistics 4.0 benefits from some technologies. Wang et al. (2016) summarizes the technical components of Logistics 4.0 as follows:

i. Automatic Identification (RFID Technology)
ii. Real-time Locating Systems (RTLS)
iii. Smart Sensing (CPS, sensors)
iv. Networking (IoT, IoS, etc.)
v. Data analyzing (Big data, data mining)
vi. Internet for business service (IoS, ERP, Marketing, CMS).

These technologies are component of Industry 4.0, therefore there is a close relation between Industry 4.0 and Logistics 4.0.

3.4 Supply Chain Management 4.0

The introduction of Industry 4.0 has caused many effects on the supply chain as in other areas. Digitalisation and automation of collaboration between suppliers, manufacturers, customers, and processes is crucial to improve the transparency of all steps from the time the order was sent to the end of the product's lifespan. After the introduction of these new technologies, it is necessary to analyze the impact of industry 4.0 on the supply chain as a whole in order to understand possible opportunities and possible threats (Tjahjono et al. 2017).

A supply chain is a network of suppliers, factories, warehouses, distribution centers, retailers, and customers throughout the purchase, conversion, production, and delivery of raw materials. Supply chain management is a series of concurrent activities that enable the integration of suppliers, manufacturers, carriers, and customers to ensure that the right product or service is delivered to the right places at the right time (Lou et al. 2011; Xu 2011).

Businesses are forced to expand their offer of products in order to remain competitive in the market. As a result, they offer a high degree of flexibility in supply chains, with a high degree of uncertainty. Failures to cope with these uncertainties can lead to fluctuations in the supply chain of the company. As a result, the right products cannot be delivered to the customer at the right time (Xu 2011). Such a situation is not only the problem of the company, but also because of the high dependence of the supply chain can affect other companies do business. The supply chain, which is a complex network, conducts these activities effectively and efficiently depends on the rapid flow of information and material (Lou et al. 2011).

It is observed that information systems are used more and more in order to manage more complex processes in supply chain management and to achieve real-time cooperation. Studies conducted on this subject show that developments in information systems should be reflected in supply chain management. It is stated that supply chain management cannot be carried out successfully today without these developments and technologies. At this point, information and communication systems that allow information sharing gain importance. In particular, information systems that allow intercompany information transfers are of significant importance for the success of supply chain management. It is stated that such information sharing capability increases the performance of the supply chain. In addition, it is demonstrated in academic studies that supply chain integration in order to increase information exchange and cooperation between companies, consequently increase supply chain performance. As a result, an appropriate information system infrastructure is indispensable for a successful supply chain management (Xu 2011).

With the development of information technology and economic globalization, it has become compulsory to carry out the supply chain management strategy for gaining comprehensive advantages in the face of challenging market competition. In addition, ensuring the smooth transmission of information between businesses in the supply chain is necessary to improve the advantages of the entire supply chain. However, in most of the existing supply chain systems, information requests and feedback information are transmitted step by step. Such vertical integration supply chain models have some disadvantages. Because top suppliers cannot get timely market information and it is very difficult for them to react correctly. This situation leads to distorted demand information.

The remarkable developments in information systems and the gradual expansion and development of the Internet in recent years have provided an important opportunity for information sharing, storage, and transfer. Integration of Internet supply chain management and information sharing provides an important infrastructure for ensuring coordination and cooperation between units. However, while information sharing occurs very quickly, material transfers in the supply chain are slower. Because while physical operations take place in real time, information exchange takes place in networks that provide very fast information exchange. It is observed that IoT technology makes significant contributions in terms of further accelerating physical movements in real time. IoT, which is an important element of Industry 4.0, minimizes the loss of time during transfers by providing automatic information exchange of objects in the supply chain. Thanks to IoT, transported materials can be monitored in real time and counted and recorded with automatic information exchange at transfer points. Records related to these transactions can be automatically transferred to a central database (Lou et al. 2011). Thanks to today's Web 3.0 technologies, it is seen that the Internet can provide visual interfaces to supply chain management that show operations dynamically. As a result, all stakeholders in the supply chain can monitor, control, plan, and optimize their business processes remotely instead of observing and managing them on site (Verdouw et al. 2016).

It has been stated above that information exchange is important in monitoring physical operations in the supply chain. Studies on this subject, harmonization of business processes and the use of information systems increase the quality of decision-making processes and supply networks. Information technology is becoming an increasingly important issue because supply networks are transforming into cyber-physical systems, one of the key elements of Industry 4.0 (Tjahjono et al. 2017). These supply networks are used not only in the production area but also in many other areas. For example, it is observed that cyber-physical systems are used in urban traffic control, emergency response units, and security control networks. Cyber-physical systems have both the information processing layer and the physical layers. These systems work in conjunction with other cyber-physical systems and take their decisions in harmony and perform their tasks in a coordinated manner. These systems are supported by information systems infrastructure and are kept in touch with the help of these systems. In addition, these systems can be reconfigured to adapt to new tasks. In other words, it has a dynamic and flexible structure to perform new tasks (Ivanov et al. 2016).

Performing identification, real-time monitoring, and tracking and control functions in supply chain management can often involve many difficulties. The main reasons for this situation include the presence of many stakeholders in the system, the heterogeneous structure of a wide variety, and the use of different technologies together. It can be stated that IoT, one of the most important elements of Industry 4.0, brings significant solutions to the problems that arise in this regard. Thanks to

IoT and cloud computing, it is possible to collect, transmit, store, and share logistics information. This opportunity provides businesses with a better collaboration and interoperability capability (Tjahjono et al. 2017). The problem discussed here is the development of cooperation between supply chain stakeholders to facilitate the flow of the process in the supply chain. Existing platforms fail to solve the problem at some key points, such as collecting data directly from time-consuming and reporting sensors, defining a common policy and communication protocol for all stakeholders, providing information from mobile devices to remotely process and update data, and managing multiple interactions between supply chain partners. In order to address the above-mentioned shortcomings, a collaborative cloud-based platform must be in place to support the data sharing, integration, and processing requirements for logistics product monitoring and tracking. The added value of this architecture is to integrate different IoT, sensor, data transfer and cloud computing storage layer, and ultimately set up the collected data for users. Therefore, it is to facilitate the exchange of information in logistics flows for traceability, cooperation, and interoperability between different actors in the supply chain (Gnimpieba et al. 2015). Use of all these new Technologies related to Industry 4.0 in the supply chain management brings the new concept of "SCM 4.0".

3.5 Meta-Analysis

In this part of the study, a meta-analysis of academic publications in the literature related to Industry 4.0 has been carried out. In this context, academic publications published since 2013, when the concept of Industry 4.0 emerged, have been analyzed and their contents have been investigated. In particular, the key technologies that form the basis of Industry 4.0 have been put forward and the most studied technologies are listed under Industry 4.0. On the other hand, publications examining the concepts of logistics and supply chain management with Industry 4.0 have been also examined. As a result, it is aimed to reveal the most commonly used Industry 4.0 technologies in the Logistics 4.0 and SCM 4.0 applications.

Although there are a large number of publications in the literature on the subject of Industry 4.0, more than 200 articles have been scanned in this study. Some of these publications have been eliminated, because of being related other aspects of Industry 4.0 such as management, marketing etc., and a total of 166 publications have been examined. A similar study in the literature has been conducted by examining 51 publications (Hermann et al. 2016). In another study, Lu (2017) examined the articles published in Industry 4.0 between the years 2011–2016. It was observed that the researcher made a meta-analysis by examining the 88 publications about Industry 4.0 (Lu 2017). Therefore, although there are more than this amount in the literature, 166 articles are considered to be sufficient to reveal the technologies and trends in the Industry 4.0.

IoT, IoS, CPS, cloud computing, smart factory, smart product, big data, and artificial intelligence technologies are frequently used when the literature related to

Industry 4.0 is reviewed. However, some of these technologies come to the forefront in the field of Industry 4.0. When Table 3.1 is examined, it is understood that IoT and CPS technologies are clearly seen as the main technologies of industry 4.0. On the other hand, smart factory, big data, smart product, and IoS technologies are also frequently used in the literature. Herman et al. (2015) also found that the most commonly used technologies are CPS, IoS, and smart factory in his study as seen in Table 3.1. Lu (2017) stated that the smart factory, IoT, CPS, and smart product technologies came to the forefront in his meta-analysis study (Fig. 3.3).

Since 2013, when the concept of Industry 4.0 emerged, untill today, there have been a great number of Industry 4.0 related publications in the literatüre. According to the166 articles that are in dataset, Fig. 3.3 shows that the number of these publications has been increasing continuously since 2013. Therefore, it is seen that the subject of Industry 4.0 is still an increasingly popular topic. Although the number of publications published in 2018 seems to have decreased, it is considered that it would be more appropriate to evaluate the 2018 data in 2019. Because there are still unpublished articles in 2018.

The relationship between Logistics 4.0 and Industry 4.0 components have been analyzed through the publications in the dataset. In this context, it has been seen that 15 articles from 166 articles deal with the subject of Logistics 4.0 together with the subject of Industry 4.0. In Fig. 3.4, it is determined that IoT and CPS technologies are commonly used in Logistics 4.0 studies. Today, it is seen that real-time systems, big data, RFID, CPS, and IoT technologies are widely used in the field of logistics (Wang et al. (2016). It has been determined that there is a focus on these technologies in the Logistics 4.0 studies conducted in the literature. It was seen that the publications about Logistics 4.0 and Industry 4.0 were published especially in 2017 (5) and 2018 (8) years. In this respect, it is seen that Logistics 4.0 is a new field of study (Fig. 3.4).

In the meta-analysis, the relationship between industry 4.0 components and SCM 4.0 has been investigated as well. In this context, there are 23 publications related to SCM 4.0 in the dataset. In terms of supply chain management, it has been seen in Fig. 3.4 that IoT, CPS, big data, and smart factory technologies are commonly used

Table 3.1 Key technologies for industry 4.0

Key technologies	Number of related Publications (This Analysis) 166 Pubs. (%)	Herman et al. (2015) 51 Pubs (%)
Internet of Things	113–68	36–70
Cyber-Physical Systems	104–62	46–90
Smart Factory	59–36	24–47
Big Data	41–25	7–14
Smart Products	20–12	10–20
Internet of Services	19–11	19–37
Machine-To-Machine Comms (M2M)	12–7	8–16

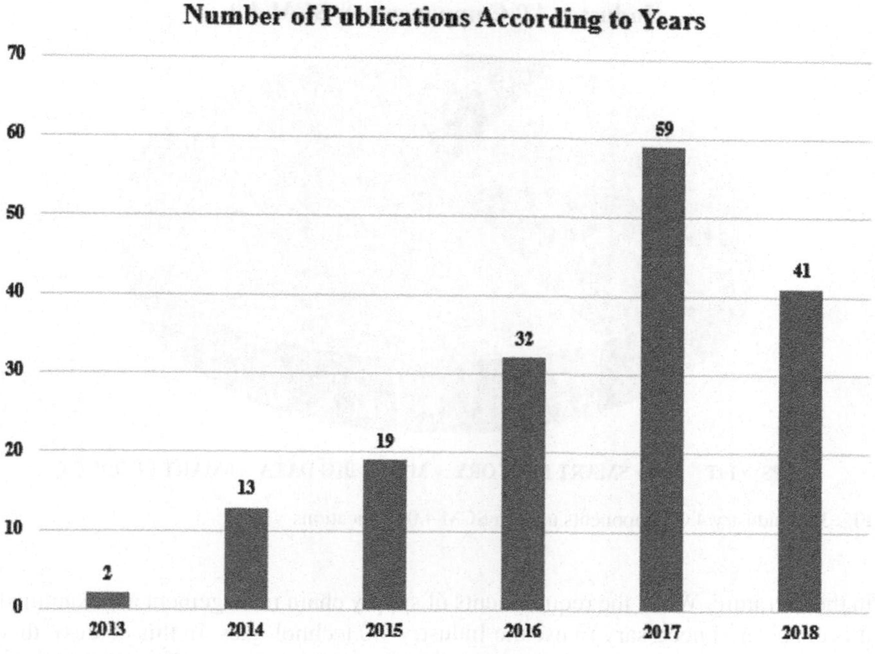

Fig. 3.3 Number of publications related to industry 4.0 between 2013–2018

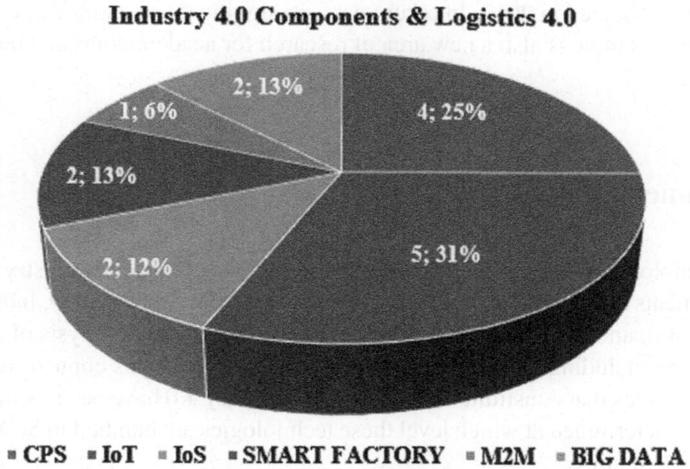

Fig. 3.4 Industry 4.0 components used in logistics 4.0 applications

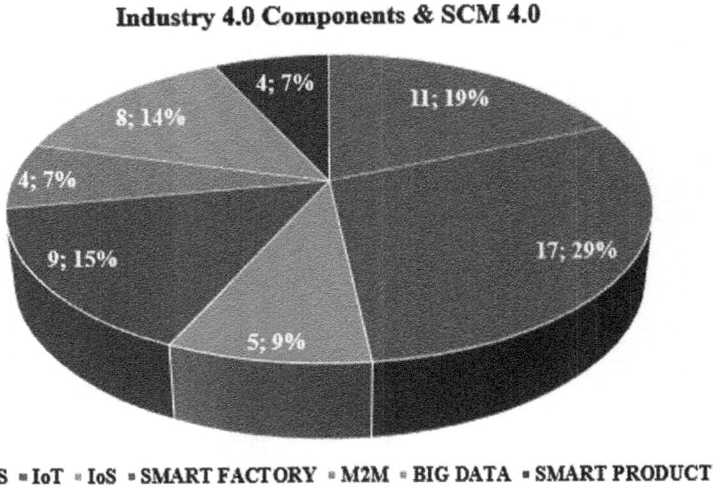

Fig. 3.5 Industry 4.0 components used in SCM 4.0 applications

in the literature. When the requirements of supply chain management are examined, it is considered necessary to use the Industry 4.0 technologies. In this context, their is a strong relationship between Industry 4.0 technlogies and SCM 4.0 When the studies on SCM 4.0 were evaluated according to years, it has been observed that the study was conducted in 2017. In other years, no significant research was conducted. Therefore, this topic is also a new area of research for academicians and researchers (Fig. 3.5).

3.6 Conclusion

New technologies emerged with the development of the concept of Industry 4.0 led to improvements in logistics and supply chain management. In this study, Industry 4.0, Logistics 4.0, and SCM 4.0 topics were discussed and a meta-analysis of academic publications including these topics has been carried out. In this context, first of all, the technologies that constitute the basis of the Industry 4.0 have been explained and it has been determined at which level these technologies are handled in SCM 4.0 and logistics 4.0 studies.

As a result of the meta-analysis study, technologies such as IoT, CPS, smart factory, big data, and smart product have been at the forefront in the Industry 4.0. On the other hand, it is seen that many different technologies such as cloud computing, artificial intelligence, and data mining have been used in the researches examined. However, in the meta-analysis, it was determined that the concepts of IoT, CPS, smart factory, big data, and smart product were discussed more in the publications in the dataset.

In the area of Logistics 4.0, it has been found that IoT and CPS technologies were the most discussed topics, while in the SCM 4.0 area, the technologies of CPS, IoT, and smart factory were the most discussed topics examined in the publications. In this study. It has been observed that the applications of Logistics 4.0 and SCM 4.0 were realized with Industry 4.0 technologies. It has been determined that the studies, especially about logistics and supply chain management, which were taken together with Industry 4.0, have been concentrated in 2017 and 2018. This shows that Logistics 4.0 and SCM 4.0 are very new fields of research.

As a result, industry 4.0 approach and technologies have been found to influence logistics and supply chain management practices. The fact that the studies carried out in this area have been done especially in the last 2 years shows that the researcher has a new research field.

References

Abbott D (2014) Applied predictive analytics: principles and techniques for the professional data analyst. Wiley, New York

Addo-Tenkorang R, Petri TH (2016) Big data applications in operations/supply-chain management: a literature review. Comput Ind Eng 101:528–543

Andersson P, Mattsson LG (2015) Service innovations enabled by the internet of things. IMP J 9(1):85–106

Atzori L, Iera A, Morabito G (2010) The internet of things: a survey. Comput Netw 54(15):2787–2805

Barros A, Oberle D (eds) (2012) Handbook of service description. Springer, New York

Chen M, Mao S, Liu Y (2014) Big data: a survey. Mob Netw Appl 19(2):171–209

Galindo LD (2016) The challenges of logistics 4.0 for the supply chain management and the information technology. Master of science in mechanical engineering, norwegian university of science and technology, department of production and quality engineering

Giannetti C, Ransing RS (2016) Risk based uncertainty quantification to improve robustness of manufacturing operations. Comput Ind Eng 101:70–80

Gnimpieba ZDR, Nait-Sidi-Moh A, Durand D, Fortin J (2015) Using internet of things technologies for a collaborative supply chain: application to tracking of pallets and containers. Proc Comput Sci 56:550–557

Hermann M, Pentek T, Otto B (2016) Design principles for industrie 4.0 scenarios. In: 2016 49th Hawaii international conference on system sciences (HICSS). IEEE, pp 3928–3937

Hofmann E, Rüsch M (2017) Industry 4.0 and the current status as well as future prospects on logistics. Comput Ind 89:23–34

Ivanov D, Dolgui A, Sokolov B, Werner F, Ivanova M (2016) A dynamic model and an algorithm for short-term supply chain scheduling in the smart factory industry 4.0. Int J Prod Res 54(2):386–402

Kagermann H (2015) Change through digitization—value creation in the age of industry 4.0. In: Management of permanent change. Springer Gabler, Wiesbaden, pp 23–45

Kang HS, Lee JY, Choi S, Kim H, Park JH, Son JY, Do Noh S (2016) Smart manufacturing: Past research, present findings, and future directions. Int J Precis Eng Manuf-Green Technol 3(1):111-128

Lasi H, Fettke P, Kemper HG, Feld T, Hoffmann M (2014) Industry 4.0. Bus Inf Syst Eng 6(4):239–242

Lee EA (2008) Cyber physical systems: design challenges. In: 11th IEEE symposium on object oriented real-time distributed computing (ISORC), pp 363–369

Loskyll M, Heck I, Schlick J, Schwarz M (2012) Context-based orchestration for control of resource-efficient manufacturing processes. Future Internet 4(3):737–761

Lou P, Liu Q, Zhou Z, Wang H (2011) Agile supply chain management over the internet of things. In: 2011 International conference on management and service science, pp 1–4

Lu Y (2017) Industry 4.0: a survey on technologies, applications and open research issues. J Ind Inf Integr 6:1–10

MacDougall W (2014) Industrie 4.0: smart manufacturing for the future (Germany Trade & Invest)

Mario H, Tobias P, Boris O (2015) Design principles for Industrie 4.0 scenarios: a literature review. Technische Universität Dortmund, working paper, (01)

Pisching MA, Junqueira F, Santos Filho DJ, Miyagi PE (2015) Service composition in the cloud-based manufacturing focused on the industry 4.0. In: Doctoral conference on computing, electrical and industrial systems. Springer, Cham, pp 65–72

Qin J, Liu Y, Grosvenor R (2016) A categorical framework of manufacturing for industry 4.0 and beyond. Procedia Cirp 52:173–178

Rüßmann M, Lorenz M, Gerbert P, Waldner M, Justus J, Engel P, Harnisch M (2015) Industry 4.0: the future of productivity and growth in manufacturing industries. In: Boston consulting group, vol 9

Shahbaz M, Gao C, Zhai L, Shahzad F, Hu Y (2019) Investigating the adoption of big data analytics in healthcare: the moderating role of resistance to change. J Big Data 6(1):6

Schmidt R (2013) Industrie 4.0 - revolution oder evolution. Wirtsch. Ostwürtt, pp 4–7

Schmidt R Möhring M (2013). Strategic alignment of cloud-based architectures for big data. In: Proceedings of the 17th IEEE international enterprise distributed object computing conference workshops (EDOCW). Vancouver, Canada, pp 136–143

Schmidt R, Möhring M, Härting RC, Reichstein C, Neumaier P, Jozinović P (2015) Industry 4.0-potentials for creating smart products: empirical research results. In: International conference on business information systems. Springer, Champp, pp 16–27

Schroth C, Janner T (2007) Web 2.0 and SOA: converging concepts enabling the internet of services. Prof IEEE Comput Soc 9(3):36–41

Schuh G, Potente T, Wesch-Potente C, Weber AR, Prote JP (2014) Collaboration mechanisms to increase productivity in the context of industrie 4.0. Procedia CIRP 19:51–56

Thames L, Schaefer D (2016) Software-defined cloud manufacturing for industry 4.0. Procedia CIRP 52:12–17

Timm IJ, Lorig F (2015) Logistics 4.0-A challenge for simulattion. In: 2015 winter simulation conference (WSC). IEEE, pp 3118–3119

Tjahjono B, Esplugues C, Ares E, Pelaez G (2017) What does industry 4.0 mean to supply chain? Procedia Manuf 13:1175–1182

Verdouw CN, Wolfert J, Beulens AJM, Rialland A (2016) Virtualization of food supply chains with the internet of things. J Food Eng 176:128–136

Wahlster W, Grallert HJ, Wess S, Friedrich H, Widenka T (eds) (2014) Towards the internet of services: the THESEUS research program. Springer, Switzerland

Wang S, Wan J, Li D, Zhang C (2016) Implementing smart factory of industrie 4.0: an outlook. Int J Distrib Sens Netw 12(1):3159805

Weyer S, Schmitt M, Ohmer M, Gorecky D (2015) Towards industry 4.0-standardization as the crucial challenge for highly modular, multi-vendor production systems. Ifac-Papersonline 48(3):579–584

Witkowski K (2017) Internet of things, big data, industry 4.0–Innovative solutions in logistics and supply chains management. Procedia Eng 182:763–769

Yıldız A (2018) Endüstri 4.0 ve akıllı fabrikalar. Sakarya Univ J Sci 22(2):548–558

Zamfirescu CB, Pirvu BC, Loskyll M, Zuehlke D (2014) Do not cancel my race with cyber-physical systems. Weyer IFAC Proc Vol 47(3):4346–4351

Zhong RY, Huang GQ, Lan S, Dai QY, Chen X, Zhang T (2015) A big data approach for logistics trajectory discovery from RFID-enabled production data. Int J Prod Econ 165:260–272

Zhong RY, Newman ST, Huang GQ, Lan S (2016) Big Data for supply chain management in the service and manufacturing sectors: challenges, opportunities, and future perspectives. Comput Ind Eng 101:572–591

Xu LD (2011) Information architecture for supply chain quality management. Int J Prod Res 49(1):183–198

Ismail İyigün has been working as a lecturer in Vocational School of Social Sciences at the Trakya University since 2006. His current research areas are marketing, logistics and supply chain management. He graduated from Ph.D. Programme on Business Administration at Gebze Technical University in 2016. His dissertation is on the "Effects of Sales and Logistics Competencies on the Relationship Continuity on Supply Chain and Corporate Governance". He received master degree from Economics, Social Graduate School at the Trakya University. He researched the commodities market of historical process and application of Uzunköprü. He graduated from Economics Department, Marmara University in 1994. Dr. İyigün teaches on supply chain management, physical distribution systems, logistics information systems, selling techniques, and basic marketing courses in Trakya University and Kadir Has University.

Chapter 4
A General View of Big Data and Machine Learning

Özhan Görçün and Hande Küçükönder

Abstract Nowadays, winds of digitalization have continuing blow in the world. No one cannot ignore these kinds of changes and is continuing to cause dramatic changes in our lives. Big data is one of the crucial parts of these changes. Nowadays, almost millions of people have become the main element providing the data flow. Furthermore, they are continuing to perform it voluntarily and willingly. As a result, companies, industries, and other stakeholders have started to encounter a very huge volume of data about their customers. However, most of the data are not be useful because they are not structured. This chapter is organized into two parts. In the first part, the big data concept was explained and its important elements were discussed. In the second stage, machine learning and its elements were presented information on some computational tools used for converting the unstructured data to structured data.

Keywords Big data · Machine learning · ANN · Genetic algorithms · Decision trees

4.1 Introduction: Big Data

Big Data is one of the most important parts of digitalized world and industry 4.0 process. In recent decades it has been crucial not only for industries but also individual users. Big data which has many components such as users, cloud systems, data processors, and so on is a term defining a system hosting a very huge volume of structured and unstructured data. Although it can host a huge volume of data, the most important issue is the quality of these data since most of the unstructured data may not be useful for industries and users. Therefore, converting these kinds of data

Ö. Görçün (✉)
Gelişim University, Istanbul, Turkey

H. Küçükönder
Department of Numerical Methods, Faculty of Economics and Administration, Bartin University, Bartın, Turkey
e-mail: hkucukonder@bartin.edu.tr

İ. İyigün and Ö. F. Görçün (eds.), *Logistics 4.0 and Future of Supply Chains*, Accounting, Finance, Sustainability, Governance & Fraud: Theory and Application,
https://doi.org/10.1007/978-981-16-5644-6_4

to structured data is a very crucial task for experts who are responsible to develop these systems as it is required for forecasting and evaluating when decision-makers take strategic decisions.

Nowadays, using big data has become very important for industries, companies, and other stakeholders to survive in a highly competitive environment. These actors want to know how the customers are thinking as well as what is important for them. As a very useful tool, big data has the potential to provide wider information about customers even their special days and daily activities, pleasures, and so on. Moreover, there are many users who provide data both about themselves and about many things for this system and it continues to increases each passing day at the present. Therefore, the data volume which flows to the system continuing to ever-increased. Today, it is possible to see the elements of the information society in all areas of life. Most people now have a smartphone in their pocket, most people have a computer at home and units that manage information technology in the back offices of all companies. However, the information itself is not so visible. However, only half a century after computers entered human life, the amount of information began to be collected in such a way as to gain a meaningful and special quality. Today, not only the amount of information has increased, but also the speed of access to information has increased. The quantitative change brought with its qualitative change. The collection of data to form a meaningful whole took place first in the field of astronomy and genetics. The concept of big data was first used in these areas, and then this concept started to be used for each area. Big data has begun to show itself in all areas of our lives. For example; an Internet search engine from Google, we find big data in every area from diagnosis and treatment of diseases to online shopping. (Mayer-et al. 2013).

According to some reports on this issue, 90% of the data used in the world has been created only in the last two years, as the information has rapidly become obsolete and technology has changed (Url1,2020). On the other hand, there are some disadvantages of big data. It is strongly expected that consisting of very large and complex datasets that are difficult to process using existing database management tools or traditional data processing applications; it causes difficulties in data acquisition, improvement, storage, searching, sharing, transfer, analysis, and visualization stages (Yılmaz et al. 2017).

As a result, although some disadvantages, Big data is started use in various fields of the economy in recent years. According to Sağıroğlu and Koç (2017), it is used in almost all industries from health to logistics. Big data consists of five components defined as 5 V and the 5 V are presented in Fig. 4.1 (Url 2,2020).

4.1.1 Volume

The volume of data is directly proportional to the size of the data and is generally expressed in various data measurement units such as gigabytes, terabytes, and bytes. as well as the quantity of generated and stored data, the size of the data determines the value and potential insight, and whether it can be considered big data or not.

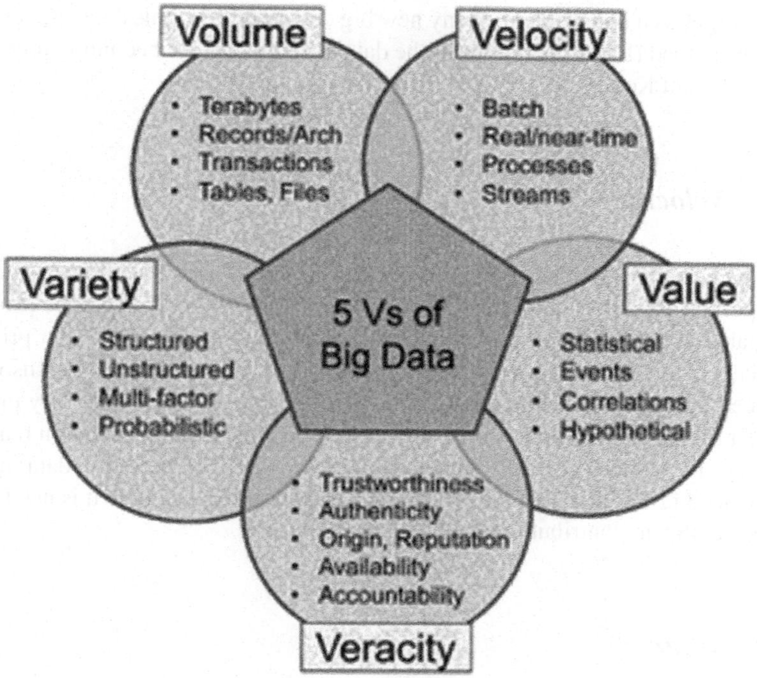

Fig. 4.1 5 V of big data system (Url 2 2020)

However, the storage of high amounts of data is among the issues that need attention. The said increases, new data processing, storage, integration, and archiving makes use of technologies necessary. But new like cloud technology with the proliferation of technologies, the storage of data can be easily and data can be used at any time (Dülger 2015).

4.1.2 Variety

The type and nature of the data help people who analyze it to effectively use the resulting insight. Not only from text, but Big data can also be drawn from images, audio, video, experts, and sources related to this issue state that the data may be either structured or unstructured or semi-structured and unfortunately, these three data types have the possibility of being replaced and only 20 percent of them are structured (Dülger 2015). The diversity of data is due to the diversity of data sources. The produced data can be seen in structural, semi-structural, or non-structural formats. Data in relational databases to structural data, semi-structural data can be given as XML, and JSON format data as non-structural data, as well as audio, video, text files. If the data is used together in these different structures, this diversity in the data emerges as a big problem. Many new big data specific to this data, especially in the convert-load (ETL) operations of the data technologies are becoming imperative (Sağıroğlu and Koç 2017).

4.1.3 Velocity

The data is in constant motion. In this context, the analysis of data flow has become an important issue for data scientists (Cyganek et al. 2016). Data can be produced at variable speeds. While large volumes of static data create a big data problem, especially phone or IoT devices that we frequently use in daily life, sensor data produced by various machines and similar data sources produce data very quickly. The fact that the data produced in a flow can be analyzed and managed in real-time is another problem of big data. (Sağıroğlu and Koç 2017). Therefore, data that can be produced faster increases the processing speed of the data where it is needed and makes significant contributions to the diversity of data.

4.1.4 Value

Analysis of the developments and innovations that may occur in the coming period, preparing for the changes that will occur, and making decisions regarding the results will be possible thanks to the value provided by the big data. Data that does not produce value has no value. Data is as valuable as the value it creates. (Venkatram and Geetha, 2017). Data that is not quickly grasped is only trash in the real business world. If they serve no purpose, they are digital waste (Spann 2017).

4.1.5 Validity

Analyzes to be made on data that cannot be confirmed for accuracy reduce the real value. There are technologies that can measure the validity of big data and the accuracy of the data. Existing data to the above properties if it has, this data is traditional value it is very difficult to produce. So, this data has a big data perspective the need to be handled and analyzed with big data technologies has arisen (Sağıroğlu and Koç 2017).

4.2 Big Data and Machine Learning

As mentioned above, without data processing processes, big data is a non-functional system. Converting data from unstructured to structured is a crucial and critical task for being able to make comprehensive analyses. For this purpose, there are many tools used to obtain meaningful information and data. Furthermore, these techniques, i.e., Artificial Neural Networks, Genetic Algorithms, and so on are ever developed each passing day, and new techniques are presented to the industries and scientific world.

4.2.1 Machine Learning

Computer technology developing in recent years has brought along collecting and processing the very huge and various types of data. As a result, big data has been very crucial for a new technological world. By contrary, the very important part of data which flow into the systems is not structured data and decision-makers who are in all side of lives such as business, public authorities and so on need to structured data to make right and suitable decisions. Therefore, they require a systematic and applicable tool for obtaining meaningful and applicable data. Based on these requirements the machine learning has become a very strong and popular research area.

In 1959, machine learning suggested by Arthur Samuel is one of the parts of artificial intelligence (Wei et al. 2019). Machine learning has a very crucial role in solving very complicated problems includes modeling and computer-based algorithms which can self-learning based on information, can make self-assessment, and embracing the patterns (Atalay and Çelik 2017). The machine learning consists of two parts. First is to solve a problem with the help of an algorithm and self-learning the model, making meaningful evaluations based on this self-learning (Şahinarslan 2019). Within this perspective, in order to solve problems, which require some functions such as forecasting, classification, and grouping, using some novel machine learning techniques (i.e., Decision trees, ANN, Bayesian networks, genetic algorithms, k the nearest neighbor algorithm (KNN)) have become common techniques as alternatives to the traditional techniques. The main procedure of the machine learning which has very strong relations with some fields such as artificial intelligence, data mining, pattern recognition, and bioinformatics can be demonstrated in six steps as shown in Fig. 4.2 (Wei et al. 2019; Zhang et al. 2020).

As is seen in Fig. 4.2, the first step of the process starts with collecting appropriate data from databases. In this step, the selection of the proper dataset is crucial and critical task for decision-makers considering the type of data, data size, quality, and form (Wei et al. 2019). Therefore, it is required to data transfer from appropriate databases by users for using more qualified data. In addition to that, the form of data can be prepared in different types considering the algorithm used in this process. The next step called future engineering includes operations of extrapolating the suitable features from raw data. Also, remain step of this process consists of some

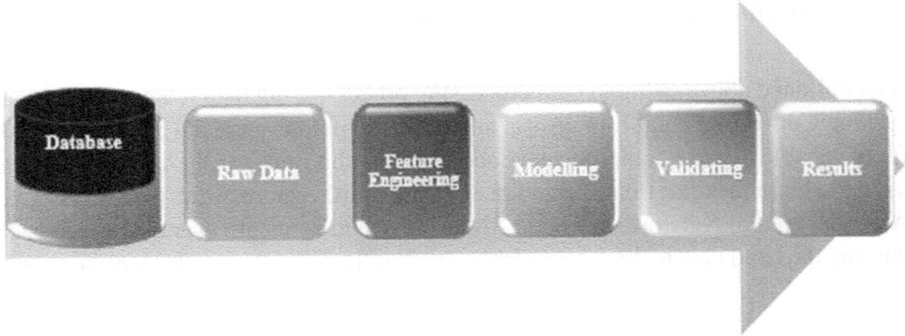

Fig. 4.2 The main procedure of machine learning approaches (Wei et al. 2019)

applications such as the selection of the proper algorithm, forming test and validation sets, modeling, and examination of validation of forecasting (Wei et al. 2019; Zhang et al. 2020). Some fields such as engineering approaches, cognitive simulation, and theoretical analysis (Şahan 2020) have been effective in developing and forming process of the machine learning which uses in many fields (i.e., medical diagnosis, education, biology, engineering, and so on.).

4.2.2 Artificial Neural Network (ANN)

Artificial Neural Network (ANN) is related to a biological neural structure of humans and it is a mathematical model which is developed by inspiring the working principles of the human brain and is effective for examining the complicate and nonlinear relations (Mjalli et al. 2007). There are four different components such as dendrite, soma, axon, and snaps. When their functions are examined briefly, snaps are spaces providing transmissions electric signals among nerve cells. Soma is the section that is to collect and process these signals. By processing these signals, nerve cells form their own signals and these formed signals are transmitted by axons to the dendrites having a view as tree branches. Also, dendrites have functions to transmit these signals to the snaps and data transfer between two cells can be provided by neurotransmitters which are in snaps (Öztemel 2003).

Artificial neural cells (process) exist in the artificial neural networks (ANN) similar to the biological nerve cells. Each process element includes five different elements such as inputs, weights, the sum function, activation function, and outputs (Aladağ 2019). Inputs collect information and data from outside and the significance and impacts of the information are determined with the help of weights. The sum function provides to compute the net input value. Using the weighting sum is a common approach for purposing to calculate the net input transmitted to the network (each input multiplied by their own weight and calculate the sum of these values) the function of activation symbolizes the functions providing to produce outputs by processing the inputs (Öztemel 2003).

Artificial Neural Network (ANN) formed many artificial neural cells, consists of many processing elements connected parallelly to each other (Maind and Wankar 2014). There are different three layers such as the input layer, interlayers, and output layers in a neural network whose processor is an artificial neuron. While the section that entering data related to the problem is the input layer, the section processing these data is a hidden layer. The section that transforms the processing information to the outputs is defined as the output layer (Ocak and Şeker 2013; Özşahin and Singer 2019). In Fig. 4.3, an ANN example, which shows these layers as is presented as diagrammatic (Maind and Wankar 2014).

ANN is a very useful tool applied to solve very complicated and nonlinear problems. It has the abilities of self-learning and generalization by considering the data and information without any theoretical assumptions or a rule base (Mjalli et al. 2007). Also, determining the fitting function is not required and it can be accepted as another advantage of the ANN. While some methods used for solving the problem provide a range of parameters that are useful for understanding and evaluating the problem, ANN cannot assess the weights of the connection for the present. Because of that, modeling of ANN is defined as a closed black box (Mjalli et al. 2007; Maind and Wankar 2014).

4.2.3 Genetic Algorithms (GA)

Genetic Algorithms (GA) is a search technique based on natural selection principles and transferring the biological and genetic development processes of lives to the computer environment (Nabiyev 2012). GA described under the mainframe of evolutional algorithms was developed by inspiring the evolution theory of Darwin. Genetic Algorithms, which was firstly introduced by J.H. Holland who are a phycologist and expert of computer sciences, is an intuitional and randomized method (Emel and Taşkın 2002; Nabiyev 2012).

GA can provide effective solutions to real-life problems has been commonly used in the literature for solving the various combinatorial optimization problems. These works can be summarized as vehicle routing (Aydemir and Karagül 2020; Ghannadpour and Zandiyeh 2020), project scheduling (Özköse and Gencer 2019) transportation (Yakıcı Ayan 2008), traveling-salesman problem (Şahin and Karagül 2019), location problems (Gülsün et al. 2009). GA has some differences than other optimization algorithms and these differences can be described as (İşçi and Korukoğlu 2003): GA deal with coded parameters set instead of single parameter and it performs to start making a search with a group of points instead of a single point. Another difference is to focus on the target function directly and it uses the stochastic passage rules. An example of a basic search procedure of the GA is presented in Fig. 4.4 (Gholamia et al. 2014).

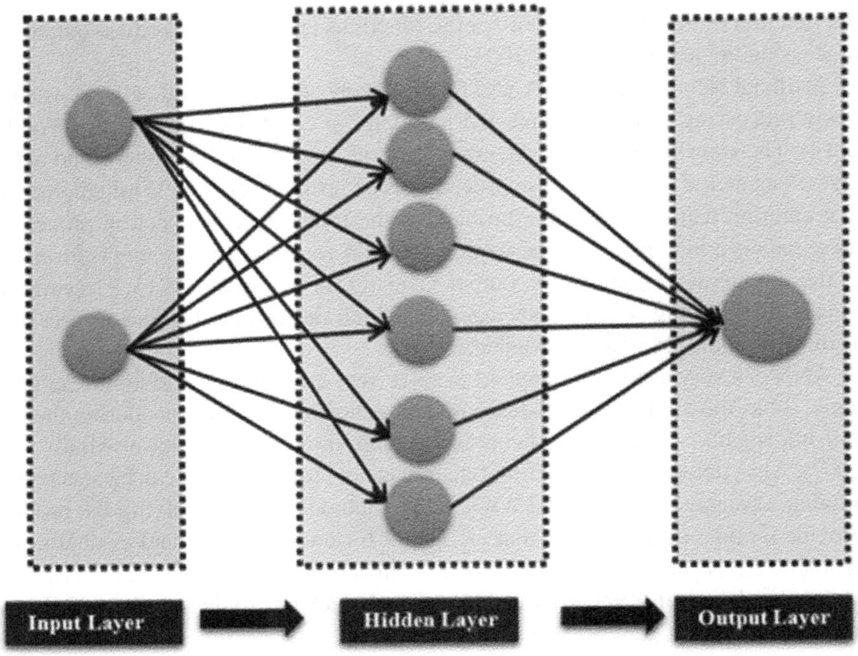

Fig. 4.3 An example of ANN (Maind and Wankar 2014)

When the operational procedures of GA given in Fig. 4.4 are examined, the operation of genetic representation of the problem which is handled is carried out with the help of genetic concepts (genes and chromosomes). After the process of coding is completed, first population examples are produced by forming an initial population for the initial implementation step. Each individual is individual has chromosomes set and the initial population can be produced in different types by using a randomized or special algorithm. In the next step, the target function is determined for individuals in the populations, and the implementation of genetic operations and the selection of the individuals are performed. In this stage, operators of reproduction, crossing, and mutation are used (Gümüşoğlu et al. 2013; Qureshi et al. 2006). The performances of the individuals in the new population after their validation values are calculated based on the validation function. These operations are continued until reach to finalization criterion determined at the beginning and when iteration is reached to this criterion the optimal solution is presented by finalizing the algorithm (Emel and Taşkın 2002; İşçi and Korukoğlu 2003; Gholamia et al. 2014).

4.2.4 Decision Trees (DT)

Decision trees are a diagram, which is viewed as a tree and repetitive diversification of a dataset to sub-groups by classifying (dichotomous classifications) (Namazkhan et al.2020). Understanding and evaluating the learning rules (if–then rules) in decision trees are very easy. In addition to that, costs are very low and integration between databases is possible and it has not non-flexible assumptions (Çalış Boyacı et al. 2014). Decision trees, which can self-learning from data with the help of the method of generalization, were introduced by J. Ross Quinlan in year 1960 (Wei et al.. 2019). In addition, algorithms of the decision trees which are accepted as a data mining technique are an effective tool for presenting complex problems in graphical form by means understandable and lean. The decision trees that have a definitive structure (continuously or categoric) are defined as classifying tree, others which can be used for solving the regression types of problems are regression tree (Alan, 2014). Within this perspective, a decision tree consists of three sections such as node, branch, and leaf. An example of a decision tree is presented in Fig. 4.5 (Aytekin et al. 2018).

When Fig. 4.5 is examined, nodes represent each self-feature. While the root is on the top, the leaf is below the tree. Branches are between roots and leaves. When a decision tree is built, the process starts with the classification of data that is in the root and asking the questions. This respective operation continues until obtaining nodes having no branches or obtaining the leaves. When any dividing is out of the question, the algorithmic process is finalized. Whether gain the generalization ability by the obtained decision tree is determined by using test data. In this process, a new dataset is used and test data is tested in the root section of the tree. Then it is progressed to the next node. This process is finished when it is reached to the leaf of the tree (Kavzoğlu and Çölkesen, 2010; Aytekin et al. 2018).

In this context, some algorithms (i.e., CHAID, CART, ID3, C4.5, and so on) are used in decision trees (Alan 2014; Aytekin et al. 2018). Obtaining the optimum

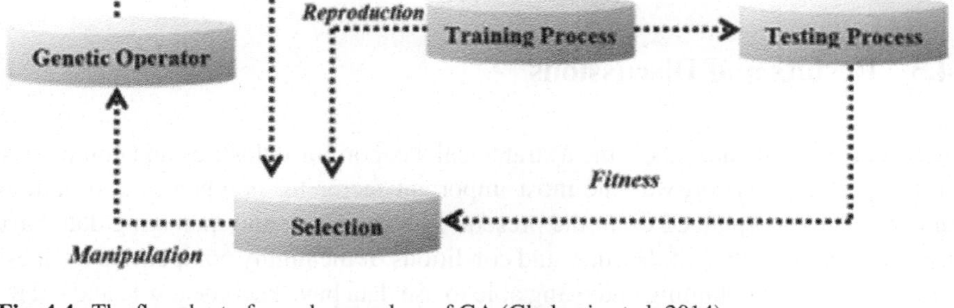

Fig. 4.4 The flowchart of search procedure of GA (Gholamia et al. 2014)

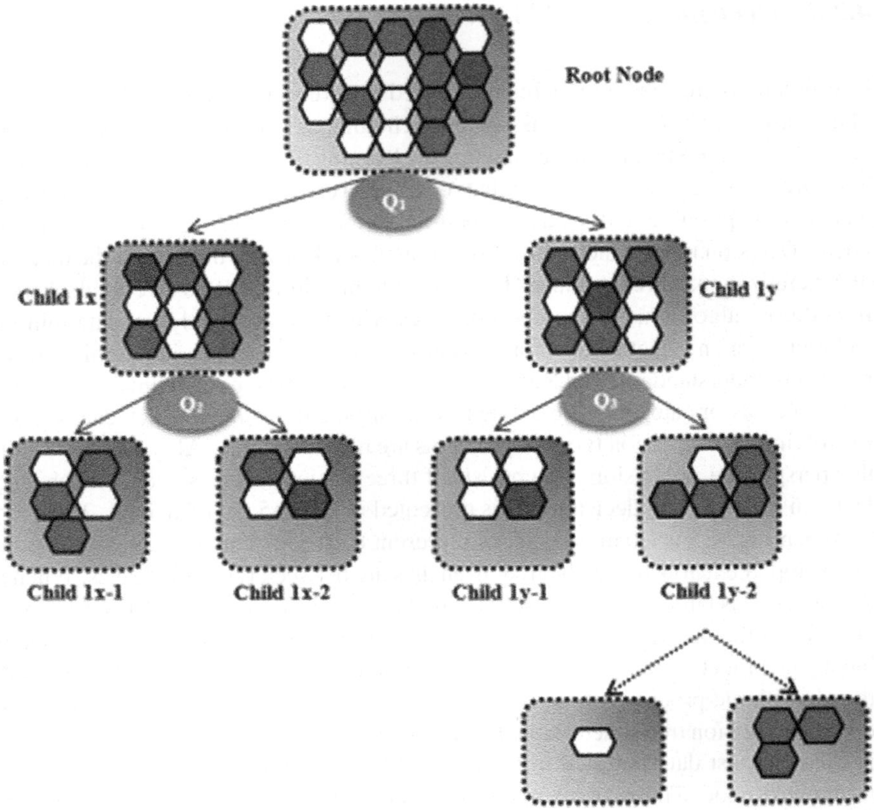

Fig. 4.5 A basic decision trees (Aytekin et al., 2018)

structure of the tree which can help to reduce to a minimum of generalization faults is one of the main aims in this process. Therefore, forming a small-sized tree can help to obtain more successful results. some measurements suggested in the literature are commonly used. Information gaining, ki-square test, entropy, and Gini index are examples (Kavzoğlu and Çölkesen 2010).

4.3 Results and Discussions

In recent years, data has become a strategical weapon for industries and companies. in the past century, oil was the most important factor for companies and nations and data have supplanted oil at the present. Actors having and managing data have become determinative of the rules and conditions of the highly competitive business environment and obtaining and being able to use data have become a sole and exclusive remedy to survive in the highly competitive for companies and other actors. As the competition hardens in global markets, businesses learn the way to control the increasing data flow, learn about big data processes, and transform the most appropriate data into information by filtering infinitely large data. Simultaneous tracking of what the manufacturer will produce and what the customer will demand over big data will be managed by big data systems. As a result, big data has increased to a size that cannot be managed with traditional processes right now. therefore, mathematical models and computational tools and may be needed to solve this problem for companies and industries to solve these kinds of problems. Finally, we are going through a new era, as one of the fundamental parts of the Big Data Industry 4.0 process, it is clearly seen that it will be the fundamental building block of digital businesses and the digital economy in the coming years.

References

Alan MA (2014) Karar Ağaçlarıyla Öğrenci Verilerinin Sınıflandırılması. Atatürk Üniversitesi İktisadi Ve İdari Bilimler Dergisi 28(4):101–112

Aladağ ÇH (2019) Architecture selection in neural networks by statistical and machine learning. Orient J Comput Sci Technol 12(3):76–89

Atalay M, Çelik E (2017) Büyük Veri Analizinde Yapay Zekâ ve Makine Öğrenmesi Uygulamaları. Mehmet Akif Ersoy Üniversitesi Sosyal Bilimler Enstitüsü Dergisi 9(22):155–172

Aydemir E, Karagül K (2020) Solving a periodic capacitated vehicle routing problem using simulated annealing algorithm for a manufacturing company. Braz J Oper Prod Manag 17(1):1–13

Aytekin Ç, Sütcü CS, Özfidan U (2018) Text classification via decision trees algorithm customer comments case. J Int Soc Res 11(55):782–792

Cyganek B, Graña M, Krawczyk B, Kasprzak A, Porwik P, Walkowiak K, Woźniak M (2016) A survey of big data issues in electronic health record analysis. Appl Artif Intell 30(6):497–520

Çalış Boyacı A, Kayapınar Kaya S, Çetinyokuş T (2014) Veri Madenciliğinde Karar Ağacı Algoritmaları ile Bilgisayar ve İnternet Güvenliği Üzerine Bir Uygulama. Endüstri Mühendisliği Dergisi 25(3–4):2–19

Dülger Ü (2015) Stratejik Büyük Veri Yönetiminin Yatırımlar Üzerindeki Etkileri. Yayınlanmamış Yüksek Lisans Tezi, İstanbul Üniversitesi Fen Bilimleri Enstitüsü

Emel AG, Taşkın Ç (2002) Genetik Algoritmalar ve Uygulama Alanları. Uludağ Üniversitesi İİBF Dergisi 21(1):129–152

Ghannadpour SF, Zandiyeh F. (2020) An adapted multi-objective genetic algorithm for solving the cash in transit vehicle routing problem with vulnerability estimation for risk quantification. Engineering applications of artificial intelligence, vol 96

Gholamia R, Moradzadehb A, Malekic S, Amiric S, Hanachid J (2014) Applications of artificial intelligence methods in prediction of permeability in hydrocarbon reservoirs. J Petrol Sci Eng 122(2014):643–656

Gülsün B, Tuzkaya G, Duman C (2009) Genetik Algoritmalar ile Tesis Yerleşimi Tasarımı ve Bir Uygulama. Doğuş Üniversitesi Dergisi 10(1):73–87

Gümüşoğlu Ş, Erboy N, Özdağoğlu G (2013) Siparişe Dayalı Üretim için Ürün Gruplarının Oluşturulmasında Genetik Algoritma Tabanlı Bir Yaklaşım. Yönetim Ve Ekonomi Dergisi 20(2):259–284

İşçi Ö, Korukoğlu S (2003) Genetik Algoritma Yaklaşımı ve Yöneylem Araştırmasında Bir Uygulama. Yönetim Ve Ekonomi 10(2):191–208

Kavzoğlu T, Çölkesen İ (2010) Karar ağaçları ile Uydu Görüntülerinin Sınıflandırılması Kocaeli Örneği. Harita Teknolojileri Elektronik Dergisi 2(1):36–45

Maind SB, Wankar P (2014) Research paper on basic of artificial neural network. Int J Recent Innov Trends Comput Commun 2(1):96–100. ISSN: 2321-8169
Mayer-Schonberger V, Cukier K (2013) Big data: a revolution that will transform how we live, work and think. Houghton Mifflin Harcourt, Boston, Massachusetts.
Mjalli FS, Al-Asheh S, Alfadala HE (2007) Use of artificial neural network black-box modeling for the prediction of wastewater treatment plants performance. J Environ Manag 83(3):329–338
Nabiyev VV (2012) Yapay Zekâ. Ankara: Seçkin Yayıncılık
Namazkhan M, Albers C, Steg L (2020) A decision tree method for explaining household gas consumption: the role of building characteristics, socio-demographic variables, psychological factors and household behaviour. Renew Sustain Energy Rev 119:109542
Ocak İ, Şeker ŞE (2013) Calculation of surface settlements caused by EPBM tunneling using artificial neural network SVM and Gaussian processes. Environ Earth Sci 70(3):1263–1276
Özköse H, Gencer C (2019) Proje Planlama ve Çizelgelemede Genetik Algoritma Tabanlı Bir Yöntem ile Kritik Yolun Proje Tamamlanma Zamanının tespiti ve Zaman Maliyet Analizi. Bartın Üniversitesi İktisadi Ve İdari Bilimler Fakültesi Dergisi 10(20):278–300
Özşahin Ş, Singer H (2019) Odun Yüzey Pürüzlülüğü Tahmininde Bir Yapay Sinir Ağı Modelinin Kullanılması. Düzce Üniversitesi Bilim Ve Teknoloji Dergisi 7(3):1764–1777
Öztemel E (2003) Yapay Sinir Ağları. İstanbul: Papatya Yayıncılık
Qureshi SA, Mirza SM, Arif M (2006) Fitness function evaluation for image reconstruction using binary genetic algorithm for parallel ray transmission tomography, emerging technologies. In: ICET'06 international conference. Islamabad, Pakistan, pp 196–201
Sağıroğlu Ş, Koç, O(eds) (2017) Büyük Veri ve Açık Veri Analitiği, Yöntemler ve Uygulamalar. Ankara: Grafiker Yayınevi
Spann T (2017) The physics of big data. Erişim: 22.05.2020. https://dzone.com/articles/the-physics-of-big-data/
Şahan AN (2020) Stratejik Yönetim Perspektifinden Sigortacılık Sektöründe Makine Öğrenmesi Algoritmaları ile Anomali tespiti. Yüksek Lisans Tezi, İstanbul Teknik Üniversitesi Fen Bilimleri Enstitüsü
Şahin Y, Karagül K (2019) Gezgin Satıcı Probleminin Melez Akışkan Genetik Algoritma MAGA Kullanarak Çözümü. Pamukkale Üniversitesi Mühendislik Bilimleri Dergisi 25(1):106–114
Şahinarslan FV (2019) Makine Öğrenmesi Algoritmaları ile Nüfus Tahmini: Türkiye Örneği. Yüksek Lisans Tezi, İstanbul Teknik Üniversitesi Sosyal Bilimler Enstitüsü.
URL 1: http://www.accaglobal.com. Access Date: 22.05.2020
URL 2: https://andressilvaa.tumblr.com/post/87206443764/big-data-refers-to-5vs-volume. Access Date: 22.05.2020
Venkatram K, Geetha MA (2017) Review on big data & analytics–concepts, philosophy, process, and applications. Cybernetics Inform Technol 17(2):3–27.
Wei J, Chu X, Sun X-Y, Xu K, Deng H.-X, Chen J, Wei Z, Lei M (2019) Machine learning in materials science. InfoMat 1:338–358. https://doi.org/10.1002/inf2.12028
Yakıcı Ayan T (2008) Sabit Maliyetli Ulaştırma Problemi için Bir Genetik Algoritma. Gazi Üniversitesi İktisadi Ve İdari Bilimler Fakültesi Dergisi 10(1):97–116
Yılmaz B, Bülbül S, Atik M (2017) Büyük Verinin Big Data Muhasebe Üzerindeki Etkisi ve Muhasebeye Sağladığı Katkıların İncelenmesi. Kara Harp Okulu Bilim Dergisi 27(1):79–112
Zhang L, He M, Shao S (2020) Machine learning for halide perovskite materials. Nano Energy 78:105380

Özhan Görçün graduated from Department of Public Administration at Abant İzzet Baysal University in 2002, Logistics and Supply Chain Management Department at Institute of Science in 2009. In 2015, Özhan started Doctorate Program on Business Administration in Istanbul Aydın University. He started his business career in the logistics industry and also taught in the Logistics program at Bahçeşehir University. He is head of Logistics Program of Adıgüzel Vocational

School. He published and contributed to many national and international books, articles and notifications published.

Hande Kucukonder (Ph.D.) is a Vice-Dean of Economics and Administrative Sciences at the Bartin University. She gives many courses in statistical data analysis, Artificial Neural Networks (ANN), Data Mining, Decision-Support Systems and so on. in addition to those, she studies about numerical analysis, statistics, multi-criteria decision-making, and data mining. She got a master's degree in the field of Biometry and Genetics and completed doctoral studies about the same area. she has numerous scientific studies about Artificial Neural Networks (ANN), Data Mining, Multi-criteria decision-making methods, statistical data analysis.

Chapter 5
3D Printing and Logistics

Ayşegül N. Bayraktar

Abstract 3D printing is a production process. 3D printing takes place among the techniques known as additive manufacturing and is used in the market as synonymous with additive manufacturing, and even commonly 3D printing is used instead. When there are three elements as digital model, feed material and 3D printer, then the needed product can be produced at any time and at any place by 3D printing. This technique has changed the principles of traditional production as producing the parts separately and then assembling. In other words, the traditional way of business will have been transformed to a new style. 3D printing is a disruptive and a growing trend in the business. It is assumed that 3D printing will cause and economic effect as 230$ billion to $550 billion per year by 2025. Also the increase in enthusiasm of companies in trying and using of this technology supports this assumed economic effect. Worldwide trademarks as Mercedez Benz, BMW, IKEA, Ford have shared their experiences, productions and savings. Experiences with 3D printing have shown that its usage is beneficial. Some of the benefits gained by 3D printing can be defined as shorter lead time, cheap production, fewer stocks, elimination of assembling requirement, acceleration of design process, etc. Shorter lead time, fewer stocks and elimination of flow for assembling requirements point out the change in supply chain and logistics market. With 3D printing, it is possible to produce closer to the customer by minimizing inventory, transportation and handling costs. As a service provider, logistics companies should review their service infrastructure in line with the needs of their customers who will adopt these strategies.

Keywords 3D printing · 3D printers · Supply chain management · Logistics · Inventory · Warehouse · Transportation · Raw material · Digital files

---

A. N. Bayraktar (✉)
Beykoz Üniversitesi, Istanbul, Turkey
e-mail: aysegulnuriyebayraktar@beykoz.edu.tr

İ. İyigün and Ö. F. Görçün (eds.), *Logistics 4.0 and Future of Supply Chains*, Accounting, Finance, Sustainability, Governance & Fraud: Theory and Application,
https://doi.org/10.1007/978-981-16-5644-6_5

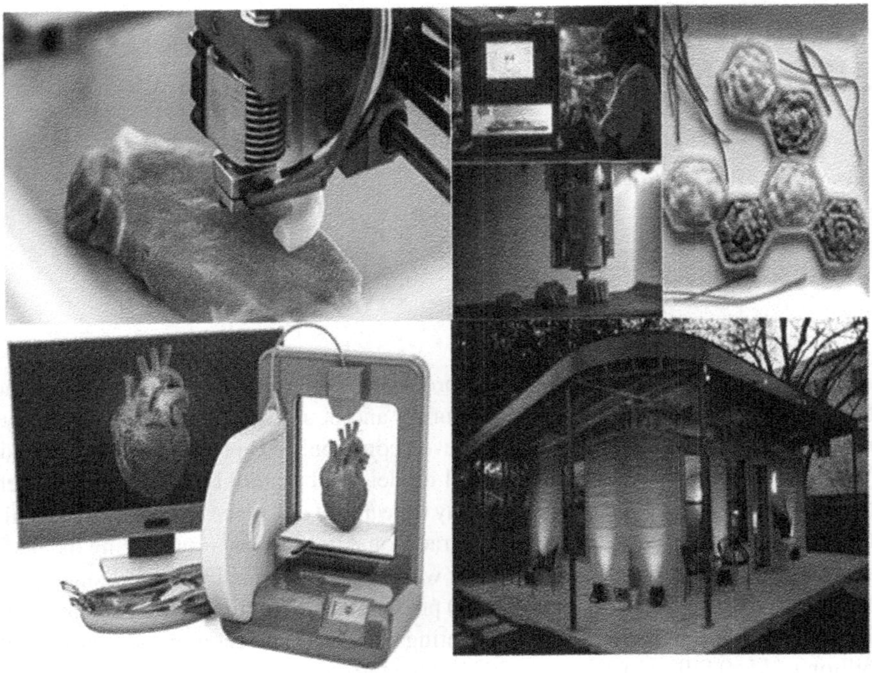

Fig. 5.1 3D printed products (Ho 2020; Carlota 2019; Danova 2019; Bendix 2019)

5.1 Introduction

When you need a product, your action is to purchase a digital CAD file, upload it to your printer at home and wait for the product you ordered to be printed for you. In a such world, you won't have to leave your house to go to any store or need anyone to deliver the product to you (Durach et al. 2017). Isn't it a beautiful dream? Not anymore. With the entrance of 3D printing to production area, science fiction thought related to 3D printing, quickly turned into a scientific reality. The technology that provides this printing is named as additive manufacturing, which is mostly known as 3D printing. Many sectors for example automotive, aerospace, health, engineering, etc., realized that 3D printing is an opportunity to do things differently (DHL 2016). Examples of 3D printed products can be seen in Fig. 5.1.

5.2 3D Printing

Although 3D printing takes place among the techniques known as additive manufacturing, is used in the market as synonymous with additive manufacturing, and even commonly 3D printing is used instead of additive manufacturing. Additive processes

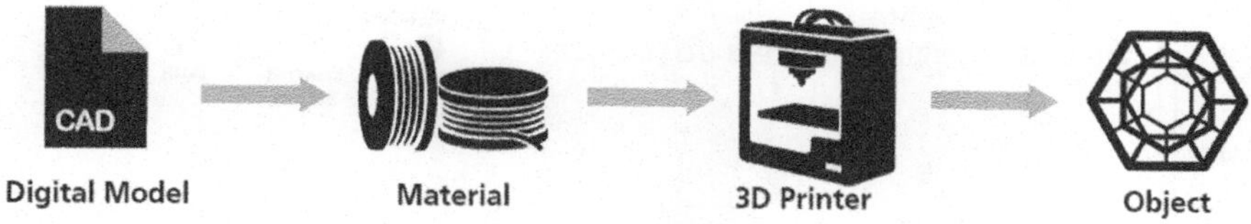

Fig. 5.2 Materials for 3D object formation (DHL 2016, p. 5)

make objects layer by layer rather than through traditional techniques as molding or subtractive techniques. For example, a computer mouse is produced by combining many different parts. In other words, with traditional production, parts are produced separately and then assembled. In 3D printing, the computer mouse is obtained as a whole printout by printing layer by layer from 3D printer (McKinsey Global Institute 2013; Leering 2017). 3D printing is a manufacturing process.

In a broad description, 3D technologies can be classified as virtual and physical. Computational models are the virtual part, technologies that produce real objects from virtual models are the physical part. To manufacture a product by using a layer by layer model, this is known as rapid prototyping and latterly named as 3D printing by media. With this technology, a prototype of an object can be rapidly and automatically created. Most industries used this to accelerate the time in the creation of new products (Silva and Rezende 2013). 3D printing's base is to build up layers of material by applying a computer-aided design. Each layer is printed until a three-dimensional object is composed; therefore, this is ascribed as an additive process (Manners-Bell and Lyon 2012). Additive manufacturing, commonly called as 3D printing, has been evolving since 1980s. Although its existence from 1980s, latterly gained importance and popularity in business world and become an effective mercantile producing system (Pour et al. 2016).

Three elementary materials are necessary for the 3D formation of an object as shown in Fig. 5.2.

1. **Digital model**—this is required for printing the product. Digital models can be formed by programs as CAD (computer-aided design), or by scanners that create 3D virtual image of the required product. For allowing 3D printer to build object from virtual model; the model is divided into horizontal layers in a number of hundreds or thousands by 3D modeling software.
2. **Feed material**—this is the material that is used to ultimately manufacture the final object. According to the Sculpteo's Survey of 2019, printing materials that are used, presented as in Fig. 5.3. Mostly the used ones are plastics, metals and resins.

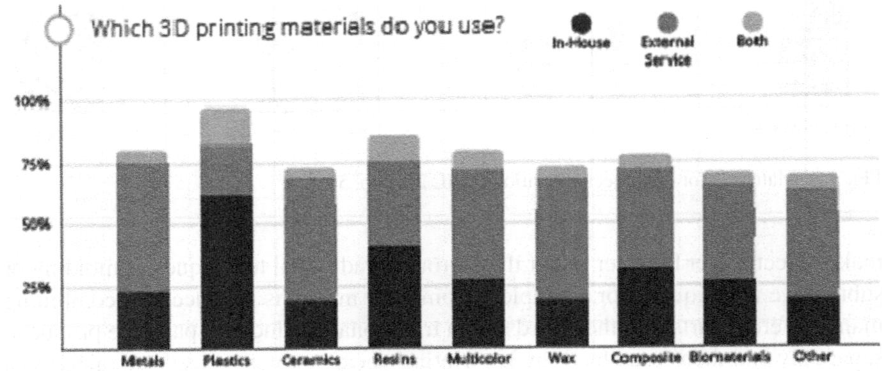

Fig. 5.3 Materials used in 3D printing (Sculpteo 2019, p. 8)

5.3 3D printer

This is the hardware used to create the solid object out of the digital model and feed material. 3D printers can differ according to the technology used as SLS, FDM, SLA, etc. The usage purpose of consumer or industry is taken into consideration in the selection of 3D printers. (DHL 2016; Sculpteo 2019) It would be useful to create a standardization in order to achieve a controlled growth because of the higher focused of the production world on the subject. With this view; a commission of American Society for Testing and Materials (ASTM-F42), which is formed in 2009 for dealing with this innovative way of manufacturing, presented a standard terminology describing a formal name as additive manufacturing (Silva and Rezende 2013). ASTM is a worldwide recognized institution in the creation and delivery of more than 12,000 voluntary consensus standards, which are used globally to increase product quality, health and safety, strengthen market access and trade and increase consumer confidence ("Detailed Overview", n.d., para.1).

5.4 Techniques and Commercial Processes of Additive Manufacturing

Progress in materials science, laser technics, computer-aided issues as design (CAD) and CNC (computer numerical control) supported the quick development of different kinds of additive manufacturing techniques. Prime seven category of these techniques, which are presented by American Society for Testing and Materials (ASTM) group "ASTM-F42 according to the set of standards that they formulated, as below;

1. Sheet lamination;
2. Vat photopolymerization;
3. Powder bed fusion;

4. Binder jetting;
5. Material Extrusion;
6. Material jetting;
7. Directed energy deposition (Dvorak et al. 2018).

Today a variety of additive manufacturing techniques are in use as commercial processes. According to the Sculpteo's State of 3D Printing report of 2019, the most commonly used printing technologies are SLS, FDM and SLA. FDM is the dominating one. This survey result presents the view of more than 1300 respondents. In Fig. 5.4, all the used technology range can be seen.

The most popular three technology is shown in Fig. 5.5.

Selective Laser Sintering (SLS); SLS takes place in powder bed fusion category of American Society for Testing and Materials (ASTM) that thermal energy is used for

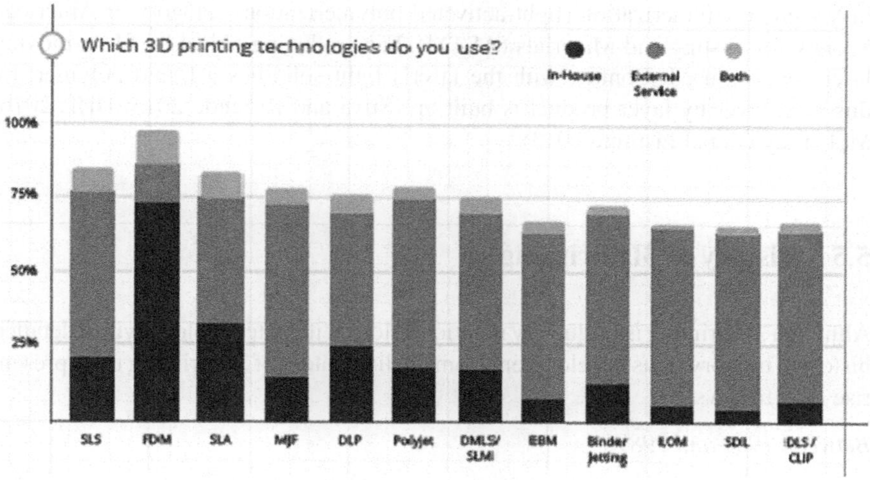

Fig. 5.4 3D printing technologies (Sculpteo 2019, p. 9)

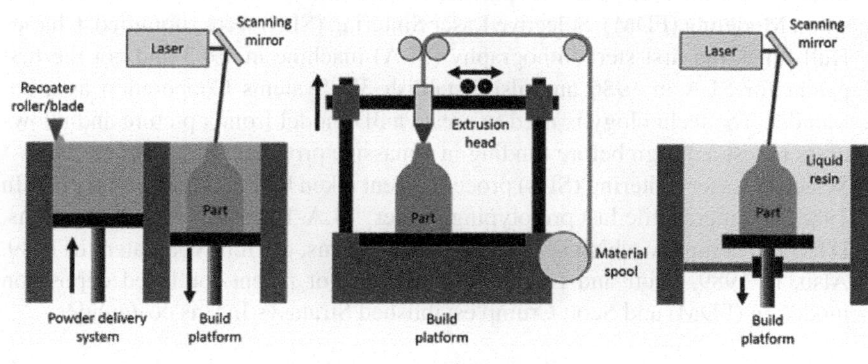

Fig. 5.5 SLS, FDM, SLA processes respectively (Thompson et al. 2017, p. 2)

melting regions of a powder bed. SLS manufactures strong models in many different materials and composites; SLS also lets metal-based materials to be printed by using lasers beams. Laser beam movement is automatically controlled by computer. Energy is transferred by laser beam into a plane, which covers a thin layer of pre-heated powder material.

Fused Deposition Modeling(FDM) is an additive manufacturing technology, which is commonly accepted, especially the most known by consumers and the user-friendly one. FDM takes place in Material extrusion category of American Society for Testing and Materials (ASTM). A filament of material (plastic, wax, etc.) is swiped through a heated nozzle to produce the final object. FDM printers are fed with seared material, which will be melted layer by layer. Low-volume production, single and multipart prototyping are the areas that FDM technology is mostly used for.

Stereolithography (SLA); SLA is commonly used for rapid prototyping and for producing complex forms with superior quality, such as jewelry. SLA takes place in Vat photopolymerization (light-activated polymerization) category of American Society for Testing and Materials (ASTM). For producing object by SLA, moving laser beam is used. Contact with the laser's light solidifies a liquid polymer; by this way, layer by layer product is built up (Silva and Rezende 2013; DHL 2016; McKinsey Global Institute 2013).

5.5 History of 3D Printing

Although 3D printing is frequently mentioned today, it is a technology with a detailed history. The story of its development from the first patent of 3D printing to its present rise is as follows:

Between 1980 and 1989;

> Hideo Kodama offered the first 3D printing patent application in 1980. This patent never had a chance to be commercialized. Till 1989 the most used Additive Manufacturing commercial processes patents as Stereolithography (SLA), Fused Deposition Modeling (FDM), Selective Laser Sintering (SLS) were submitted. Charles Hull funds the first stereolithography (SLA) machine in 1983 and got the first patent for SLA in 1986 and also establish 3D Systems Cooperation as a co-founder. The technology is used to create a 3D model from a picture and allows users to test a design before funding in a massive producing program in 1983.
> Selective Laser Sintering (SLS) process patent taken by Carl Deckard in 1987. In 1988, first mercantile fast prototyping printer "SLA-1" was sold by 3D Systems. DTM is a company which is bought by 3D Systems, got this SLS patent in 1989. Also, in 1989, Scott and Lisa Crump applied for patent for fused deposition modeling (FDM) and Scott Crump established Stratasys Inc. as co-founder.

Between 1997 and 1999;

In 1997, the first 3D-printed metal treatment by laser additive manufacturing (LAM) is manufactured by AeroMat.
In 1999, urinary bladder as the first 3D-printed organ is produced by Wake Forest Institute of Regenerative Medicine for transplant. This organ is produced by using 3D synthetic scaffold coated with patient's own cells, by this way there is little to no risk of rejection of the organ.

Between 2005 and 2019;

In 2005, Dr. Adrian Bowyer's RepRap concept provided the creation of several new 3D printers as an open-source initiative. By this way, it was possible to make a 3D printer that can print most of its own pieces. In 2008, the first 3D printer named as Darwin was commercially available, which is created under the RepRap concept.
In 2006, the first SLS machine was available to use. With this printer, mass customization and on-demand producing became possible for manufacturing industrial parts and prostheses. A 3D Printing System and material supplier presents a machine that produces a part with a range of densities and material specifications.
In 2009, The Stratasys's FDM patent expired and this caused decrease FDM 3D printer prices from $10.000 to under $1000. Makerbot presented do-it-yourself kits that ever body can make their 3D printers. And also, Makerbot establishes a file library that 3D printable files are submitted and downloaded.
In 2011, the first unmanned 3D-printed aircraft is printed by the design of University of Southampton and a prototype car with a 3D-printed body expressed by Kor Ecologic in a conference.
In 2012, Kickstarter campaigns that introduced different 3D printing process as DLP technology and stereolithography were launched by B9Creator and Form 1.
In 2013, Makerbot was bought by Stratasys
In 2015, a Swedish Company Cellink presented bio-ink, which is the first standardized trading product and later in the same year 3D printer for bioprinting named as INKREDIBLE 3D. Body parts can be printed with this bio-ink made of a seaweed material

After the expiration of patents and issue of open source projects, the number of 3D system producers' number reached over 170 worldwide in 2019. 3D Systems, Stratasys, Fusion3 and Voxel8 companies are some of them (González 2020; Grynol 2014).

In this historical flow, 3D printing has a three-stage evolutionary process. In the first stage, 3D printing technologies were used for making prototypes or models of new designs. In the second stage, 3D printers were used in the manufacturing of final products mentioned as direct digital manufacturing. Finally, these stages include end-users who have and use 3D printers in their homes (Berman 2012).

5.6 The Usage Area of 3D Printing

Manufacturing companies prefer to use 3D printing in all processes, from product design to after sales support. Table 5.1 summarizes the industries that use 3D printing according to the rates of their purchasing about 3D printers (Leering 2017).

When time passed, companies become more enthusiastic to try and use 3D printing. For example, Mercedes Benz trucks declared in 2016 that spare parts that were printed by 3D printing were used in their vehicles. And also, BMW shared their experience, after using 3D-printed parts, which are approximately one hundred thousand parts. This experience result is, BMW provided more profit by shortened manufacturing with 3D printing (Wieczorek 2017). Boeing has used in plane model of 787 Dreamliner 30 printed parts and another application of Boeing was a printed cabin. Ford has been using 3D printing for a long time, one of their application is printed engine cover for new Ford Mustang. When Ford compared this technology to the traditional way, Ford presented its cost-saving as 497.000$ and production time saving as 4 days instead of 4 months. Also, GE presented a comparison from a different view as production quality. When GE used 3D printing instead of traditional assembly, GE had able to get more stronger and lighter products. Nike's 2014 Super Bowl cleats and American Pearl Jewelry company products were examples of different sectors that 3D printing technology was used (Gilpin 2014). Sugar producer Hershey produces customized pralines with 3D printing, IKEA conducts experiments with 3D technology in home-related designs. Siemens and Airbus use 3D printing to

Table 5.1 Industries using 3D printing (Leering 2017, p. 5)

Fields of application	Share in sales of 3D printers (%)	Examples of what is made
Industrial machinery	19	Production of tools like jigs and fixtures
Aerospace	18	Geometrically complex and lightweight parts
Automotive	15	Functional prototypes, small and complex parts for luxury and antique cars. Mainly non-mass production of specific tools and parts and for prototyping
Consumer products	13	Micro-electromechanical systems, microwave circuits fabricated on paper substrates, radio frequency identification devices inside solid metallic objects (RFID), polymer-based, three-dimensional, grippers
Medical and dental devices	11	Hips, knees, dental aligners, hearing devices, digital prostheses, etc
Other	24	

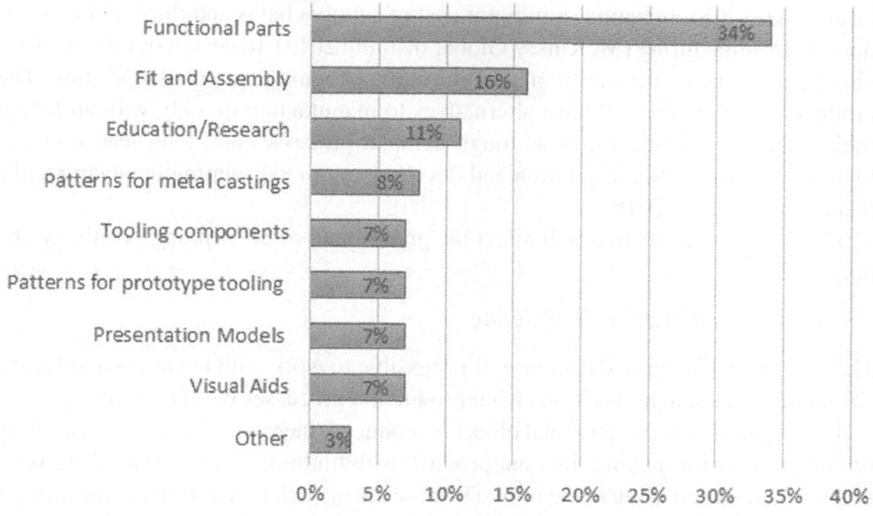

Fig. 5.6 How 3D printers are utilized (Leering 2017, p. 4)

produce gas turbine components and aircraft parts (Mohr and Khan 2015). Figure 5.6 shows the rates of how 3D printers are utilized.

Experiences with 3D printing have shown that its usage is beneficial. Some of the benefits gained by 3D printing production can be defined as shorter lead time, cheap production, fewer stocks, elimination of assembling requirement, reduction in processing times, digitalization of manufacturing process, acceleration of design process, etc (Leering 2017). The advantages of 3D printing support the rise in usage of 3D printing technology in the market. Many examples of large global companies especially that are in the manufacturing sector are accepted as the proof of success of 3D printing in market (Mohr and Khan 2015).

5.7 Growth of 3D Printing Technology in the Market

According to Grynol assumption, 3D printing will expand by 300% from 2012 to 2020 from 1.3 billion $ to 5.2 billion $ (Grynol 2014). As supporting this assumption, Mckinsey Global Institute presented an economic effect assumption as $230 billion to $550 billion per year by 2025, which will be caused from 3D printing. They link this impact to 3D printing application types as consumer uses, direct manufacturing and the use of 3D printing in forming equipment and molds. Although these predictions show that 3D printing will continue to grow, from a different point, 3D printers become already ordinary for those who uses like to form object designs and prototypes. And also, 3D printing attracts attention for direct manufacturing of equipment and final object. The usage of 3D printing could cause an unexampled level of mass customization, minimum cost of supply chains and the "democratization" of manufacturing (McKinsey Global Institute 2013) To design or produce final objects and reduce obstacles to make innovation become easier by 3D printing. The product manufacturer will have alternatives to manufacture quickly without facing traditional bureaucratic forms as tough demand processes and long lead times of logistics processes. So, innovation and development in 3D printing democratize the manufacturing (EY 2016).

Here are the factors that will affect the growth rate of 3D printing positively and negatively.

Positive factors that drive 3D Printing

The main effect is the cost decrease. It is possible to work with lower costs since the elimination of some transactions from production processes by 3D printing.

By 3D printing, a directly final object is produced, there is no need to an assembly of components for making the final product as in traditional flow. Also, there is no need to plan and coordinate the flow. The absence of such transactions in the process lowers the cost of workforce and also provides minimum human mistakes. Fewer people take place in the process, this causes minimum error, there will be no need for correction transaction. Since there is no need for an additional process, time and cost are saved. As direct manufacturing of the final object without assembly is possible with 3D printing, there will be no need to store semi-products and also transport them to the manufacturing area. Only raw material is required. 3D printers consume raw materials in right quantity, by this way, there is no wastage. This capability reduces the raw material amount also. As a result, inventory, transportation and raw material costs will decrease.

Besides these, it will be possible to produce prototypes in a cheap way by 3D instead of time-consuming and workforce-intensive techniques (Leering 2017).

Negative factors that hold 3D Printing back

Some factors are related to monetary issues and some of them are with shortcomings in industry and quality. The quality of 3D printed products is insufficient in some cases, for example, printed metal products contain defects, there are standards required by the products that 3D printers still cannot meet yet.

It is hard to find talented designers to write digital models required for 3D printers. In addition, copyright law is not enough. Besides these, business world does not have enough information about 3D printing so this slows the acceptance and awareness of business world.

The traditional production method is cheaper than 3D printing when economy of scale is considered. Unavailability of usage of 3D printing for mass production causes a risk. In addition, traditional production has high investment costs. Companies do not want to switch to 3D printing without amortizing this cost. Another monetary issue that increases the cost is the fact that the raw material market is under the monopoly of some suppliers (Mohr and Khan 2015; Leering 2017).

5.8 Success Factors for 3D Printing

The global expansion of 3D printing will depend on the realization of five key success factors as material technology, process speed and quality, warranty and liability, intellectual property and cost of printer and material.

As ***material technology***; 3D printers need to handle a variety of materials in a single object. Current printers in the market are available to process up to three different materials in one single print run although there is a wide range of available materials. Some of the institutions are working on prototypes for the feasibility of using up to 10 materials in a single print run, which will be a main effect in expanding the range of printable products. To provide an ongoing success, 3D printing process must be a faster option to buying from store or ordering from online, with a continuous quality. With an FDM printer, a golf ball can be printed in 4–5 h and more complex product of the same shape needs almost 9 h for being printed. To overcome this challenge, there are new developments as HP's Multi Jet Fusion Technology and Carbon 3D's liquid 3D printing that can provide 10 to 100 times faster printing with required quality. Increasing the speed by providing also the quality is the required integration in the scope of ***process speed and quality***. The current regulatory infrastructure has to be renewed with 3D printing consideration related with ***warranty and insurance issues.*** For example, where will be the responsibility related with broken 3D-printed parts? Insurance solutions will be critical about the potential of the liability claims. Otherwise, nobody wants to be a consumer or user in processes in which the rights are not developed. Another subject to be considered as a complementary subject is ***intellectual property challenges.*** This subject looks like the issue that media companies are dealing with pirating digital music and video files. Digital files of 3D printers provide the competitive benefit. If these files are not protected, then this will let possible the replication of original printed products by third parties. To create a rely on 3D printers and security of digital files, intellectual property has to be organized. ***3D printer, material and scan costs are other important factors***. Minimum costs will be effective in the future of 3D printing. There are 3D printers that their costs are around 3500 USD and also there are others that cost only a few dollars. There occurred a decrease of approximately %30 in prices between 2010 and 2016. Also, a decrease as %6 is expected annually. This is an indicator of 3D printers more purchasable in future. With the cost of printers, also the cost of materials and scanning technology (required for converting the object into a CAD file) have to be considered. For minimizing the scanning technologies, mobile devices are taken into consideration. By this way cost that changes between several hundred dollars and several thousand dollars can be converted to affordable ranges. For example, a company presented a low-cost and high-quality laser scanner, which is powered via smartphone (DHL 2016).

5.9 Future of Printing Technology

A new concept as 4D printing has started to be discussed in the sector as an advanced form of 3D printing. 3D printing creates three-dimensional objects, 4D printing is adding a fourth dimension as time to the 3D dimensions. 4D printing takes into account the influence of external factors as light, heat, etc. to show the change in the printed object over time. That's why time-dependent deformations of the objects have to be used in the preprogramming of 4D printed structures. Wise nanocomposites, shape memory alloys, etc. as smart materials are the basic elements of 4D printing. 4D printing draws attention; because of this attention, market assumptions have occurred. Although 4D printing is at the first stage of its evolution, the assumption for growth rate target is a yearly increase of 42.5%. In 2022 expectation of 3D printing to reach 30.19 million $ while 4D printing's possibility is $ 537.8 million (Quanjin et al. 2020).

Although the expectations for revenue growth are like this, it is uncertain to what extent 3D printing will take place in the market in the future, but wide acceptance of this technology could have an impact in many areas. Their impact domains can be summarized as manufacturing, trade, customs, supply chain, laws, environment and sustainability etc. While affecting trade volumes, it can shorten the times in the supply chain. Using 3D printing in production can minimize the workforce requirement. Since the transmission and copying of digital models in 3D printing are easy, an increase in intellectual property and legal difficulties can be expected. This will necessitate reviewing existing laws or revising them or creating new regulations. There can be challenges but besides, it will also provide the opportunity to establish sustainable production systems that take the environment into account (World Economic Forum 2020).

5.9.1 Effects of 3D printing on Logistics Management

Hessman (2013) presented Ed Morris's speech in his article, "You can print on demand, meaning you don't have to have the finished product stacked on shelves or stacked in warehouses anymore. Whenever you need a product, you just make it. And that collapses the supply chain down to its simplest parts, adding new efficiencies to the system."—Ed Morris, director of NAMII (National Additive Manufacturing Innovation Institute).

The statement of Ed Morris draws attention to the change in the activities of the supply chain, which we are used to by 3D printing. In other words, traditional way of business will be transformed into a new style. In new style, some of the logistics activity's area will be narrowed or redesigned or will disappear completely.

The flow within the supply chain that we are used to or traditionally applied is managed towards meeting the needs of mass production. If we need to summarize this flow briefly, first of all, companies prepare a production plan according to customer orders and demand increase in the market. The required amount of material and labor is determined according to the production plan. Determining the amount of required materials as raw material, semi-finished product, packaging material, etc., will form the basis of the company's purchasing, storage and transportation plans. Before production stage, domestic or foreign purchases are organized according to the production plan, the custom clearance process is completed for foreign purchases and the materials are stored until they are used. The products that will be sent to the production are picked according to the work orders and shipped from the warehouse to the production point. After production, different materials are brought together by assembling and the final product is obtained. Final products are shipped to the warehouse for storage, then selected according to customer orders, if required value-added services are organized as labeling, special packaging, adding promotional material etc. Products that are ready for delivery are shipped to inland or international as FTL (Full Truck Load)/FCL (Full Container Load) or LTL (Less than Truck Load)/LCL (Less than Container Load) and delivered to customers. But now, there is an opportunity that the customer can print the product with 3D printing technology as a complete final product from a printer, at any time, in any place with desired features; this new manufacturing structure will cause a change in supply chain management. Because with 3D printing, by using only digital files, raw material and printers, final product can be obtained close to the customer. So, in this respect, it will be no longer efficient to work for traditional way, for example, there will be no need to ship the products worldwide to deliver the customer anymore. It will be possible to say that there will be no need for some of the services anymore with the impact of 3D printing. This customized manufacturing structure will decrease the stock amount, square meter of required warehouse area and transportation number.

3D printing is a destructive trend. The growing requirement for mass customization and on-demand printing will force the roles of the supply chain participants like manufacturers, wholesalers, retailers and logistics providers (Heutger 2020). But many manufacturers and logistics service providers think that they have time to prepare for the effects of 3D printing. The research result of DHL presented by John Manners-Bell and Ken Lyon (2014) in their article is an indication of existing of this thinking. The distribution of the answers is shown in Fig. 5.7.

According to the answers, %22 of the respondents was confident about the adoption of the technology in the next 3 years; %20 of the respondents believed this adoption will take a little bit longer than they assume 3D will come real in the next 5 years. %20 of the respondents was suspicious about their own adoption, so they need to search more about 3D flow. %8 of respondents assumed that it is already in their supply chain that industries like aerospace, automotive, mobile telecoms and healthcare are using and adopted the methods of 3D (Manners-Bell and Lyon 2014).

When the results are evaluated; respondents other than the 8% believe that they still have time to act. That's why they don't pay enough attention to the matter. But every year, the fact that "time still exists" loses its validity. However, in order to be able to stay as a player in the sector and to get the future value, it is necessary to start investing or pilot studies immediately (Roucolle and Boilard 2017).

Fig. 5.7 Question and the distribution of the answers for the interaction of the 3D printing and supply chain (Manners-Bell and Lyon 2014)

The pandemic period that the world lives, has also shown also that everything should transform faster than predicted. The benefit provided by 3D printing during the pandemic period will attract more attention to 3D printing technology. Because 3D printers helped to fill critical gaps in the supply chain, with the epidemic that affected the whole world. Instead of waiting for the products needed by the health units to be mass-produced abroad and shipped from abroad, 3D technology enabled the production of the needs on demand and in a close place to the demand. 3D printing technology enabled the design of masks, ventilator pieces and even for tools for nasal testing (Attaran 2020). In the crisis environment during the pandemic period, there was a huge contribution of 3D printing in the production of these critical products for health units. In this period when there is a struggle to survive and the usual living conditions are interrupted, the effect of this production technology, which creates significant benefits, will be more striking in the market.

Recent evaluations reveal that 3D printing will represent a $ 400 billion production market by 2030 (Roucolle and Boilard 2017) According to presented data in the report of Leering (2017); annual growth of investment in 3D printing is 3 times bigger than traditional machinery over the past 5 years (%29 for 3D printing; %9, 7 global investment growth in traditional machines). Based on these data, two scenarios regarding years 2040 and 2060 are worked to guess how the share of 3D production might be. If there is the same continuity in the growth rate of investment in 3D printers, %50 of the world production output will be produced by 3D printers in 2060. If there is a doubling in investment, then this result as %50 will be reached in 2040 (Leering 2017). With this assumption, the estimated reduction in world trade is as one-quarter. Another estimation from Mckinsey Global Institute is a reduction in physical trade by 1–2% by 2030 (World Economic Forum 2020). Both assumptions pointed out that 3D printing will slow down the trade. Manufacture phases closer to the customer would reduce cross-border trade and also impact the form of global value chains, driven by manufacture scalability, logistics costs and workforce input (Leering 2017; World Economic Forum 2020).

Spare part logistics will be affected when parts' manufacturer moves closer to the consumer by the effect of 3D printing. This will decrease the delivery distance and impact their current delivery organization. Current organizations are based according to long-distance deliveries worldwide and able to manage stocks for their customers (Roucolle and Boilard 2017). In this respect, changes depend a lot on raw materials and where they can be manufactured. While scarce materials might still need to be delivered around the world, mass market materials might be available locally. Even 3D printers do not have to be shipped anymore, since they can be produced by other 3D printers (Suberg 2018). On-demand manufacture at point of use to customers' specifications, reduce the movement of the goods. There will be still transportation but the type of the moving good will differ. There will be a shift from final and intermediary products to raw materials along an increase in electronic transmissions in delivery. There will be no need for reverse flow as 3D printing produces customers' specifications so delivery of unwanted or damaged products will be eliminated, except wrong raw material. (Mckinon 2018; World Economic Forum 2020). Also, this opportunity reduces the nonessential distances to be covered for transportation. The reduced transport process will also reduce carbon dioxide emissions (Mantey 2017). So, there will be changes in transportation related with volume and flow. And also, stock levels are likely to decrease (Heutger 2020).

5.9.2 What Should the Logistics Companies Do?

Logistic companies have to adapt their structure to new market demands, by this way 3D printing threat can be changed to opportunity. UPS clearly understands the developing trend and established 3D printing in some of their customer-facing stores and try to apply 3D printing for very important and time-sensitive parts (Roucolle and Boilard 2017).

UPS, one of the world's largest package delivery companies, developed two approaches for providing 3D printing for their customers. In both of the two approaches, UPS chose to work with partners as SAP, Fast Radius and etc. First off, all UPS evaluates the suitability of 3D printing applications for a particular product. UPS tries to find out which parts make sense for 3D printing and then confirm suppliers and certify the parts. This is UPS's pre-production approach. According to this approach, UPS decides the next step. If there is a confirmation then UPS takes place in process from manufacturing through transportation, otherwise UPS does not involve in process. UPS invested in a 3D printing company (Fast Radius) for using its experience and capabilities. By this way, UPS integrated its logistics network to the manufacturing, so that 3D-printed products can be handled from production to the final destination of the part. UPS has more than 60 3D printers, which are called as 3D printing factories, in U.S stores. They got low-volume production orders as prototypes, service parts and etc. TNT, an international courier delivery service provider started to imagine 3D printing offices that customers will come and select the products which will be ready in a few hours. The products can be taken by the customers or delivered to them. That's why TNT started to work with a 3D printing software and service provider as a partner. UPS, TNT and also FedEx began to consider becoming a manufacturer to integrate with their logistics competencies to cope with the effects of 3D printing (Mantey 2017).

DHL presented a report in 2012 including five future scenarios to set an insight for the Logistics in 2050. One of the scenarios is about customization of consumer demands. In that situation, consumers are able to compose, design and modify their own products, which is the result of 3D printing. According to the DHL scenario evaluation related with customized life style, there will be positive effect on World GDP (Gross Domestic Product) development and new business alternatives for logistics. There will be negative impact for quantity of global stream and standard logistics growth potential. 3D printers change the manufacturing technologies that caused speeding up in customization of consumers and manufacturing patterns. This will cause an increase in local trade flow, with only raw materials and digital data still streaming globally. Long-distance delivery of final and semi-final products will decrease. Raw material and energy consumption will increase because of the increase in customization process. The effect of this energy consumption is assumed as a heat increase as 3.5 °C by the end of this century. The other insights of DHL about the future logistics under the impact of 3D are:

- Decrease in long-distance delivery is expected because of the re-localization of the value chain, this decrease will be observed in air transportation.
- Logistics companies have to provide online service beside the physical. Online section must have a trustworthy infrastructure. Online service as reliable data transfer and data retails will be carried via online shops. Raw material transportation and reverse logistics will be fulfilled physically. Mostly required raw materials as metal dusts or plastics will be delivered from worldwide to the target markets terminals/seaports and then transported to the local producing and printing sites by rail or land transportation.
- Logistics providers will have to compete with Information Communication Technology companies. Because logistics suppliers will deliver coded data flows required for the transfer of construction and design plans as they have to act as a systems integrator between product design companies, contract producers and local transportation suppliers. Besides they continue to deliver the total physical product chain.
- Regional logistics capabilities will be a highly important success factor (Deutsche Post/DHL 2012)

Logistics companies are trying to understand the future by evaluating the extreme conditions of no manufacturing in another country or no shipping from another country or from far point of the same country. Their main question is if there is no need for a logistics company, then what will happen? (Mantey 2017). They have tried to shed light on the required transformation of the supply chain by sharing their indications and scenarios. As presented above, UPS, TNT and DHL are trying to determine the necessary actions for surviving in line with the market conditions that 3D printing will change. In this respect, beside the current physical services, they are organizing or planning some of the actions as integrating logistics activities with manufacturing, providing online shops, operating virtual warehouses for digital files. With virtual warehouses, it will be easy to provide localized printing according to demand and then deliver.

3D printing technology transforms companies, so companies have to understand this transformation to manage it. Two strategies could be adopted as efficiency and growth. In the efficiency strategy, the main aim is to apply 3D printing in the current supply chain, main issue is improving the process of final product creation rather than the final product. This strategy includes actions that reduce the time in all the processes of final product to market. In growth strategy, the main focus is to obtain a new or redesigned 3D printed product. 3D printing will be used directly for production technology to improve the design and to add extra customer value. As these developments can provide new market alternatives, thus this strategy is called as growth. With 3D printing, it is possible to produce closer to the customer by minimizing inventory, transportation and handling costs (EY 2016). As a service provider, logistics companies should review their service infrastructure in line with the needs of their customers who will adopt these strategies. Both strategies will affect the logistics activities. They will minimize the required transportation frequencies, warehousing area and warehousing activities. There has to be an efficient integration between 3D printing requirements and current business operations. This transformation period should be planned systematically and should be handled with a project management perspective.

A four-phased approach is suggested to provide more value while integrating the current process with the 3D printing application. The first phase is forming organizational awareness about the 3D printing capability. The second phase is performing a 3D printing identification. In this step, the main issue is to define the areas that are suitable for the application and also able to provide higher value by the application. In the third phase, a transformation plan is developed for sustainability of the 3D printing. For this sustainability, building new capabilities and redesign of the organization can be required, that's why this phase requires a time period. The last phase is the implementation, in other words, the realization of every planned detail in the application (EY 2016).

This approach will be useful both for companies that want to convert their production to 3D printing and for logistics companies that will feel the effects of this transformation. In fact, it will create more benefit for the manufacturer to carry out a project in which the stakeholders of the entire supply chain will be included in order to realize the transformation more successfully. For this realization, 6 Sigma Project Management methodology can be considered. There are two approaches in 6 Sigma as DMAIC and DMEDI. DMAIC is used for improvement projects, DMEDI is used for designing a new service, product or process. DMEDI is a useful tool for this transformation and includes five phases as Define, Measure, Explore, Develop and Implement. When we look into the issue in terms of logistics activities, logistics companies will be able to establish their new structure systematically by taking into account their customers' and their own needs and evaluating their risks in every phase. In define phase project scope, objectives, resources, constraints and risks will be determined. Also, customer' requirements will be collected as Critical Customer Requirement (CCR). CCR has to be measurable, otherwise targets cannot be organized and then performance cannot be tracked. In measure phase, House of Quality will be prepared with the customers" requirements, company's technical competencies, competitor information and required targets of each CCR. Within the Explore phase, alternative concepts for service will be determined and according to the selected concept, high-level service/process/product design is planned including process description, methods, human resources, information systems, facilities, equipment. In Develop phase, details of the high-level design are defined as detailed process map, layouts, job and task description, database, system interfaces, performance details (targets, measurement frequencies and responsible) etc. In Implement phase, pilot studies are carried and before implementation, required improvements will be completed and then new process/service/product will be implemented. And before every phase completion, risk evaluation will be repeated. Projects like transformation of physical warehouses to virtual ones or to develop new business opportunities from idle transport and storage resources (for example manufacturing trucks can be an alternative solution) can be handled with this kind of systematic tool.

Since change is inevitable, it is important to be able to manage correctly and realize this change systemically according to the new requirements, because it will not be possible to continue with traditional structures within the 3D printing world.

5.9.3 Conclusion

As a result, consumers and businesses compared to the past, they want to have their demands immediately without waiting. They are becoming more demanding in this regard. This is a sign that the demand-driven economy will not slow down. As 3D printing technology continues to develop, demand-driven economy will increase related to the reduction of technology and material costs. With 3D printing, the needed product can be produced at any time and at any place. In this scope instead of sending the parts to the warehouses or distribution centers after production, all needed is the transmission of the digital files, which need to be printed on the 3D printer. This will affect the supply chain and logistics market. Suppliers will decrease, merge or even be eliminated in some cases. Such a structure will reduce the need for warehouses and consequently warehouse expenses, because printing on demand minimizes the need to carry inventory. Also, this opportunity reduces the nonessential distances to be covered for transportation. Thereby shortening the length of the supply chain, which reduces transportation time, logistics and inventory costs (Mantey 2017).

3D printing is a disruptive trend, nothing will ever be the same. It is necessary to be prepared for the new business style.

References

Attaran M (2020) 3D printing role in filling the critical gap in the medical supply chain during COVID-19 pandemic. Am J Ind Bus Manage 10(05):988

Bendix A (2019) These 3D-printed homes can be built for less than $4000 in just 24 hours. *Businessinsider*. https://www.businessinsider.com/3d-homes-that-take-24-hours-and-less-than-4000-to-print-2018-9. Accessed 22 Aug 2020

Berman B (2012) 3-D printing: the new industrial revolution. Bus Horiz 55(2):155–162.

Carlota V (2019) A guide to 3D printed food—revolution in the kitchen?. 3Dnatives https://www.3dnatives.com/en/3d-printing-food-a-new-revolution-in-cooking/. Accessed 22 Aug 2020

Danova I (2019) 3D printing revolutionizing Medtech. PegusDigital. https://pegus.digital/3d-printing-revolutionizing-medtech/. Accessed 22 Aug 2020

Detailed Overview (n.d) In About Section. https://www.astm.org/ABOUT/full_overview.html. Accessed 22 Aug 2020

Deutsche Post A G/DHL (2012) Delivering tomorrow: logistics 2050—a scenario study. Bonn, Germany: Deutsche Post AG. https://www.dhl.com/content/dam/Local_Images/g0/aboutus/SpecialInterest/Logistics2050/szenario_study_logistics_2050.pdf. Accessed 21 June 2020

DHL (2016) Printing and the future of supply chains. A DHL perspective on the state of 3D printing and implications for logistics. https://www.dhl.com/content/dam/downloads/g0/about_us/logistics_insights/dhl_trendreport_3dprinting.pdf. Accessed 31 May 2020

Durach CF, Kurpjuweit S, Wagner SM (2017) The impact of additive manufacturing on supply chains. Int J Phys Distrib Logist Manage 47(10):954–971.

Dvorak F, Micali M, Mathieug M (2018) Planning and scheduling in additive manufacturing. Intel Artif 21(62):40–52

EY (2016) If 3D printing has changed the industries of tomorrow, how can your organization get ready today? https://www.ey.com/Publication/vwLUAssets/ey-3d-printing-report/$FILE/ey-3d-printing-report.pdf. Accessed 1 Aug 2020

Gilpin L (2014) 3D printing: 10 companies using it in ground-breaking ways. TechRepublic. https://www.techrepublic.com/article/3d-printing-10-companies-using-it-in-ground-breaking-ways/. Accessed 22 Augu 2020

González CM (2020) Infographic: the history of 3D printing. Mech Eng Mag ASME. https://www.asme.org/topics-resources/content/infographic-the-history-of-3d-printing. Accessed 20 June 2020

Grynol B (2014) Disruptive manufacturing: the effects of 3D printing. Deloitte. https://www2.deloitte.com/content/dam/Deloitte/ca/Documents/insights-and-issues/ca-en-insights-issues-disruptive-manufacturing.pdf. Accessed 31 May 2020

Hessman TM (2013) 3-D printing the supply chain. IndustryWeek. https://www.industryweek.com/technology-and-iiot/article/21961492/3d-printing-the-supply-chain. Accessed 22 Aug 2020

Heutger M (2020) DHL: how 3D printing is distrupting the logistics industry. Supply Chain. https://www.supplychaindigital.com/logistics/dhl-how-3d-printing-disrupting-logistics-industry. Accessed 22 Aug 2020

Ho S (2020) China dominates 3D printed meat market despite covid-19 crisis. Greenqueen. https://www.greenqueen.com.hk/china-dominates-3d-printed-meat-market-despite-covid-19-crisis/. Accessed 22 Aug 2020

Leering R (2017) 3D printing: a threat to global trade. ING. https://think.ing.com/reports/3d-printing-a-threat-to-global-trade/. Accessed 21 June 2020

Manners-Bell J, Lyon K (2012) The implications of 3d printing for the global logistics industry. Transport Intelligence, 1–5. http://www.reloaderitalia.it/wp-content/uploads/2013/01/3d_printing_impact_on_global_logistics_industry.pdf. Accessed 30 May 2020

Manners-Bell J, Lyon K (2014) The implications of 3D printing for the global logistics industry. Supply Chain 247. https://www.supplychain247.com/article/the_implications_of_3d_printing_for_the_global_logistics_industry. Accessed 22 Aug 2020

Mantey C (2017) Virtual inventory in the virtual warehouse: will 3D printing kill the traditional supply chain? Supply Demand Chain Executive 18(2):34–37. https://www.sdcexec.com/warehousing/article/12329351/virtual-inventory-in-the-virtual-warehouse-will-3d-printing-kill-the-traditional-supply-chain. Accessed 18 June 2020

Mckinon A (2018) 3D printing: decarbonising silver bullet? Focus 20(10):28–29. https://ciltuk.org.uk/Portals/0/Images/Focus%20Archive/Focus%202018/FOCUS_Oct18.pdf?ver=2020-05-12-165507-430. Accessed 18 June 2020

McKinsey Global Institute (2013) Disruptive technologies: advances that will transform life, business, and the global economy. https://www.mckinsey.com/~/media/McKinsey/Business%20Functions/McKinsey%20Digital/Our%20Insights/Disruptive%20technologies/MGI_Disruptive_technologies_Full_report_May2013.ashx. Accessed 30 May 2020

Mohr S, Khan O (2015) 3D printing and supply chains of the future. In: Proceedings of the Hamburg International Conference of Logistics (HICL), vol 20, pp 147–174. http://hdl.handle.net/10419/209254. Accessed 30 May 2020

Pour MA, Zanardini M, Bacchetti A, Zanoni S (2016) Additive manufacturing impacts on productions and logistics systems. IFAC-PapersOnLine 49(12):1679–1684

Quanjin M, Rejab MRM, Idris MS, Kumar NM, Abdullah MH, Reddy GR (2020) Recent 3D and 4D intelligent printing technologies: a comparative review and future perspective. Proc Comput Sci 167:1210–1219

Roucolle G, Boilard M, Forbes (2017) 3D printing is already starting to threaten the traditional spare parts supply chain. https://www.forbes.com/sites/oliverwyman/2017/03/06/3d-printing-is-already-starting-to-threaten-the-traditional-spare-parts-supply-chain/?sh=3adc0e972646. Accessed on June 15th, 2020

Sculpteo (2019) The state of 3D printing 2019. https://www.sculpteo.com/fr/ebooks/the-state-of-3d-printing-report-2019/. Accessed 1 Aug 2020

Silva JVL, Rezende RA (2013) Additive manufacturing and its future impact in logistics. IFAC Proc Vol 46(24):277–282

Suberg T (2018) The future of logistics tech: print or ship? Predict. https://medium.com/predict/the-future-of-logistics-tech-print-or-ship-7b1e3fb2fc8e. Accessed 22 Aug 2020

Thompson A, McNally D, Maskery I, Leach RK (2017) X-ray computed tomography and additive manufacturing in medicine: a review. Int J Metrol Qual Eng 8:17

Wieczorek A (2017) Impact of 3D printing on logistics. Res Logist Prod 7(5):443–450

World Economic Forum (2020) 3. rs [White paper]. http://www3.weforum.org/docs/WEF_Impacts_3D_Printing_on_Trade_Supply_Chains_Toolkit.pdf. Accessed 31 May 2020

Ayşegül N. Bayraktar works as an Assistant Professor in Beykoz University since 2017. She is the manager of Graduate Programs Institute in Beykoz University. She graduated from Istanbul Technical University, Department of Industrial Engineering in 1999. She continued her graduate and doctorate education at Marmara University and got her PhD in Management and Organization program at the Institute of Social Sciences in 2014. She worked in a logistics company from 2000 to 2016 and in sales, customer relations and project management fields. Since 2017 she lectures for associate degree and graduate classes in Beykoz University. Among the courses she is responsible for are Logistics Management, Project Management, Information Systems in Logistics, Free Zones and Foreign Trade Incentives and Organizational Behavior.

Chapter 6
Autonomous Robots and Utilization in Logistics Process

Ömer Faruk Görçün

Abstract The autonomous robots are required in our logistics operations? It is one of the crucial questions that asked to him by logistics managers. Depending on technological developments, some requirements, which can be defined as speed, lower cost, and high quality, have become more important to stay in a highly competitive environment for companies and supply chains compared with the past. Autonomous robotic systems can help to reach these aims. They can provide a more safe working environment in warehouses, distribution centers, and other logistics facilities. İn addition to that, they can provide opportunities for carrying out lower costly, speedy, and quality logistics operations. Furthermore, robotic systems continue to improve, and creating more specific robotics that will be used in the field of logistics will be possible in the near future. Moreover, robotic systems becoming smart and autonomous thanks to artificial intelligence, developed sensors, and trainability. In the future, requirements for human force may reduce in logistics operations since robotics may carry out logistics activities autonomously depending on these technological improvements. Finally, autonomous robots can provide opportunities for carrying out more quality and excellent logistics operations to logistics managers and staff in the near future.

Keywords Autonomous robots · Logistics process · Logistics operations

6.1 Introduction

Although usage of robotics in daily lives goes back a long way, we have met with modern robotic systems in more recent times. Likely, they will have become the most important elements of our near future. Even, these systems have been an indispensable part of industries such as automotive and white appliances already (Schwab 2016). The first industrial robot called UNIMATE was installed in the general motors' factory in 1961. From this year to present, robotic systems have recorded remarkable

Ö. F. Görçün (✉)
Department of International Trade and Logistics, Kadir Has University, Istanbul, Turkey
e-mail: omer.gorcun@khas.edu.tr

İ. İyigün and Ö. F. Görçün (eds.), *Logistics 4.0 and Future of Supply Chains*,
Accounting, Finance, Sustainability, Governance & Fraud: Theory and Application,
https://doi.org/10.1007/978-981-16-5644-6_6

developments. There are some discussions about can these systems be substituted perfectly instead of human force. İt is seen that these discussions will continue in the near future.

More special and functional algorithms developed depending on technological advancements can provide to opportunity to become self-learning organisms and for robotic systems. In addition to that, these robots can also gain autonomous features thanks to these improved algorithms. In the near future, it is foreseen that these systems may make optimal decisions and they can self-implement these decisions without the human factor.

It is a fact accepted by everybody that the error probability of advanced technologies including robotic systems is low compared with humans. Furthermore, dangerously and obtrusive jobs can uncomplainingly be performed by robotics. İn addition to that, as the robotic systems can be functional whenever and wherever operators want, almost all logistics activities can become more flexible. Many restrictions such as organizing human force, meeting the customer orders only in the hour of work, and so on can be removed by these systems. Also, as well as deviation values, the performance and efficiency of robotics can be monitored, and operators can intervene in operational errors very quickly.

Although the usage of robotic systems in logistics activities has increased depending on the technological advancements, expectations from these systems could not meet right now. Many logistics and supply chain managers expecting that these systems can help to reduce operational costs in addition to increase the speed, quality, and effectivity.

It is a fact accepted by everybody that the robotics can only be operated thanks to algorithms created by programmers. But this idea has started to change because these robotic systems have become autonomously trainable elements, and they have gained some specifications such as self-learning. On the other hand, robotic systems may not carry out successful operations in the field of logistics compared with industrial activities. In fact, creating an algorithm for industrial activities is easy than an algorithm created for logistics activities because there are standard operations in industrial activities. Operators expect the same movement from machines and robots in each operation and deviation occurring in industrial activity is an undesired result. By contrast, there are many variations in logistics activities, and operations can be carried out by using many and different ways.

Indeed, there are many options to conduct successful logistics operations, and happening a change in logistics operations is possible at any moment because there are many factors and variables affecting the processes. Therefore, the more variations and options in logistics activities are available, the more carrying out successful logistics operations by robotic systems may difficult.

6.2 The Terms and Scope of Robotics

Due to they are programmable, the robotics can be defined as machines that can perform a defined job semi or fully autonomously. Therefore, they have the ability to be able to make the same movement at each operation to fulfill their duty. Therefore, if a mechanical device can be programmed to perform a wide variety of applications, it is probably an industrial robot (Dyson 2018). The main difference between classical machines and robotics is the versatility of the robot that it is provided with tools of different types and has a large workspace compared with the volume of the robot itself (Wallen 2008). In fact, the robotic systems are the composition of different sciences such as mathematics, mechanical and electronic engineering, and computer programming. In order to obtain the desired movement, designing algorithms is the main task for programmers. İn addition to that, these systems need sensors, which are designed by electronic engineers to detect the variations and changes in external situations.

According to the Robot Institute of America, a robot is a reprogrammable multi-functional manipulator designed to move materials, parts, tools, or specialized devices through variable programmed motions for the performance of a variety of tasks. In addition to that, the International Standard Organisation has made a definition for robots in a wider perspective:

> Manipulating industrial robot is an automatically controlled, reprogrammable, multi-purpose, a manipulative machine with several degrees of freedom, which may be either fixed in place or mobile for use in industrial automation applications. (Stock and Selinger 2016, p. 25)
>
> Manipulator is a machine, the mechanism of which usually consists of a series of segments jointed or sliding relative to one another, for the purpose of grasping and/or moving objects (pieces or tools) usually in several degrees of freedom. (Stock and Selinger 2016, p. 25)

6.3 History of Robotic Systems

The idea of automation can be accepted as the motivators of designing robotic systems. From the first industrial revolution to present, getting automation in production processes has been one of the main targets of the engineers and managers for reducing production costs (Görçün 2016). Nevertheless, these attempts are valuable, they are not sufficient to install the full-automated production systems. Engineers and managers have been obliged to wait for the technological revolution for that. With the help of technological developments such as computers, data processors, data collecting, integrated circuits, and designing algorithms, they have obtained an opportunity for constructing full-automated production systems. Even though computer science and electronics are the basis of contemporary robotic systems, the inspiration of the initial robots is mechanical science. The usage of the mechanical forces to obtain continuous movement was the prior approach for engineers. They thought that if full mechanization is possible, the construction of.automatized production systems is also possible. Automatization and mechanization were synonyms for them (Banger 2016).

In later years, mechanization was provided in a wider perspective but machines had no autonomous features, and it is required to repeat the mechanical programming in each operation. After the third industrial revolution, developments in the field of computer and data processing technologies provided remarkable opportunities for designing autonomous robotic systems (Badaraite 2016).

Ford Automobile Company has been recorded as the first institution that done an attempt to construct the automatized production system, which using industrial robotic systems (Öztuna 2017). But there was no full success, which was obtained in the result of this attempt because the technological development level in this date was not sufficient to reach the desired result. In later years, the second attempt that resulted in success was done by General Motors in 1966. General Motors installed 66 robotic systems to use in the production processes, and it was a milestone for the future of industrial robotics. Industry explored the power and potential of these technological instruments in these years. Especially, in automotive industry, robotic systems have become an indispensable part of the industry in later years. İn these production systems, the usage of human force has been reduced sharply, and any job that can be performed by industrial robotics is not performed by using human force.

The year 1977 that is the date of producing the robots using in the houses can be accepted as the starting point for today's contemporary and modern robotics technology. These robots were started to produce by benjamin sokora in this year. Later, remarkable developments have been recorded in robotics technology and autonomous robots called irobot is a milestone in the process of development of the robotics technology. İn addition, robotic arms designed by Engelberger in 2004 have been started to use in various fields such as space research, medicine, various industries, and this year is the milestone for producing and supply chain management.

Over the past 30 years, companies have faced intensive and backbreaking competition with the effect of the globalization process. İt is difficult to give an answer to the question that technological advancement is the cause or result of the globalization process. As a result, the globalization process continues and it leads to an increase in the severity of global competition. At the same time, determinative actors of the traditional supply chain such as producers and suppliers have lost their positions, and customers' role has become more important and determinative in recent years. As a result of that, flows and operations carried out in the supply chains have dramatically changed. Nowadays, customers want to buy more specialized and customized products, and they have started to ignore the concept of fashion. Moreover, both their tolerance level is low, and meeting their demand at the high-quality service level is the priority for this kind of customer. Therefore, carrying out logistics activities at a high level of efficiency, performance, and speed has become a requirement for the supply chains and they have started to seek new ways and methods to meet these kinds of requirements.

As a result of difficulties in matching the abilities and qualifications of human force and requirements of supply chains, managers are in quest of different and new ways. Correspondingly, they consider that robotics is an instrument that can solve these problems, and they believe that this technology will be alternate to human force completely in the near future. Although there are some attempts to use this technology in the field of logistics, these attempts have not provided desired impacts up to now. Even though robots have provided benefits at a limited scale, they have been stayed underperform in the field of logistics different from their performance that shows in the field of production. Its reasons have been explained in the section of Introduction.

This situation requires using autonomous, having the ability to make the evaluation, and can self-operated systems in logistics activities. Therefore, ın order to the usage of robotic systems in the fields of logistics and its subfields, it is required that these systems should have these specifications and functions. Although these machines can perform a range of mechanic works if these machines have no features such as being autonomous and can self-operated, they can not remove the requirement to programming by humans. İn addition, a well-functioned robotic may lose its functions due to continuously occurring changes over time. For this reason, robotics should be a machine that is self-learning and easily updates.

In fact, it is very important that a robot should be an autonomous machine, which can give a quick response to all likely requirements in different situations with optimal decisions. Furthermore, a non-autonomous robot can not give to respond to requirements in very complicated business processes because it should be programmed by an operator in each different situation. As a result of that, occurring a range of retards and interruptions in the business environment is likely (Celiktas et al. 2015).

6.4 The Potential of Robotics in the Field of Logistics

As has been mentioned earlier, although some partial successes have been obtained, some restrictions making it difficult to use them in logistic operations affect the usage of the robotic systems in the field of logistics. On the other hand, improvements occurring in the field of technology are promising in this issue. İt can be foreseen that when we evaluate the developments that happen in the present if these machines become smart machines by equipping with artificial intelligence, usage of robotics in the field of logistics will boom in the near future.

Nowadays, robotics are used in warehouse operations that are an important part of logistics. Since routine activities are performed in warehouses in general, the usage of static and nonautonomous or semi-autonomous robotics is enough to run warehouse operations. On the other hand, almost all robots used in the warehouses have not autonomous features and they may become functionless without an operators' intervene (Göçmen and Erol 2018). On the other hand, more specific and special logistics activities have also started to begin to carry out in warehouses.

In previous times, while the logistics activities were limited with receiving, placement, and dispatch, many operations such as sorting, consolidation, complementations, and so on, which can be carried out in modern warehouses, have been added to these traditional operations. As logistics operations have gained a more specific character, the requirement for autonomous robotics is increasing. Therefore, warehouses have gained mechanized characteristics recently. An important verge has been passed thanks to the usage of the RFID systems, sensors, and automatic shelving systems in warehouse operation, even though the usage of autonomous and smart robotics just limited right now (Çağlar 2014).

In fact, warehousing operations and activities are the highest costly operations in logistics processes. The main reason for that is warehousing operations need human force intensively, even if their mechanization level has already started to increase (Şekkeli and Zümrüt 2018). A range of warehousing operations such as order picking, product sorting, and preparing the dispatch that proper to customer orders requires to use of intensive human force already. From this point of view, supply chains and companies should increase the usage of smart and autonomous robotic systems in the warehousing operations to keep their competitive power.

The most important requirement is to reach a high standardization level of warehousing operations. Pallets and containers can be accepted as important instruments to reach high standardization in warehousing operations. Both of them have helped to increase the logistics flow rate as well as the standardization of these operations. As a result of those, logistics costs have started to ever-decrease. In this situation, increasing logistics speed has required to re-overview all operational processes that carry out in the warehouses. Especially, interruptions and other likely problems lead to re-think on how these operations can be conducted with different instruments and systems that can help to remove these disorders. It can be seen that autonomous and intelligent robotic systems have the potential that can help to eliminate these kinds of problems and can reduce losses (Pesti and Nick 2016).

As shown in Fig. 6.1, the robotics used for handling operations consists of telescopic conveyors, integrated sensors, remote sensing systems, electronic arms, and other elements providing movements. These systems can also be defined as handling robots. These robots can do the right operations expected by operators by scanning the rfıd tags. They can collect or put products that are wanted only thanks to these advanced technological elements.

One of the difficult and complicated applications of logistics activities is order-picking operations. Traditionally, an operator collects products by visiting shelves after he takes work order and he may visit distance shelves each other. It may cause to use of intensive human force and loss of time that is already limited. Robotics designed to solve these problems can be considered as a promising factor to carry out more effective and lower costly logistics operations. The working principle of these order-picking robots depends on shelves gain the ability of movement instead of their static features. These kinds of robotics can visit these shelves and can carry to the desired collecting point by lifting them. By this means, the requirement for visiting distant shelves can be removed. Order-picking robots can perform their duties by determining the optimal routes at high-performance and efficiency levels. These systems can perform order-picking operations both visiting shelves and carrying shelves to collecting points.

Fig. 6.1 Typical material handling robots in warehouses

It is foreseen that the robots used for order picking can reduce human force and loss of time on a vast scale. These systems need some components such as electronic says, mobile shelves as well as carrier robotics. Completely constructing these systems in a logistics process may make necessary to re-design the logistics systems and flows. In addition to those, these technological elements may cause disrupt ordinary logistics flows because they do not compatible with traditional warehousing operations. Moreover, since the installation costs are at a high level at present, companies stay undecided to make investments in these systems. Although the usage of these systems is staying limited in the logistics industry, some companies such as alibaba, amazon, and etc. that have intensive logistics flows seeking the best ways to use these autonomous and smart robotic systems more effectively. Within the near future, order-picking robots will be used in almost all parts of logistics processes as independent of the scale of supply chains.

Another robotics that can be used in logistics operations are mobile collector robots (Fig. 6.2). Though they have similar features with order-picking robotics, the main difference is to collect products by walking around shelves in the warehouse freely without need to static movement ways. Nevertheless, they have not been commercially used in the logistics processes at a sufficient level, they are systems that have the potential to perform very important duties.

These robots can reach the shelves and can detect ordered products by scanning the shelves and products and they can carry them to the collecting points. The most

Fig. 6.2 Mobile shelves and carrier robots

Fig. 6.3 Autonomous robotics in warehouse

important advantage of the system is to remove the requirement to the static electronic ways and mobile shelves. Correspondingly, installation costs may be reduced. Mobile collector robots have equipped three-dimensional (3d) cameras and sensors to be able to detect the positions of products on shelves and pick the right products. When they reach the shelf, detect to right products and can show different and sensitive behaviors in accordance with the specifications of products (Fig. 6.3).

Also, another robotic system that can use in the logistics process is the robots performing some duties such as packaging, assembly, decomposition. It may require that packing, labeling, and packaging of each product have become different depending on the customization of customer orders. Customers don't want to take similar products and services at present and they have different expectations of their own requirements and personal characteristics. As a result of those, slowing in logistics flow rate, complexities, and a range of problems may occur in the logistics processes. Furthermore, the human force used in these processes can cause more serious errors. On the other hand, these intelligent robots can perform their duties with almost zero mistakes.

6.5 The Future of Robotic Systems in Logistics

When the robotic systems are evaluated in general, it can be seen that they have not been sufficiently used and their contributions making in logistics operations are low. At present, 80% of logistics activities conducted in global logistics systems have been run without full automation, they are carried out by using traditional or semiautonomous systems. As has been mentioned before, autonomous and intelligent robotic systems will play an important and key role in the logistics processes, and these systems will be able to take on many tasks carried out in logistics.

Taking more tactical and strategical tasks by these systems in the logistics operations may lead to changes and transformation dramatically in not only logistics flows but also almost all sides of our lives. Also, this kind of development may cause occur new paradigms and cause to remove existing paradigms. İn the near future, it is possible to mention on the logistics system will very different from the existing logistics system at present. İn the near future, re-defining the role and functions of supply chain actors such as suppliers, manufacturers, distributors, logistics service providers may be required.

When we evaluate at the micro-scale, it is possible to be that distribution centers will have more specific and technologically improved features. In these areas, the usage of human force will ever-decrease depending on the development level of technology and technological instruments such as autonomous robotics and intelliğent systems will completely supersede the human force. It is expected that these trends will also accelerate when the installation costs of these systems are reduced and these systems are more advanced. In addition, logistics activities will be organized as work parts at a micro-level, and each work will be performed by autonomous and intelligent robots working at high performance and efficiency. Each robotics will be related to each other by means of the Internet of Things (ıot), they will have features that can send commands motiving other robots and systems autonomously thanks to the cloud systems.

At the same time, because intelligent robots can plan and optimize their maintenance and repair processes, they can organize these processes in they are idle. As a result of that, they can reduce the impacts of some negative situations such as reducing production and loss of time.

6.6 Conclusions and Discussions

In this chapter, both the existing and potential impacts of robotic systems on the logistics systems as well as our daily lives have been evaluated. At present, robots are available even if they are semi-autonomous features. Whereas they are not an important part of our daily lives, they are in the position of a crucial part of production systems of factories that make mass production. They will also become an important part of humans' daily lives in the near future. One day, it is likely that an intelligent robot can knock at the door as a delivery officer. Moreover, a robot that is a delivery officer will not be a surprise for too many people in the future.

The majority of projection on logistics and supply chains prognosticates that the majority of customers will shop in the electronic environment instead of face to face shopping. Correspondingly, the importance and role of retailers in the supply chains will decrease and supply chains will directly send products to their customers through distribution channels. Concordantly, because customers want small and unitized products, while the number of daily logistics operations will increase, the product range will also increase. As a result of those, the logistics flow rate will increase at an unprecedented pace. Due to these kinds of transformations occurring in logistics processes, carrying out logistics operations will not be possible depending on human abilities and human intelligence. Many research and studies have shown that robotics have capabilities and abilities to respond to these kinds of requirements and they will become a very important part of logistics systems in the near future.

References

Badaraite Z (2016) Welcome to the future: industry 4.0. https://www.hylasoft.com/en/posts/welcome-to-the-future-industry-4-0. Cited 5 May 2018

Banger G (2016) Endüstri 4.0 ve Akıllı İşletme. Ankara: Dorlion Publications

Çağlar BM (2014) Lojistik İşletmelerinde Bilişim Teknolojilerinin Kullanımı, Müşteri Memnuniyeti ve İşletme Performansı İlişkisi: Bir Araştırma. Selcuk University J Inst Soc Sci 32:41–55

Çeliktaş MS, Sonlu G, Özgel S, Atalay Y (2015) "Endüstriyel Devrimin Son Sürümünde Mühendisliğin Yol Haritası. TMMOB Makina Mühendisleri Odası Mühendis Ve Makine 56(662):24–34

Dyson S (2018) Mind the gap – industry 4.0 and the future of manufacturing

Görçün FÖ (2016) Dördüncü Sanayi Devrimi Endüstri 4.0. İstanbul: Beta Basım

Göçmen E, Erol R (2018) The transition to industry 4.0 in one of the Turkısh logistics company. Int J 3D Print Technol Digit Ind 2(1):76–85

Öztuna B (2017). Endüstri 4.0 (Dördüncü Sanayi Devrimi ile Çalışma Yaşamının Geleceği. Ankara: Gece Kitaplığı

Pesti I, Nick AG (2016) Industry 4.0 from the aspect of logistics innovations. society for regional science and policy. In: 6th winter seminar of regional science, 9–12 March 2016, Slovakia
Schwab K (2016) Dördüncü Sanayi Devrimi. Zülfü Dicleli (translated). İstanbul: Optimist Publications
Şekkeli H, Zümrüt Bİ (2018) Endüstri 4.0'ın Etkisiyle: Lojistik 4.0 (By The effect of the industry 4.0 on logistics 4.0). J Life Econ 5(2):17–36
Stock T, Selinger G (2016) Opportunities of sustainable manufacturing in industry 4.0. In: 13th global conference on sustainable manufacturing - decoupling growth from resource use. https://www.core.ac.uk/download/pdf/82023214.pdf. Accessed 12 May 2020
Wallen J (2008) The history of industrial robots. Technical reports from the Automatic Control group in Linköping, Linkoping University

Ömer Faruk Görçün who is currently a Faculty Member of the Faculty of Business Administration at Kadir Has University is also a member of transportation council of Ministry of Transportation of Turkey. In addition to that, he has performed as president and member in the committees of the Research & Development and the railway regulation. He acted as the member of the board in Rail Systems Association and was advisor for the Crane Operator Association and Heavy Transport Association. In aspect of his academic studies and works, he works on optimisation and decision-making techniques in the fields of logistics, supply chain management. In addition to many studies published in international and national reputable journals, he wrote many scientific books, which published by international and national publishers, such as Industry 4.0, Integrated Logistics Management, Supply Chain Management, Warehouse and Inventory Management, Railway Transportation. Currently, Görçün continues his academic life as scientific committee member, editor, technical committee member in addition to his managerial and academic duties at the university. He is married and has two wonderful daughters.

Chapter 7
Warehousing 4.0 in Logistics 4.0

Mahmut Tutam

Abstract Warehousing is an important component of logistics and plays a critical role in contributing to a company's success. With the advent of e-commerce, warehouses are becoming increasingly more important than ever in logistics. Typically, warehouses receive, store, preserve, retrieve and deliver millions of products on a daily or weekly basis. Therefore, warehouse systems are being continuously revised to accommodate a continuous flow of products in logistics. In recent years, a large and growing number of studies related to the development of warehouses have been conducted. Accordingly, mobile, autonomous, compact, or collaborative systems are increasingly applied in warehouses (called Warehousing 4.0). This chapter presents the state of advancement of Warehousing 4.0 by evaluating the current situation and considers future prospects by including emerging technologies.

Keywords Logistics 4.0 · Warehousing 4.0 · Robotic mobile fulfillment system · Autonomous vehicle storage and retrieval system · Compact storage and retrieval system · Collaborative robot system

7.1 Introduction

With the fourth industrial revolution (known as Industry 4.0 and abbreviated to I4), all fields are restructured, including but not limited to production, logistics, and warehousing. Integrating information and communication technologies into industry, I4 simply refers to an industrial transformation from traditional to digital. The development of information technologies (informatization) is the beginning of this transformation. Using information technologies, non-digital data (represented in analog or physical format) are converted to digital data (represented as bits or bytes), which is called digitization. Digitized information is then used in adopting information technologies across all industrial fields (digitalization).

M. Tutam (✉)
Department of Industrial Engineering, College of Engineering, Erzurum Technical University, Erzurum, Turkey
e-mail: mahmut.tutam@erzurum.edu.tr

İ. İyigün and Ö. F. Görçün (eds.), *Logistics 4.0 and Future of Supply Chains*, Accounting, Finance, Sustainability, Governance & Fraud: Theory and Application,
https://doi.org/10.1007/978-981-16-5644-6_7

The advent of internet has resulted in increasing the interdependence of nations (globalization). As a result of globalization, companies are forced to improve their organizations and compete with international companies in addition to their national competitors. Staying competitive in business, companies have started restructuring their organizations by including computer technologies in their systems (computerization). With the adoption of automatic identification and communication technologies, operations are performed by computer-based automated systems with a minimal human intervention (automatization). More importantly, computerization and automatization of industry must be optimized as computers take charge of systems without human intervention (autonomization) and connect to each other (communicatization).

The ultimate goals of companies can be accomplished with the help of autonomous and interconnected computer systems by minimizing production, storage, and delivery costs (minimizing time can also be considered an objective) while increasing the quality and quantity of products. Therefore, I4 has recently received a big boost from academia and industry. Hence, the pace of transformation to I4 has been accelerated. Academicians focus on finding an effective way to integrate I4-driven technologies into industry; whereas, industry leaders look for opportunities to implement the principles of I4.

I4 not only restructures production systems but also reorganizes all activities from production to delivery of products to customers. Logistics consists of all processes for the transportation of products from production to customers and all services in between. Moreover, it has a profound effect on total costs, especially on storage and delivery costs. Therefore, companies must incorporate I4 technologies in their logistics systems and deliver their products much cheaper and faster than in earlier times. As expected, a cheaper price and a faster delivery can be accomplished by taking advantage of autonomous and interconnected computer technologies in logistics (known as Logistics 4.0). Therefore, companies can have an important advantage over their national or international competitors by increasing the efficiency of their organizations and the satisfaction of their customers, resulting in an increase in the number of customers. For a comprehensive framework of Logistics 4.0, the interested reader is referred to Winkelhaus and Grosse (2020), Abdirad and Krishnan (2020) and Holubčík et al. (2021).

Logistics 4.0 compels all businesses to revisit and optimize their logistics systems. By doing so, most companies tend to focus on their warehousing systems, which require large space for facilities and high investment for materials and equipment, most importantly skilled and expensive labor are required. Recent technological innovations created an incentive for warehouse entrepreneurs or managers to invest in mobile, autonomous, compact, or collaborative systems and reduce escalating labor costs by increasing the productivity of their systems or workforces. Therefore, the fourth revolution has become inevitable in the warehousing industry (called Warehousing 4.0).

As with I4 and shown in Fig. 7.1, the warehouse revolution is classified into four stages. Transition to mechanical systems is considered the first revolution in warehousing (Warehousing 1.0). With the invention of pallets and the use of mechanical

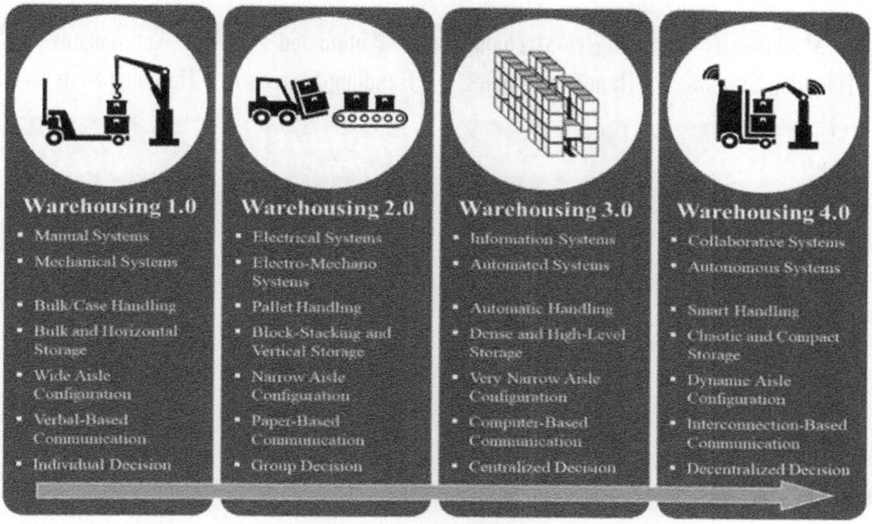

Fig. 7.1 The four stages of the warehousing revolution

or electrical power, the second revolution is materialized based on electro-mechano systems (Warehousing 2.0). The third revolution (Warehousing 3.0) is marked by the development of information, computer, and automation systems. The final revolution (Warehousing 4.0) redesigns all earlier technology with the incorporation of autonomous systems.

The integration of chatbots, artificial intelligent assistants, or social media platforms into e-commerce gives online shopping a big boost going forward. Moreover, the COVID-19 pandemic changed shopping behaviors and daily routines of customers dramatically. During the enforcement of quarantine, lockdown and social distancing, customers turned toward meeting their needs by online shopping, even for food or grocery products. This resulted in a spike in Business-to-Consumer (B2C) sales. Therefore, thousands of different products were being ordered by online daily. In response, companies embraced technological disruptions to stay in business by maintaining profitability, satisfying the demand of customers, and increasing customer satisfaction. At the end of the day, enhancing existing automated systems with technologies inherent in I4, Warehousing 4.0 paved the way for the transformation of warehouses into more adaptive, flexible, and cost-effective facilities.

Although there are different types of warehouses (e.g. raw-material warehouse, finished-good warehouse, distribution warehouse, fulfillment warehouse, etc.), we are not aware of a taxonomy being created for warehouse types. However, the classification of warehouses can be based on specific features, such as design (traditional, non-traditional, etc.), type of operations performed (unit-load, order-picking, etc.), and type of technology used (manual, mechanical, etc.). Interested readers may refer

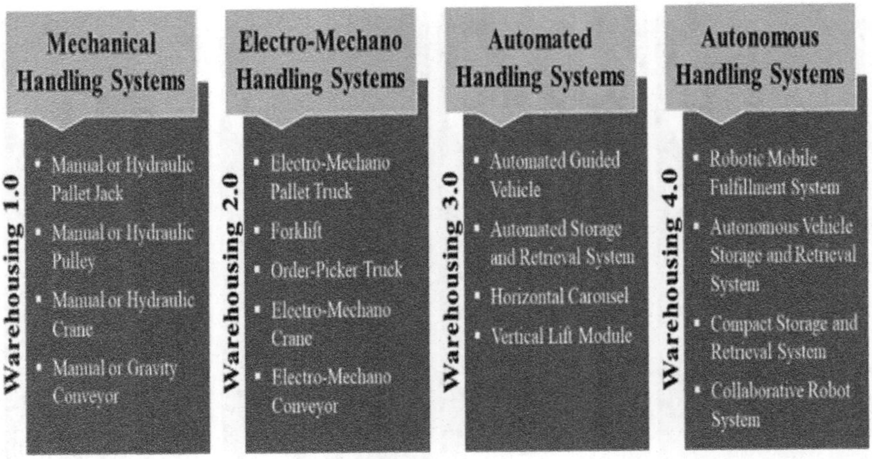

Fig. 7.2 The four levels of warehouse handling systems

to review papers such as Rouwenhorst et al. (2000), De Koster et al. (2007), Gu et al. (2007, 2010), Shah and Khanzode (2017), Kembro et al. (2018) and Masae et al. (2020).

Because the handling system has been the driver of change in warehouses, an evaluation of warehousing can be related to the handling systems used. Moreover, technology has a significant impact in transforming handling systems. Depending on the revolution of warehousing, four levels of technology-based handling can be distinguished: mechanical, electro-mechano, automated and autonomous. Figure 7.2 represents important examples of four levels of handling systems and their relations to the warehousing revolutions.

With different characteristics and the rapid adoption of I4 technologies to warehouse systems, warehouse entrepreneurs or managers are challenged to make the right design, operation, and control decisions. A Robotic Mobile Fulfillment System (RMFS) increases the productivity of pickers by bringing racks to workstations; whereas, an Autonomous Vehicle Storage and Retrieval System (AVS/RS) allows a robot to roam a storage tote independently, thereby increasing the accuracy and speed of operations. On the other hand, high-density storage is provided by a Compact Storage and Retrieval System (CS/RS, also called as a Robot-Based Compact Storage and Retrieval System, RCSRS). Last, but not least, a Collaborative Robot System (CRS) provides maximum efficiency for robots and a safer environment for workers by allowing robots to collaborate with humans on tasks.

This chapter is organized into seven sections. In the following section, the history and transition of warehousing are discussed including the advancement from Warehousing 1.0 to Warehousing 4.0. The following three sections provide brief background information on mechanical, electro-mechano, and automated systems. The sixth section focuses on autonomous systems by considering different types of RMFS, AVS/RS, CS/RS, and CRS. The last section discusses the future of warehousing systems.

7.2 History of Warehousing

The need for storing or preserving products has its roots in early civilizations. Clearly, early humans used caves to store foods such as grains, fruits, and vegetables for difficult times (drought, famine, etc.) or to protect goods such as artifacts, spears, and arrows from invaders. Fisher–hunter–gatherers began to live in permanent villages and cultivate domestic plants; as well as explore nearby regions for fishing, hunting, or/and gathering. With the transition to residential life, humans built storehouses (they can be considered the earliest versions of warehouses) to save their surpluses in case of countering seasonal fluctuations or risky situations. Looking for better opportunities, humans started traveling long distances to barter or trade whatever was caught, hunted, gathered, or/and harvested. Therefore, new trading regions and routes were discovered between cities or countries. With the passage of time, temporary trading regions became permanent trading centers (alias bazaars) in which permanent warehouses (known as granaries) were established.

The Porticus Aamelia, the first emporium constructed by the Roman Empire, is credited as the first example of public warehousing where the distribution of grains to humans took place. Later, the Porticus Minucia became a great storage and distribution center (Patrich 1996). At the end of the Roman Empire, there were almost 300 horrea (public warehouses), which were used mainly for warehousing and distribution (Lampe 2003). The Horrea Galbea was the largest public warehouse with 140 rooms and a total area of 225,000 ft^2. It was used to store the public grain supply, olive oil, wine, foodstuff, clothing, and marble (Potter and Mattingly 1999).

An increase in population size in trading centers resulted in establishing colonies and discovering new trading routes between continents (see the history of amber, incense, silk, and spice routes). In particular, trade between continents was accomplished by transporting products using ships. Therefore, canal hubs became important trading centers and storage locations (called canal warehouses). The Duke's warehouse, built in 1769, is reportedly the first canal warehouse with a shipping hole allowing loading/unloading products from/to barges (Parkinson-Bailey 2000; Nevell and Walker 2001; Nevell 2013).

The development of railroads expanded transportation over lands and enabled humans to transfer products more cheaply and efficiently between canal hubs and interior regions. Therefore, rail warehouses were situated alongside railroads. Liverpool Road Station Warehouse (built in 1830) was regarded as the first railroad warehouse including four floors: a basement for storage (called a cellar), a ground floor used to load/unload trucks, a first floor accessed by railroad for loading/unloading trains and a second floor for lifting products through the warehouse by gravity.

In the early stage of warehousing, storage operations were performed manually, as shown in Fig. 7.3. Typically, unpacked products were stored in single-deep, back-to-back storage rooms in the early warehouses. Moreover, different types of storage rooms were used for different products. Storing products in bags and bales made of cotton, linen, or plastic, etc., the stacking height in warehouses reached at most

Fig. 7.3 Bulk storage

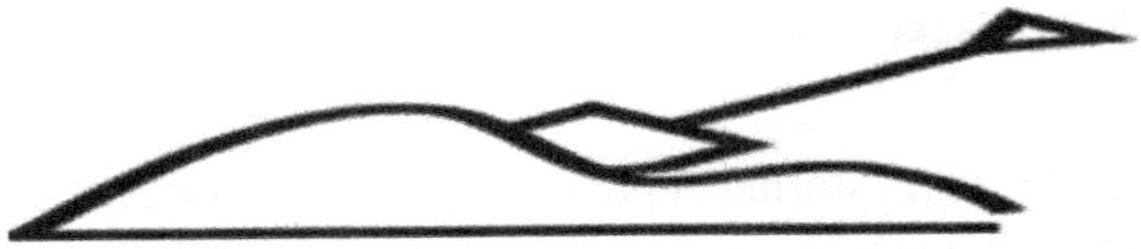

twice human height (6–10 feet). Therefore, storage rooms were extended horizontally (resulting in the size of The Horrea Galbea) and occupied large areas of primitive cities.

With the advent of technology, warehouses were no longer just storage rooms or locations. Some value-added operations were performed such as receiving, inspecting, repackaging, kitting, labeling, storing, preserving, order-picking, retrieving, sorting and delivering, etc. As a result, mechanical systems were increasingly deployed in warehouses to respond to customers' orders on a timely basis and to minimize labor costs. Handling a large volume of orders due to mass-production, warehouses first evolved from mechanical systems to electro-mechano systems by adding motors to the equipment, thereby increasing speed and vertical storage capabilities. Then, an increase in product variety, coupled with reduced volume and the development of the semiconductor, resulted in a transformation to automated systems. During the last decade, warehouses became more sophisticated, diverse, and complex because of customized or personalized orders; evolution was directed toward autonomous systems.

7.3 Warehousing 1.0

The first revolution of warehousing systems began with a structural change by using mechanical systems. With the emergence of Industry 1.0, mechanization contributed to an increase in production capacities of companies. However, storing and handling products were very laborious, time-consuming, and expensive. Therefore, companies began to package their products by using boxes for easier, faster, and cheaper transportation. Although moving boxes was easy, it was neither fast nor cheap. As a result, mechanical and powered systems were incorporated in warehousing operations. Therefore, instead of moving boxes individually, a number of boxes piled or stacked on a skid were transferred easily by using mechanical systems (see Fig. 7.4).

A hand truck (also known as a two-wheel truck or a sack truck) is the simplest and least expensive equipment used in warehouses to move bulky or heavy materials manually by leaning back on two wheels. Accordingly, a four-wheel truck (called a hand cart) is used to transfer products between locations easily by applying manual force (pushing or pulling). Although moving the four-wheel truck requires less power than the two-wheel hand truck, extra effort is required to load and unload product on the flat platform above the ground. Getting under and lifting the skid, a manual or hydraulic pallet jack enables one worker to move products from one location to another. It is simple and it helps immensely in moving multiple boxes or bales.

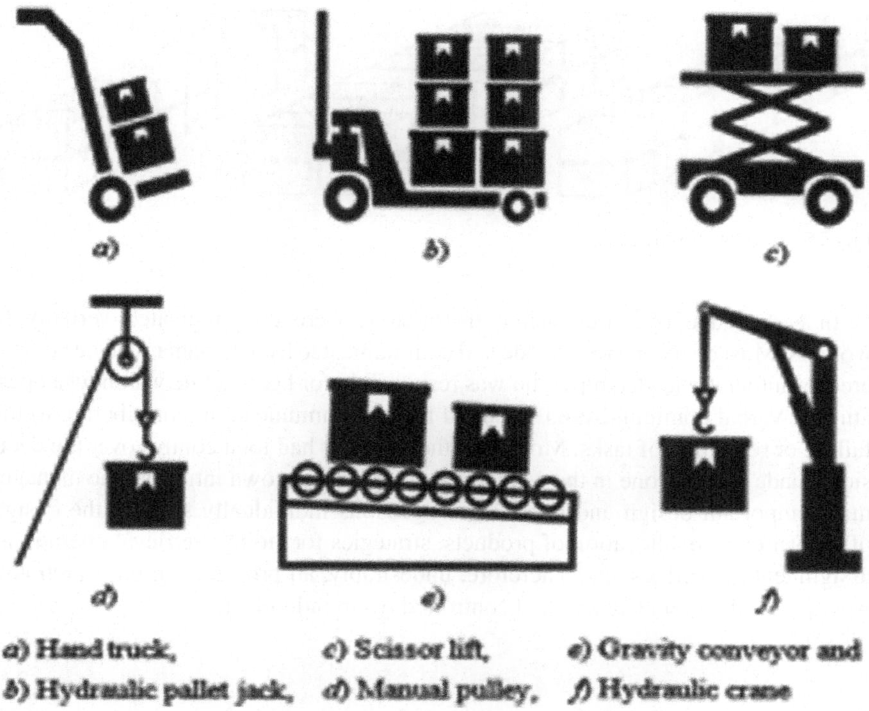

Fig. 7.4 Mechanical system examples

Stacking large or heavy products on the platform, scissor lifts allow humans to access upper storage levels and store products vertically.

Loading or unloading trucks, the pulley is used in warehouses to lift heavy products by changing the direction of force. Because it requires no human power, a gravity conveyor is a preferred-handling equipment. It allows products on rollers to be conveyed by using the force of gravity. Moreover, an unpowered roller conveyor allows products to be moved manually without much effort. In addition to moving products vertically, the crane enables products to move horizontally. Besides, the hydraulic pulley and hydraulic jib crane are also used to lift and moveproducts too heavy for a human.

As stated in the previous section, space utilization was not effective in early stages of warehousing because products could not be stored vertically. With the assistance of hydraulic systems and skids carrying boxes or bales, it was possible to have at most two or three storage levels in warehouses by stacking products on top of each other or storing on shelves. Therefore, the length of warehouses was reduced by a half or a third, as shown in Fig. 7.5. Moreover, aisles were designed wide enough to offer ample space for handling equipment.

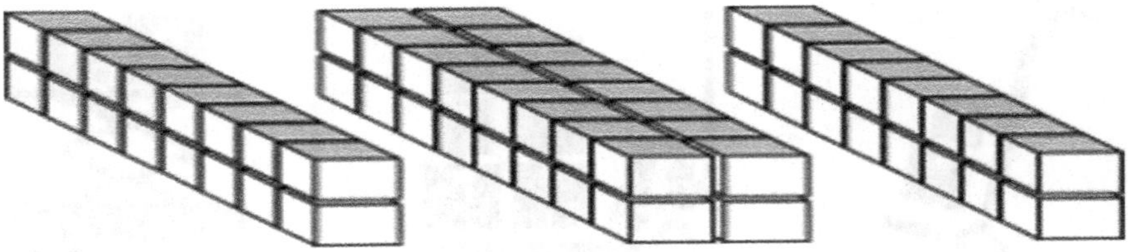

Fig. 7.5 Horizontal storage

In early stage of warehousing, instructions were communicated verbally to workers. Most decisions were made and communicated by a manager (can be considered as autocratic leadership) who was responsible for keeping the warehouse operational. Verbal communication often led to miscommunication, contributing to the failure or repetition of tasks. Moreover, the manager had total control over the decisions made by everyone in the warehouse. Using his/her own intuition, the manager made important design and operational decisions individually such as the design of the warehouse, allocation of products, strategies for storage/retrieval operations, assignment of workers, etc. Therefore, undesirably, all processes in the warehouse were planned, organized, led, and controlled by an individual.

7.4 Warehousing 2.0

The second revolution in warehousing was characterized by the transformation of mechanical systems to electro-mechano systems incorporating an internal combustion engine or electric battery. As a result of electrification and mass-production in Industry 2.0, the production capacity of companies increased dramatically. Therefore, the storage operation for a large number of items was necessary. The invention of wooden pallets resulted in stacking products on pallets, hence, loading, transferring, and unloading products faster and cheaper by using electro-mechano systems (see Fig. 7.6). Moreover, pallets were stacked on top of each other in blocks separated by aisles depending on the stackability of products. As a result, pallets became a vital part of warehouses. However, conventional racks were used to store large volumes of products. Therefore, both transportation and storage systems were improved in warehouses. Consequently, powered conveyor systems were used to speed up the transportation of products.

The powered pallet truck extended the lifting and maneuvering capability of the manual pallet jack; it also allowed the worker to ride, rather than walk. Propelling the hand cart throughout a fixed path, the towline conveyor was used to provide power to four-wheel trucks. Forklifts allowed workers to efficiently load, transfer, and unload pallets regardless of the weight of pallets.

Storing products above floor level of conventional racks, order-picker trucks were used to pick orders on upper levels. Powering a series of pulleys by engines, conveyors

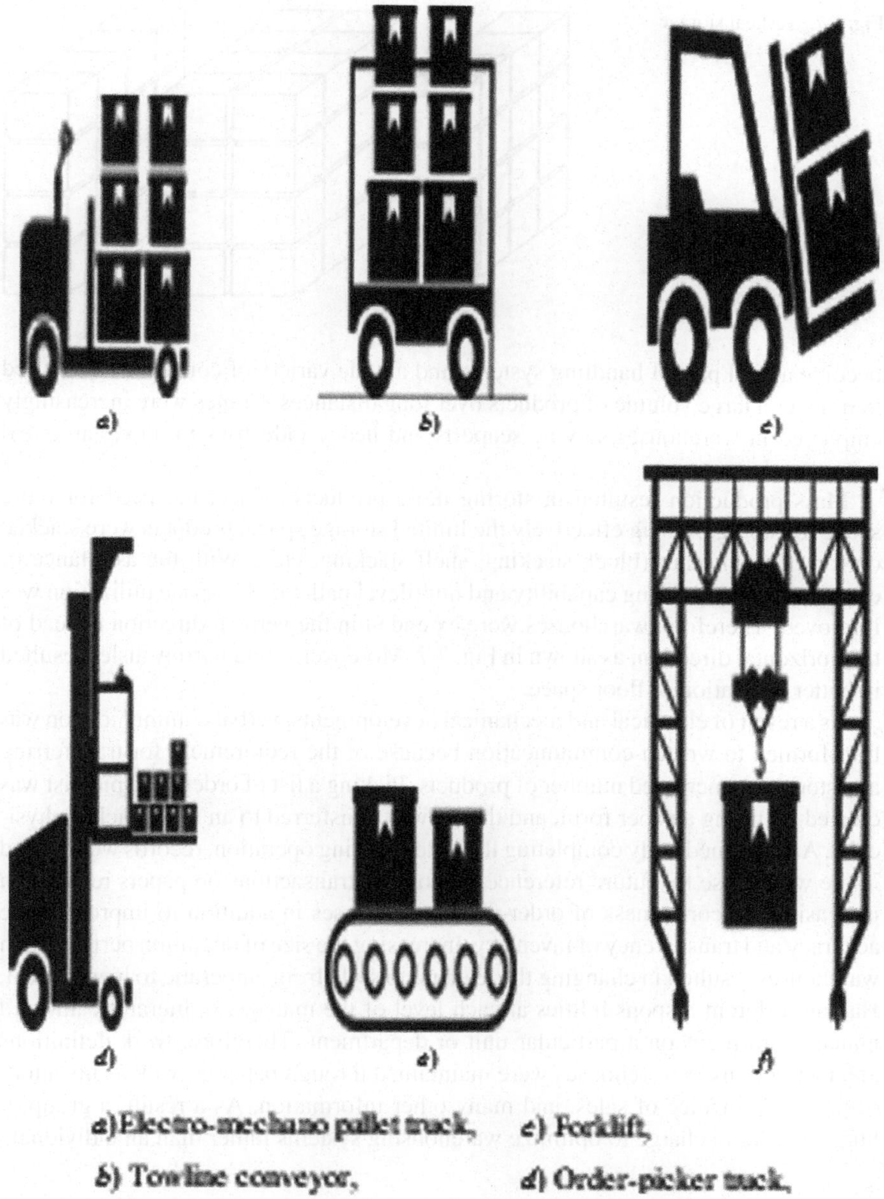

Fig. 7.6 Electro-mechano system examples

Fig. 7.7 Vertical storage

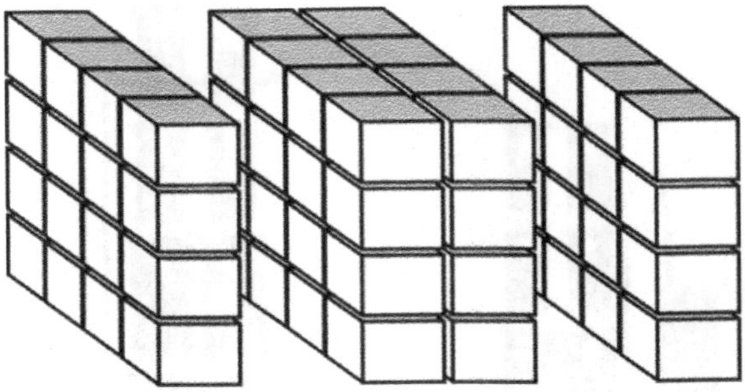

become a vital part of handling systems and a wide variety of conveyors were used to transfer a large volume of products over long distances. Cranes were increasingly employed in warehouses serving seaports and heavy industries to move large and heavy products.

Mass production resulted in storing more products, hence, the need for more space for storage. Using effectively the limited storage space, products were stacked on top of each other (block-stacking, shelf stacking, etc.). With the assistance of equipment having lifting capability and multilevel pallet racks, space utilization was improved. Therefore, warehouses were extended in the vertical direction instead of the horizontal direction, as shown in Fig. 7.7. Moreover, using narrow aisles resulted in better utilization of floor space.

As a result of electrical and mechanical developments, verbal communication was transformed to written communication because of the requirement for transferring and storing an increased number of products. Picking a list of orders, the picklist was created by filling a paper form, and then it was transferred to an order-picker physically. After immediately completing the order-picking operation, records were stored in the warehouse for future reference. Recording transactions on papers resulted in increasing the correctness of order-picking processes in addition to improving the accuracy and transparency of inventory. Increasing the size of operation performed in warehouses resulted in changing the leadership style from autocratic to hierarchical. Having different responsibilities at each level of the managerial hierarchy allowed managers to focus on a particular unit or department. Therefore, work definitions and instructions in warehouses were maintained through paper, as well as inventory of products, invoice of sales, and many other information. As a result, a group of humans were in charge to optimize warehousing systems rather than an individual.

7.5 Warehousing 3.0

The third revolution in warehousing was triggered by the improvement in information, computer and automation systems. Using computer-based automated systems was one of the most important disruptions in warehousing. The way decisions were made, products were handled, and warehouses were designed changed dramatically.

Automated vehicles began to abrogate the necessity of human power and human control in the warehouse environment. With the help of automated vehicles controlled by a computer, automated systems increased the throughput and efficiency of warehouses by reducing human intervention, hence warehouses became more centralized (see Fig. 7.8).

Dealing with higher labor costs and higher demands, products were transferred easily and coherently in a warehouse by using Automated Guided Vehicles (AGVs). An alternative to an AGV system is an Automated Storage and Retrieval System (AS/RS) traveling horizontal and vertical directions simultaneously. Consistent with the digitalization of facilities, AS/RSs changed the process of orders by increasing storage capacity, utilization of the warehouse footprint, labor productivity, safety, and security. However, requiring a high initial capital investment and locking the warehouse into an inflexible design, decisions to use an AS/RS were made carefully. Different AS/RS versions were used in industry, such as unit-load AS/RS, mini-load AS/RS, micro-load AS/RS, man-on-board AS/RS, etc. Rotating bins, horizontal carousels allowed products to be brought to order-pickers, thereby increasing the productivity of order-pickers. Different from horizontal carousels, Vertical Lift Models (VLMs) improved the ergonomic conditions for order-pickers and space utilization.

Using automated systems resulted in warehouses being taller than ever before (AS/RSs and VLMs). Additionally, storage density was increased with the usage of very narrow aisle configurations. Therefore, storage efficiency was improved in both horizontal and vertical directions as illustrated in Fig. 7.9.

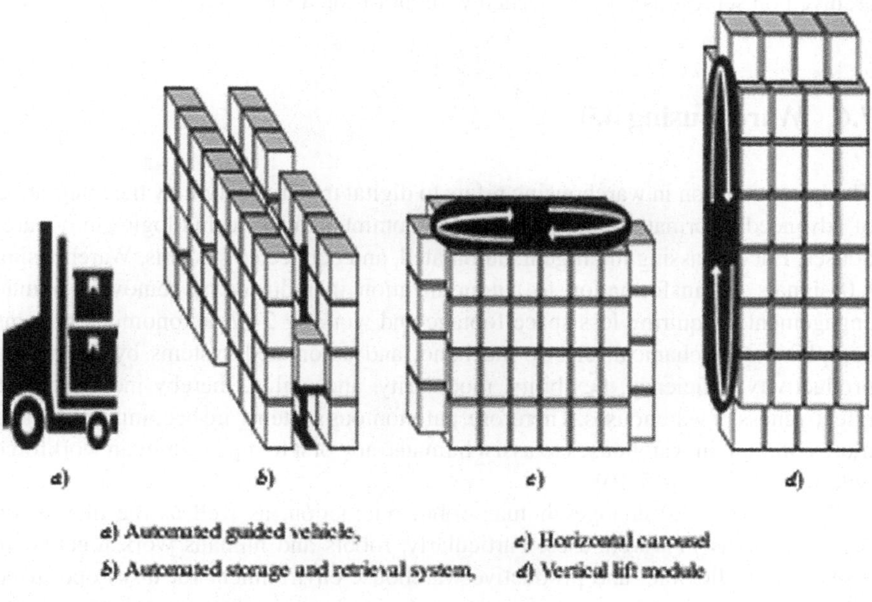

Fig. 7.8 Automated system examples

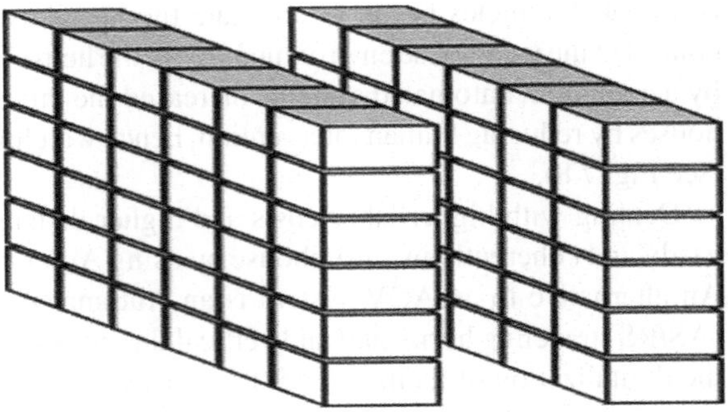

Fig. 7.9 Dense and high-level storage

With the advent of computer technology, Warehouse Management Systems (WMSs) emerged. Therefore, transition from a paper-based communication to a computer-based communication resulted in a paperless and automated warehouse environment. As a result of computerization and automatization, orders were batched in an optimal fashion and order-pickers were instructed by a central computer. Therefore, the visibility of products and the productivity of order-pickers were increased; whereas order-picking costs were decreased by batching orders and instructing order-pickers to follow optimal routes. Additionally, automatic identification (barcode, optical character recognition system, radio frequency identification tag, etc.) and communication (voice headset, pick-to-light system and pick-to-voice system, etc.) technologies enabled automated data collection systems and allowed real-time tracking of changes in the inventory. Electronic records were stored and archived on servers as long as needed without losing a single record.

7.6 Warehousing 4.0

The last revolution in warehousing refers to digital transformation by the integration of advanced information, automation, and communication technologies into warehouses. Encompassing intelligent, automated, and connected systems, Warehousing 4.0 signals a transformation to autonomization in industry by removing human engagement. Requiring less space footage and working 24/7, autonomous systems complement mechanical, electro-mechano, and automated systems by improving productivity, efficiency, flexibility, modularity, and agility, thereby increasing the effectiveness of warehouses. Therefore, autonomous systems are becoming more and more common in warehouses to assist humans, at worst to replace human workforce with robots (see Fig. 7.10).

Warehousing 4.0 changes human–robot interaction, as well as the interaction between robots in warehouses. Particularly, robots and humans work together in a safer, more flexible, and productive warehouse environment for tasks optimized by interconnected on-board computers. Reducing human effort and injuries, heavy,

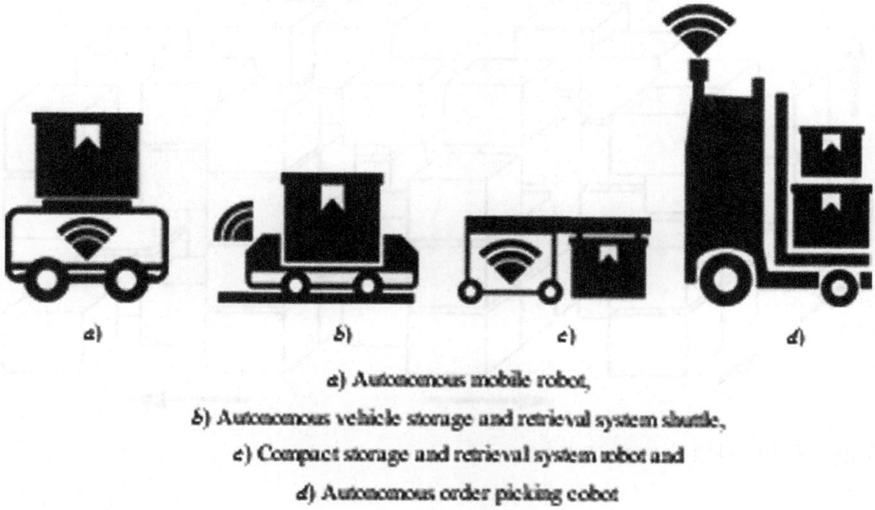

Fig. 7.10 Autonomous system examples

dangerous, or/and risky tasks are assigned to robots. By the end of the day, warehouses are expected to be transformed into intelligent warehouses with decentralized intelligent decision-making processes, innovative autonomous robotic technologies, and interconnected communication systems. We refer interested readers to Azadeh et al. (2019), Boysen et al. (2019), Jaghbeer et al. (2020), Da Costa Barros and Nascimento (2021) and Kumar et al. (2021) for information on relevant literature and additional research opportunities.

Using autonomous technologies in warehouses allows a new breed of warehousing systems. Reducing a constant need for humans, autonomous systems operate in complex warehouse environments freely and independently. Rather than directing an order-picker to a pick location throughout the warehouse (person-to-goods), the RMFS provides a boost in order-picking performance by bringing products directly to the order-picker's workstation (goods-to-person). Instead of using fixed-path cranes, an AVS/RS enables access to any storage/retrieval location at any level and aisle. Therefore, denser storage can be obtained with a flexible and modular design. Achieving a much higher storage density and maximizing space utilization, the CS/RS allows products to be stored in bins stacked on top of each other. Allowing robots to collaborate with humans on tasks, the CRS maximizes the joint performance with an optimized task assignment by maximizing the efficient use of robots and minimizing discomfort to the human body.

Autonomous systems reduce unnecessary aisle space in warehouses because autonomousvehicles require less aisle space than a manual, mechanical, electromechano, or automated vehicle (see Fig. 7.11). Moreover, storing products randomly in available locations results in optimizing floor space utilization.

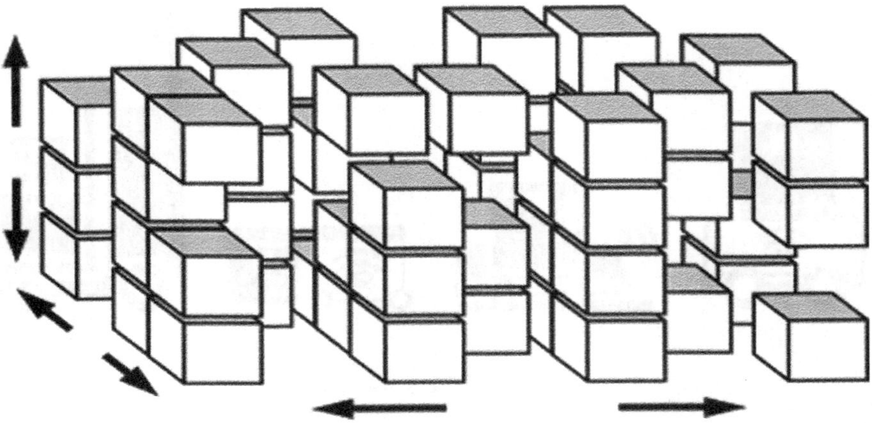

Fig. 7.11 Chaotic and compact storage

Although employing autonomous vehicles demands a high level of control, a decentralized control (an on-board computer system) of interconnected autonomous vehicles will provide a great opportunity to make better operational decisions such as routing, batching, etc. Therefore, energy consumption will be minimized. Moreover, advances in robotic and sensor technologies will improve the safety and security of warehouses. Likewise, inventory can be tracked by using autonomous tag reader robots or drones, rather than physical counts of inventory requiring days for large warehouses. Therefore, considerable money and time can be saved by counting and tracking inventory continuously, as well as increasing the visibility of products and accuracy of inventory.

7.6.1 Robotic Mobile Fulfillment System (RMFS)

An RMFS enhances the capability of an Autonomous Mobile Robot (AMR) to lift and transfer a mobile rack from/to a storage area to/from a workstation based on instructions from a WMS without human interference (see Fig. 7.12). Therefore, the RMFS allows a human order-picker to remain stationary while AMRs are moving products to the workstation. Eliminating the walk distance for the order-picker and increasing the number of picks per hour, the RMFS greatly improves the productivity of the order-picker by assigning non-value added and repetitive tasks to AMRs. Rather than following a fixed predetermined path, the AMR interprets its environment and performs given tasks efficiently with the assistance of an on-board computer system and innovative sensors. Moreover, the RMFS allows warehouses to be dynamically designed and optimized by arranging the number of AMRs employed.

Typically, an RMFS consists of three components: AMRs, mobile racks, and workstations. The AMR is instructed to transfer a mobile rack between a storage location

Fig. 7.12 Robotic mobile fulfillment system

and a workstation by an on-board computer system or a central WMS depending on the decision-making process. Handling a range of payloads from 100 to 1500 kgs, an AMR moves underneath a mobile rack, lifts, and transports the mobile rack to a workstation (rotating about its center if it is necessary). After waiting until its turn in a buffer area, order(s) will be picked at a workstation by an order-picker. After immediately completing the order-picking process, the mobile rack will be transferred to a storage location with the location determined by a decision-making system depending on the frequency of demand for products stored in the mobile rack.

There are a growing number of studies focused on the RMFS including storage area design, storage policy, performance estimation and charging strategies, etc. For further details, the reader is referred to Boysen et al. (2017), Lamballais et al. (2017), Zou et al. (2017, 2018), Lee et al. (2019), Merschformann et al. (2019), Roy et al. (2019), Gong et al. (2020), Wu et al. (2020) and Xie et al. (2021).

Offering a new approach to goods-to-person picking, the first RMFS was patented by KIVA Systems Inc. (Mountz et al. 2008). Amazon acquired the company and rebranded it as Amazon Robotics™ LLC (Amazon Robotics). Since then, Amazon stopped selling AMRs and employed thousands of AMRs internally. Closing the gap in the market, new startups emerged: for instance, Boby™ (Scallog), CarryPick™ (Swisslog), Freight500™ and Freight1500™ (Fetch Robotics), GoFer-Bot™ (Conveyco), Grenzebach L600™ (Grenzebach), MiR100™, MiR500™ and MiR1000™ (Mobile Industrial Robots), Open Shuttle™ (Knapp), OTTO100™, OTTO750™ and OTTO1500™ (Clearpath Robotics), Racrew™ (Hitachi), Ranger™ GTP (GreyOrange), and others.

RMFS manufacturers claim the RMFS will reduce travel time by 5 times, improve the productivity of order-pickers, improves speed, increase pick accuracy, lower operating costs, empower workers, reduce complexity, increase flexibility, increase modularity, and ease integration into existing environments (Conveyco, Grenzebach, GreyOrange, and Swisslog).

7.6.2 Autonomous Vehicle Storage and Retrieval System (AVS/RS)

Requiring an autonomous robot to store and retrieve products, an AVS/RS improves traditional AS/RSs by allowing the number of robots to be increased or decreased depending on the desired throughput capacity (see Fig. 7.13). Moreover, recent developments in movement patterns contribute flexibility and modularity of the AVS/RS. Therefore, the AVS/RS is more flexible and extendable than a traditional AS/RS. Unlike storage/retrieval cranes in AS/RS, the autonomous vehicle or shuttle can access any storage/retrieval location at any level in any aisle. Therefore, the AVS/RS increases the storage density of a warehouse by using storage space more efficiently.

The AVS/RS can be classified on the basis of three categories, depending on the movement in the system: horizontal, vertical, and diagonal systems (Azadeh et al. 2019). A horizontal AVS/RS requires autonomous vehicles to move products horizontally using rails and lifts to move autonomous shuttles vertically. Eliminating the necessity for using lifts, a vertical AVS/RS transfers products by allowing autonomous vehicles to travel in both vertical and horizontal directions (not simultaneously, first vertically then horizontally, or vice versa). Recently, a diagonal AVR/RS

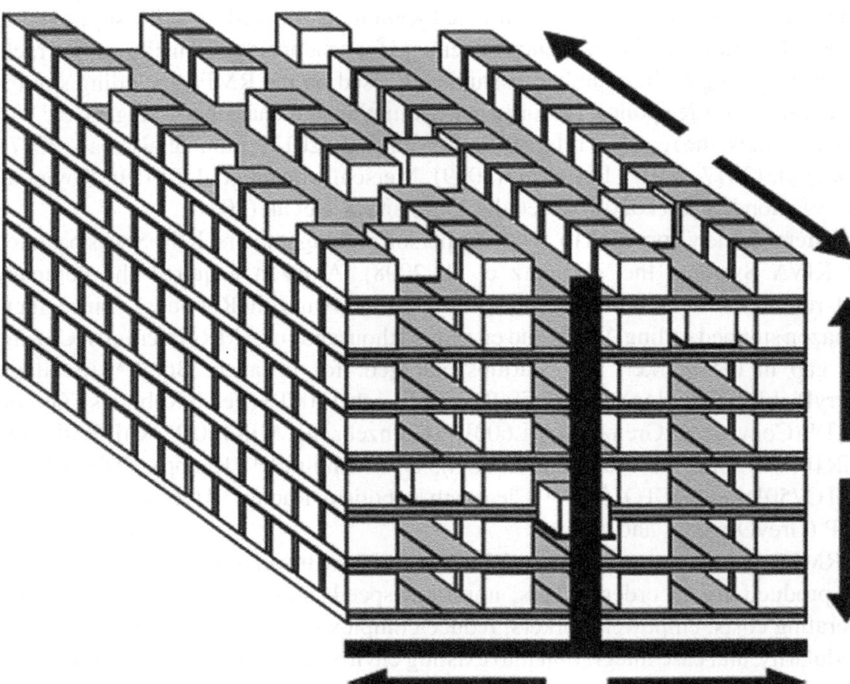

Fig. 7.13 Autonomous vehicle storage and retrieval system

was introduced by allowing the autonomous robot to travel in both horizontal and diagonal directions simultaneously.

A large number of studies have been conducted related to the AVS/RS. Malmborg (2002) is credited with developing a tool for modeling the expected performance of the AVS/RS. Following this study, considerable research has been devoted to different versions of the AVS/RS. We refer the interested reader to recent studies on the AVS/RS, including but not limited to, D'Antonio et al. (2018), Ekren et al. (2018), Lerher (2018), Ha and Chae (2019), Xu et al. (2020), Jerman et al. (2021), Ekren (2021) and Rahim et al. (2021).

Savoye Logistics is credited with the first introduction of the AVS/RS (Azadeh et al. 2019). Later on, different types of AVS/RS have been introduced by different companies, such as Adapto™ (Vanderlande), CycloneCarrier™ (Swisslog), iBot™ (Opex), OSR Shuttle™ (Knapp), RackRacer™ (Fraunhofer IML), Skypod™ (Exotec), etc.

AVS/RS manufacturers claim that the AVS/RS will minimize running cost, reduce storage cost, increase throughput capacity to handle peak demands efficiently, increase picking quality by reducing failures, and increase flexibility (Knapp, Swisslog, and Vanderlande).

The CS/RS expands storage capacity by optimizing space utilization. Storing products in bins and storing bins on top of each other, a storage grid is formed (see Fig. 7.14). Therefore, autonomous vehicles operate on the top of the storage grid by storing/retrieving bins. A very high storage density can be accomplished because the CS/RS does not require aisle structure. Moreover, the CS/RS brings more flexibility to warehousing systems by allowing the storage grid to expand with more bins or by employing additional robots or/and consignment modules. Also, storage locations can be designed to be dynamic by moving bins on a shuttle system (Azadeh et al. 2019).

Four parts of a CS/RS are: the storage grid, bins, autonomous robots, and consignment modules. Supporting the CS/RS, the storage grid offers space for bins to be stacked on top of each other. Products are stored and retrieved in bins randomly. Moving horizontally on top of the storage grid, autonomous robots lift bins and transport them to consignment modules. Any robot can move any carrier and transport it to any consignment module. Depending on the system specifications, consignment modules can be installed on one face or multiple faces of the CS/RS. Moreover, the CS/RS learns the order frequency of bins and stores them on the top layer of the storage grid.

The CS/RS is only considered by Zou et al. (2016) so far and described in the review papers by Azadeh et al. (2019) and Boysen et al. (2019, 2021). The first implementation of the CS/RS is AutoStore™ by Hatteland (Azadeh et al. 2019). Instead of selling the CS/RS to customers directly, Hatteland launched a global integrators program. Therefore, AutoStore™ is distributed, designed, installed, and serviced by integrator partners (AutoStore). Integrator partners of AutoStore™ include Bastian Solutions, Dematic, Fortna, Hörmann, Pulse and Swisslog, among others.

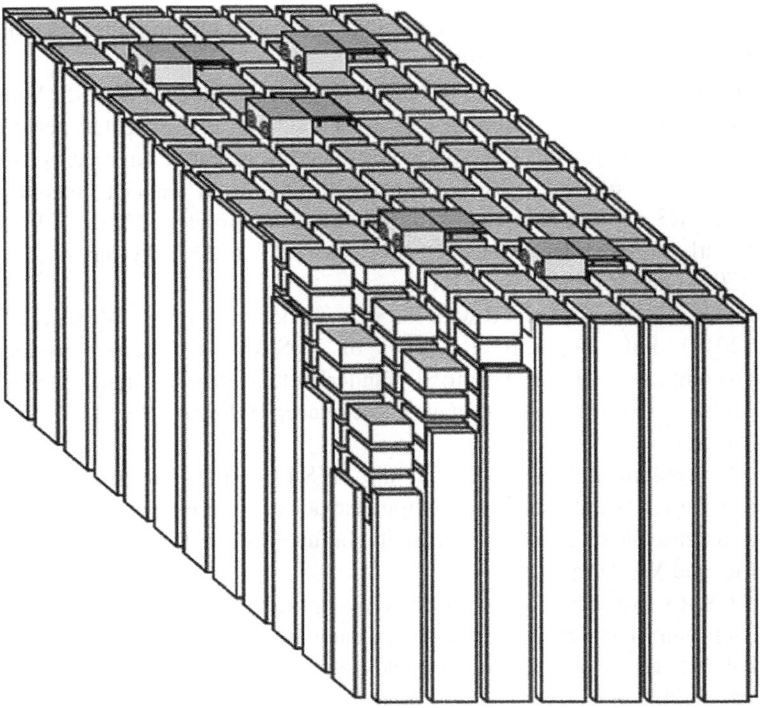

Fig. 7.14 Compact storage and retrieval system (AutoStore)

AutoStore claims that the AutoStore™ will increase flexibility, increase throughput, increase storage capacity, increase modularity, be relatively inexpensive compared with other automated systems, reduce travel distance with a compact storage space and shorten response times.

An important system related to AutoStore is GridStore (see Fig. 7.15), a rectangular grid ofsquare conveyor modules that features decentralized control to move items in the four cardinal directions (Gue 2006; Gue and Kim 2007; Alfieri et al. 2010; Gue et al. 2014; Mirzaei et al. 2017 and Yalcin et al. 2019). Communicating with four neighbors of a conveyor module, items are moved from one side of the system to the opposite side. Extending GridStore system, new grid-based storage systems are developed such as GridPick (Uludağ 2014), GridSorter (Seibold 2015), Live-Cube Storage System (Zaerpour et al. 2017a, b, c), GridHub (Shirazi 2018), and GridPick+ (Shekari et al. 2021).

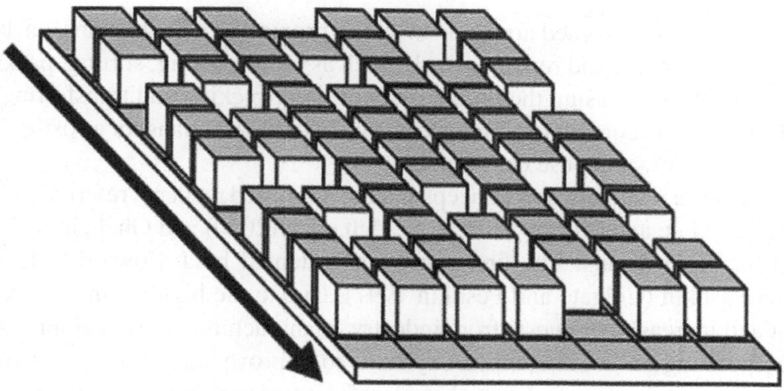

Fig. 7.15 Compact storage and retrieval system (GridStore)

7.6.3 Collaborative Robot System (CRS)

Although autonomous systems are being used increasingly in industry, humans still play important roles in completing tasks. Human and robot interactions have received a big boost in recent years. In a shared region, humans and robots can work together on tasks. For example, the collaboration between humans and robots has become popular for order-picking processes because of the joint benefit of each partner (see Fig. 7.16). Moving pallets, bins, totes, containers etc. CRSs can assist humans by following or leading them. After completion of a list of orders or fulfilling the container, collaborative robots (cobots) can transfer products to workstations. Therefore, order-pickers can continue order-picking with another cobot. Different from traditional industrial robots, the cobot is equipped with sophisticated sensors to detect the locations of humans and stop, if necessary. Moreover, cobots take on heavy and dangerous work, thereby protecting humans from injury.

A CRS includes two important parts: human and cobot. Humans are always involved in complex warehousing systems. Combining human skills with powerful

Fig. 7.16 Collaborative robot system

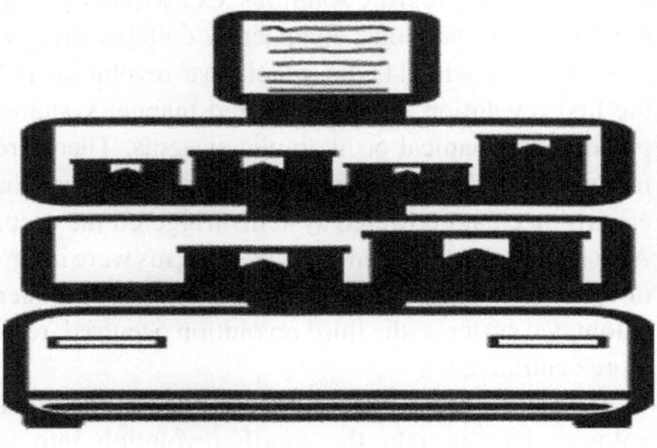

CRSs, cobots are designed to work in close proximity with humans on boring, tedious, monotonous, and repetitive tasks such as order-picking, sorting, packaging, tag reading, etc. Increasing the productivity of order-picking and transferring products at a faster rate, cobots are employed to follow or lead humans for a cost-effective, safe, and flexible warehouse environment.

The CRS is a relatively new concept and the interested reader is referred to Löffler et al. (2018), Lee and Murray (2019), Azadeh et al. (2020) and Ghelichi and Kilaru (2021) for more details. The first cobot was patented by J. Edward Colgate and Michael Peshkin (Colgate and Peshkin 1997). Due to the highly competitive robot market and increasing requests from industry, manufacturers are developing cobots by incorporating on-board computer systems to improve autonomy and integrating innovative vision systems to detect obstacles: for instance, Chuks™ (6 River System), HMIShelf™ (Fetch Robotics), LocusBots™ (Locus Robotics), Pick-n-Go™ (Kollmorgen), Ranger™ Mobile Sort and Ranger™ Pick (GreyOrange), SOTO™ and TORU™ (Magazino), Stacker-Bot™ (Conveyco), Swift™ (IamRobotics), etc.

CBS manufacturers claim that the CBS will optimize the order-picking process, provide a safe working environment for order-pickers, and ensure error-free order-picking operations, work collaboratively with human teams or autonomously 24/7, improve accuracy and minimize order-picking walking distance (GreyOrange, Locus Robotics, and Magazino)

7.7 Summary and Conclusion

This chapter presented the state of advancement of Warehousing 4.0 and provided an overview of emerging technologies in warehousing systems. As a result of technological and scientific developments, behaviors of customers change frequently. Additionally, the COVID-19 pandemic resulted in increased demand for online shopping systems and faster customer delivery. As a result, people are getting used to shopping online and ordering millions of products daily or weekly. Responding to customers' 2-day or same-day delivery expectations, companies must revisit their service levels and change their logistic solutions. Consequently, warehouse companies are tending to invest in autonomous, compact, or collaborative systems.

Consistent with I4, the warehouse revolution is classified into four stages. In the first revolution, human-powered manual systems were transformed to human-powered mechanical or hydraulic systems. Therefore, instead of moving products individually, a number of products were stacked and handled easily. The invention of electro-mechano powered systems triggered the second revolution in warehousing. Accordingly, handling and storage systems were improved in warehouses. As a result of the invention of the semiconductor and consequent advances in computers and automated systems, the third revolution occurred, resulting in warehouses becoming more centralized.

The integration of advanced information, automation and communication technologies has brought the fourth revolution into warehouses. Warehousing 4.0

improves human–robot interaction and results in a safer, more flexible, and more productive warehouse environment. Bringing racks to stationary workstations, the productivity of order-pickers is improved by the RMFS. The AVS/RS increases the accuracy and speed of operations. Storing bins on top of each other, the CS/RS provides a much higher storage density. Combining human skills with powerful CRSs, the efficiency of robots is increased by providing a safer warehouse environment for workers.

With the emergence of new autonomous technologies, unaddressed research questions arise. Given the wealth of unanswered questions, opportunities for new studies exist. Through research and evaluations of new and emerging technologies, improvements in the performance of autonomous systems will lead to Warehousing 5.0 in the not-too-distant future.

Acknowledgements The author wishes to thank Dr. John A. White for his valuable comments regarding the revolution of warehousing and constructive criticism improving the chapter significantly. This work was supported by the Scientific and Technological Research Council of Turkey (TUBITAK) (Grant Number. 1059B191900637).

References

Abdirad M, Krishnan K (2020) Industry 4.0 in logistics and supply chain management: a systematic literature review. Eng Manage J. https://doi.org/10.1080/10429247.2020.1783935

Amazon Robotics: We reimagine now. http://www.amazonrobotics.com. Accessed 14 July 2020

AutoStore: Space redefined. https://autostoresystem.com. Accessed 14 July 2020

Azadeh K, De Koster R, Roy D (2019) Robotized and automated warehouse systems: review and recent developments. Transp Sci 53(4):917–945

Azadeh K, Roy D, De Koster R (2020) Dynamic human-robot collaborative picking strategies. SSRN Electron J. https://doi.org/10.2139/ssrn.3585396

Boysen N, Briskorn D, Emde S (2017) Parts-to-picker based order processing in a rack-moving mobile robots environment. Eur J Oper Res 262(2):550–562

Boysen N, De Koster R, Füßler D (2021) The forgotten sons: warehousing systems for brick-and-mortar retail chains. Eur J Oper Res 288(2):361–381

Boysen N, De Koster R, Weidinger F (2019) Warehousing in the e-commerce era: a survey. Eur J Oper Res 277(2):396–411

Clearpath Robotics: Modern enterprises need modern material handling. https://ottomotors.com/amrs. Accessed 14 July 2020

Colgate JE, Peshkin MA (1997) Cobots. U.S. Patent No. US5952796A

Conveyco: AMR flow rack shelving replenishment robotic stacker-bot. https://www.conveyco.com/technology/autonomous-mobile-robots-amrs/flow-rack-stacker-bot. Accessed 14 July 2020

Da Costa Barros ÍR, Nascimento TP (2021) Robotic mobile fulfillment systems: a survey on recent developments and research opportunities. Robot Auton Syst 137. https://doi.org/10.1016/j.robot.2021.103729

De Koster R, Le-Duc T, Roodbergen KJ (2007) Design and control of warehouse order picking: a literature review. Eur J Oper Res 182(2):481–501

D'Antonio G, Maddis MD, Bedolla JS (2018) Analytical models for the evaluation of deep-lane autonomous vehicle storage and retrieval system performance. Int J Adv Manuf Technol 94:1811–1824

Ekren BY (2021) A multi-objective optimisation study for the design of an AVS/RS warehouse. Int J Prod Res 59(4):1107–1126

Ekren BY, Akpunar A, Sari Z, Lerher T (2018) A tool for time, variance and energy related performance estimations in a shuttle-based storage and retrieval system. Appl Math Model 63:109–127

Exotec: Skypod system changes the rules of intralogistics. https://www.exotec.com/en/solutions. Accessed 14 July 2020

Fetch Robotics: The fetch robotics virtual conveyor. https://fetchrobotics.com/products-technology/virtualconveyor. Accessed 14 July 2020

Fraunhofer: Rackracer versatile and flexible rack feeder for container racks. https://www.iml.fraunhofer.de/content/dam/iml/en/documents/OE140/Produktblatt_RACKRACER_EN_2014.pdf. Accessed 14 July 2020

Ghelichi Z, Kilaru S (2021) Analytical models for collaborative autonomous mobile robot solutions in fulfillment centers. Appl Math Model 91:438–457

Gu JX, Goetschalckx M, McGinnis LF (2007) Research on warehouse operation: a comprehensive review. Eur J Oper Res 177(1):1–21

Gu JX, Goetschalckx M, McGinnis LF (2010) Research on warehouse design and performance evaluation: a comprehensive review. Eur J Oper Res 203(3):539–549

Gong Y, Jin M, Yuan Z (2020) Robotic mobile fulfilment systems considering customer classes. Int J Prod Res. https://doi.org/10.1080/00207543.2020.1779370

Grenzebach: Transforming how you do intralogistics. https://www.grenzebach.com/products-markets/intralogistics/goods-to-person-solutions. Accessed 14 July 2020

GreyOrange: The Ranger™ series of intelligent fulfillment robots. https://www.greyorange.com/fos-overview/ranger-robot-series. Accessed 14 July 2020

Gue KR (2006) Very high density storage systems. IIE Trans 38:93–104

Gue KR, Furmans K, Seibold Z, Uludağ O (2014) GridStore: a puzzle-based storage system with decentralized control. IEEE Trans Autom Sci Eng 11(2):429–438

Gue KR, Kim BS (2007) Puzzle-based storage systems. Nav Res Logist 54(5):556–567

Ha Y, Chae J (2019) A decision model to determine the number of shuttles in a tier-to-tier SBS/RS. Int J Prod Res 57(4):963–984

Hitachi: Autonomous driving without the floor marker guiding system. https://www.hitachi.com/New/cnews/month/2015/08/150804a.pdf. Accessed 14 July 2020

Holubčík M, Koman G, Soviar J (2021) Industry 4.0 in logistics operations. Transp Res Proc 53:282–288. https://doi.org/10.1016/j.trpro.2021.02.040

IamRobotcs: I AM Swift. https://www.iamrobotics.com/products/swift. Accessed 14 July 2020

Jaghbeer Y, Hanson R, Johansson MI (2020) Automated order picking systems and the links between design and performance: a systematic literature review. Int J Prod Res 58(15):4489–4505. https://doi.org/10.1080/00207543.2020.1788734

Jerman B, Ekren BY, Küçükyaşar M, Lerher T (2021) Simulation-based performance analysis for a novel avs/rs technology with movable lifts. Appl Sci (switzerland) 11(5):1–16

Kembro JH, Norrman A, Eriksson E (2018) Adapting warehouse operations and design to omni-channel logistics: a literature review and research agenda. Int J Phys Distrib Logist Manage 48(9):890–912. https://doi.org/10.1108/IJPDLM-01-2017-0052

Kollmorgen: Increase order picking productivity up to 100% and decrease operational costs up to 40% with Pick-n-Go® from Kollmorgen. https://www.kollmorgen.com/en-us/company/newsroom/press-releases/2009/increase-order-picking-productivity-up-to-100-and-decrease-operational-costs-up-to-40. Accessed 14 July 2020

Kumar S, Narkhede BE, Jain K (2021) Revisiting the warehouse research through an evolutionary lens: a review from 1990 to 2019. Int J Prod Res (online). https://doi.org/10.1080/00207543.2020.1867923

Lamballais T, Roy D, De Koster R (2017) Estimating performance in a robotic mobile fulfillment system. Eur J Oper Res 256:976–990

Lampe P (2003) Christians at Rome in the first two centuries: from Paul to Valentinus. Fortress Press, Minneapolis

Lee CKM, Lin B, Ng KKH, Lv Y, Tai WC (2019) Smart robotic mobile fulfillment system with dynamic conflict-free strategies considering cyber-physical integration. Adv Eng Inform 42, Article 100998

Lee HY, Murray CC (2019) Robotics in order picking: evaluating warehouse layouts for pick, place, and transport vehicle routing systems. Int J Prod Res 57(18):5821–5841

Lerher T (2018) Aisle changing shuttle carriers in autonomous vehicle storage and retrieval systems. Int J Prod Res 56(11):3859–3879

Locus Robotics: Robots empowering people. https://locusrobotics.com. Accessed 14 July 2020

Löffler M, Boysen N, Glock C, Schneider M (2018) Picker routing in AGV-assisted order picking systems. Working paper, DPO-2018–01 (version 7, 23.10.2020), 01:1–43

Magazino: Robot revolution intelligent and mobile automation technology for warehousing and intralogistics. https://www.magazino.eu/products/?lang=en. Accessed 14 July 2020

Malmborg CJ (2002) Conceptualizing tools for autonomous vehicle storage and retrieval systems. Int J Prod Res 40(8):1807–1822

Masae M, Glock CH, Grosse EH (2020) Order picker routing in warehouses: a systematic literature review. Int J Prod Econ 224. https://doi.org/10.1016/j.ijpe.2019.107564

Merschformann M, Lamballais T, De Koster R, Suhl L (2019) Decision rules for robotic mobile fulfillment systems. Oper Res Perspect 6:100128

Mirzaei M, De Koster R, Zaerpour N (2017) Modelling load retrievals in puzzle-based storage systems. Int J Prod Res 55(21):6423–6435

Mobile industrial robots: optimize your workflows with the mobile robots from MiR. https://www.mobile-industrial-robots.com/en/solutions/robots. Accessed 14 July 2020

Mountz MC, D'Andrea R, LaPlante JA, Lyons DP, Mansfield PK, Amsbury BW (2008) Inventory system with mobile drive. U.S. Patent No. US7402018B2

Nevell M (2013) Bridgewater: the archaeology of the first arterial industrial canal. Ind Archaeol Rev 35(1):1–21

Nevell M, Walker J (2001) Portland basin and the archaeology of the canal warehouse. University of Manchester Archaeological Unit

Opex: A unique approach to automated order fulfillment for small- to mid-sized ecommerce businesses.https://www.opex.com/case-studies/a-unique-approach-to-automated-order-fulfillment-for-small-to-mid-sized-eco. Accessed 14 July 2020

Parkinson-Bailey JJ (2000) Manchester: an architectural history. Manchester University Press

Patrich J (1996) Warehouses and granaries in caesarea maritima. In: Raban A, Holum KG (eds) Caesarea Maritima: a retrospective after two millennia, pp 146–176. E.J.Brill, Leiden

Potter DS, Mattingly DJ (1999) Life, death, and entertainment in the Roman empire. University of Michigan Press

Rahim M, Javed MA, Alvi AN, Imran M (2021) An efficient caching policy for content retrieval in autonomous connected vehicles. Transp Res Part A Policy Pract 140:142–152

Rouwenhorst B, Reuter B, Stockrahm V, van Houtum G, Mantel R, Zijm W (2000) Warehouse design and control: framework and literature review. Eur J Oper Res 122(3):515–533

Roy D, Nigam S, De Koster R, Adan I, Resing J (2019) Robot-storage zone assignment strategies in mobile fulfillment systems. Transp Res Part E Logist Transp Rev 122:119–142

Scallog: mobile boby robot. https://www.scallog.com/en/mobile-boby-robot. Accessed 14 July 2020

Seibold Z (2015) Logical time in decentralized controlled material handling systems. Doctoral dissertation, Karlsruher Instituts für Technologie (KIT), Karlsruhe, Baden-Wurttemberg, Germany

Shah B, Khanzode V (2017) A comprehensive review of warehouse operational issues. Int J Logist Syst Manage 26(3):346–378

Shekari Ashgzari M, Gue KR (2021) A puzzle-based material handling system for order picking. Int Trans Oper Res 28(4):1821–1846

Shirazi E (2018) Flexible high-density puzzle storage systems. Master's thesis, West Virginia University, Morgantown, West Virginia
Swisslog: Automated warehouse systems from a single source. https://www.swisslog.com/en-us/products-systems-solutions. Accessed 14 July 2020
Uludağ O (2014) GridPick: A high density puzzle based order picking system with decentralized control. Doctoral dissertation, Auburn University, Auburn, Alabama
Vanderlande: Warehousing I Storage (AS/RS) ADAPTO. https://www.vanderlande.com/warehousing/innovative-systems/storage-asrs/adapto. Accessed 14 July 2020
Winkelhaus S, Grosse EH (2020) Logistics 4.0: a systematic review towards a new logistics system. Int J Prod Res 58(1):18–43. https://doi.org/10.1080/00207543.2019.1612964
Wu S, Chi C, Wang W, Wu Y (2020) Research of the layout optimization in robotic mobile fulfillment systems. Int J Adv Rob Syst 17(6):1–14
Xie L, Thieme N, Krenzler R, Li H (2021) Introducing split orders and optimizing operational policies in robotic mobile fulfillment systems. Eur J Oper Res 288(1):80–97
Xu X, Zhaoa X, Zoub B, Gongc Y, Wanga H (2020) Travel time models for a three-dimensional compact AS/RS considering different I/O point policies. Int J Prod Res (online). https://doi.org/10.1080/00207543.2019.1659519
Yalcin A, Koberstein A, Schocke K (2019) An optimal and a heuristic algorithm for the single-item retrieval problem in puzzle-based storage systems with multiple escorts. Int J Prod Res 57(1):143–165
Zaerpour N, Yu Y, De Koster R (2017) Optimal two-class-based storage in a live-cube compact storage system. IISE Trans 49(7):653–668. https://doi.org/10.1080/24725854.2016.1273564
Zaerpour N, Yu Y, De Koster R (2017) Response time analysis of a live-cube compact storage system with two storage classes. IISE Trans 49(5):461–480
Zaerpour N, Yu Y, De Koster R (2017) Small is beautiful: a framework for evaluating and optimizing live-cube compact storage systems. Transp Sci 51(1):34–51
Zou B, De Koster R, Xu X (2016) Evaluating dedicated and shared storage policies in robot-based compact storage and retrieval systems. ERIM report series research in management Erasmus Research Institute of Management
Zou B, Gong YY, Xu X, Yuan Z (2017) Assignment rules in robotic mobile fulfilment systems for online retailers. Int J Prod Res 55(20):6175–6192
Zou B, Xu X, De Koster R (2018) Evaluating battery charging and swapping strategies in a robotic mobile fulfillment system. Eur J Oper Res 267(2):733–753

Mahmut Tutam is an assistant professor in the Department of Industrial Engineering at Erzurum Technical University in Turkey and holds a BSIE degree from Gazi University in Turkey. In 2012, he was awarded a full scholarship from the Ministry of National Education of Turkey to study in the United States. He received his MSIE and Ph.D. from the University of Arkansas (UA) in 2015 and 2018, respectively. He performed research on Warehouse Design problems. A recipient of the 2015 ISERC Best Track Paper in Facility Logistics award, he has given multiple presentations and served as a Session Chair at IISE and INFORMS annual meetings. Following the postdoctoral fellowship at Erasmus University Rotterdam in the Netherlands, he pursues his academic career at Erzurum Technical University. His research interests include, but not limited to, supply chain management, facility logistics, warehousing, collaborative robots and artificial intelligence.

Chapter 8
Cloud Information Systems

Evrencan Özcan and Tamer Eren

Abstract With the development of information technologies, the needs such as transferring and storing data, and providing quick access to the data have arisen. Cloud service providers have emerged on these needs. Businesses can stream and store data, and access it at any time over a fixed portal thanks to cloud service providers which have advantages such as low storage costs, working with cross-platforms, fast data flow and access facilities at any time. In this context, considering the importance of data in the effective management of commercial organizations, these advantages of cloud computing systems offer great opportunities to businesses. From this point of view, it can be said that an analytical evaluation is of great importance in the selection of cloud service providers. Within this scope, in this study, the criteria such as performance, satisfaction, data transfer rate, suitability and network latency, etc.) that businesses should consider renting the most suitable cloud service provider were examined and cloud service providers were compared under these criteria by suitable multi-criteria decision-making algorithms.

Keywords Cloud information systems · Cloud service provider · Network latency · Cloud service provider

8.1 Introduction

Nowadays, the increase in the Internet usage rate, the transfer of data to digital media, and the desire to access to the data quickly have found out the need for virtual storage. With virtual storage, storage, access, and authorization activities can be carried out in addition to the classification of the multitude of data owned by companies. Developing technology also changes the size of the competition between companies. This change has enabled companies to turn to cloud service in order to be able to use their data and access it quickly (Uslu et al. 2019a).

E. Özcan · T. Eren (✉)
Department of Industrial Engineering, Kırıkkale University, Kırıkkale, Turkey

İ. İyigün and Ö. F. Görçün (eds.), *Logistics 4.0 and Future of Supply Chains*, Accounting, Finance, Sustainability, Governance & Fraud: Theory and Application, https://doi.org/10.1007/978-981-16-5644-6_8

The fact that companies started to exist in digital media has also brought a new level of competition between companies to digital media. Companies actively are used cloud service systems to provide fast and easy access/storage to their data. This led to an increase in cloud service companies and strengthened the competition among themselves. This competition caused cloud service companies to serve on many factors such as performance, cost, capacity, security, data processing rate (Uslu et al. 2019a).

Service differences were offered by cloud service providers make it quite difficult for companies to make decisions. Companies faces uncertainties, user-induced factors, company requirements, and many other factors in the decision-making process. Such situations are led to wrong decisions and negative consequences for the company (Uslu et al. 2019b, c, Arpacı 2019, Gireesha et al. 2020).

Companies' selection of a cloud service provider that is not suitable for them will affect negatively in terms of cost and time. Therefore, companies that want to choose a cloud service provider should first determine the importance of the criteria that are effective in selecting a cloud service provider. They should then evaluate the cloud service providers they have identified based on these criteria. The fact that the problem has a complex structure and is affected by many factors also makes the decision-making process difficult. In such complex and difficult decision-making problems, multi-criteria decision-making methods are used.

Through multi-criteria decision-making methods, while the problem becomes more understandable, time is also saved. Within the scope of the study, firstly, the factors affecting the selection of middle tier companies cloud service providers were determined. Subsequently, the cloud service provider selection problem was addressed by 7 experts. Due to the determination of the factors affecting the cloud service provider selection and the relationship between the determined factors, the ANP method was used, which is a multi-criteria decision-making method. TOPSIS and COPRAS methods were used in the ranking of the cloud service providers.

8.2 Cloud Computing

Advances in information technology have revealed concepts such as Industry 4.0, the Internet of Things, and Big data. The emergence of these concepts has increased the importance of not only the product but also the information. The fact that the markets started to go digital, also affected the companies. The importance of information and information privacy in companies has led to data diversity and has revealed the need to classify data. Industry 4.0 and its accompanying concepts have made it possible to transmit data in digital media, to flow data in a virtual platform, to store data, and to have an environment where they can access data at all times. This virtual platform is called cloud computing (Uslu et al. 2019c, Gireesha et al. 2020, Wang et al. 2019).

Cloud computing is a technology network where users can perform transactions and store their data at any time within their access authorization through the Internet portal. Owing to cloud computing technology, data flow, data storage, and uninterrupted access can be provided over a common network (Uslu et al. 2019b; Arunarani et al. 2019).

Cloud service providers have cloud computing types and models to serve users according to their needs. Cloud computing has four types of categories and four models (Kumar and Desai 2019; Colajanni and Daniele 2019).

Cloud computing service types:

- *Public cloud service type*
  It is a type of service offered by cloud service providers for general purposes. The entire service system is structured on the resources offered to the users by the company. This computing model has a free or pay-as-you-go system up to a certain capacity by users. Emails in our daily life are examples of this model (Küçüksille et al. 2013; Shawish and Salama 2014).
- *Private cloud service type*
  It is a cloud service type where private or important information is kept in the company and is generally preferred by large-scale companies. All information is accessible to authorized persons. In this model, access security and privacy are high compared to the public cloud service. This service is more costly than others because it contains company-specific information and security is higher than other models (Küçüksille et al. 2013).
- *Hybrid cloud service type*
  The hybrid cloud service model is a cloud service model that occurs as a result of the combination of the public and private cloud service models. Companies can separate their public or private data according to their use. This situation may differ in combination rates depending on the volume of companies (Küçüksille et al. 2013; Garber et al. 2013; Zhou et al. 2013).
- *Community cloud service type*
  Community cloud service is the sharing of a cloud platform with a certain group of communities. Companies using this model use the common area. Therefore, companies with the same structure prefer this model, although they are less costly than the public cloud (Küçüksille et al. 2013; Condori 2013).

Cloud computing service models:

- *Infrastructure as a service model(IaaS)*
  Cloud computing service models are similar to a pyramid. There is an infrastructure service at the bottom of this pyramid and software service at the top. Infrastructure service is the most basic cloud service model. Cloud service providers offer servers as physical or virtual machines. The infrastructure service model generally provides an advantage in hardware and software development of large-scale companies (Şengül and Bostan 2013; Sowmya et al. 2014; Rani and Ranjan 2014).

- *Platform as a service model(PaaS)*
  The platform service model is the software and hardware area offered to application developers to work on the project. In this service model, there are many platforms such as operating systems, programming languages, management systems, database. Users using the platform service model do not lose extra time and cost in operations such as software installation, maintenance and licensing, and database matching (Şengül and Bostan 2013; Sowmya et al. 2014; Rani and Ranjan 2014).
- *Software as a service (SaaS)*
  Owing to the software service model, users can directly access CRM, ERP, and various software programs without the need for software or hardware. Even if the user can access with different devices, user can benefit from the software service. Thanks to this model, accessing shelf ware is an advantage in terms of time and cost (Şengül and Bostan 2013; Sowmya et al. 2014; Rani and Ranjan 2014).

8.3 Methods

It is quite difficult to solve problems that are encountered in daily life and which have more than one factor. Multi-criteria decision-making methods are recommended for such problems. Problems can be understood with multi-criteria decision-making methods and solutions can be reached in a shorter time[4]. This study was used in ANP, TOPSIS, and COPRAS methods, which are multi-criteria decision-making methods.

There are many studies on multi-criteria decision-making in the literature. In addition to studies such as energy (Özcan et al. 2017), health (Gür et al. 2018; Korkusuz et al. 2020), and personnel selection (Esen et al. 2019; Sert et al. 2020), there are also studies on Industry 4.0 (Uslu et al. 2019d; Şahin 2019; Yalçıner and Çaylak 2020), Internet of Things (Uslu et al. 2019c; Nadhira and Dachyar 2020), and Cloud (Ur et al. 2012; Sun et al. 2016; Kumar et al. 2018; Çakır and Karabıyık 2017) related things in recent years.

8.3.1 ANP

It is a method that is effective in forming the hierarchical structure of some complex problems and determining the relationship between the factors that are effective in the decision-making process. Saaty (1980) proposed the ANP method, which is a multi-criteria decision-making method for the problem of such internal or external dependencies between alternatives and criteria (Gür et al. 2018, Nadhira and Dachyar 2020). ANP steps are presented in Fig. 8.1.

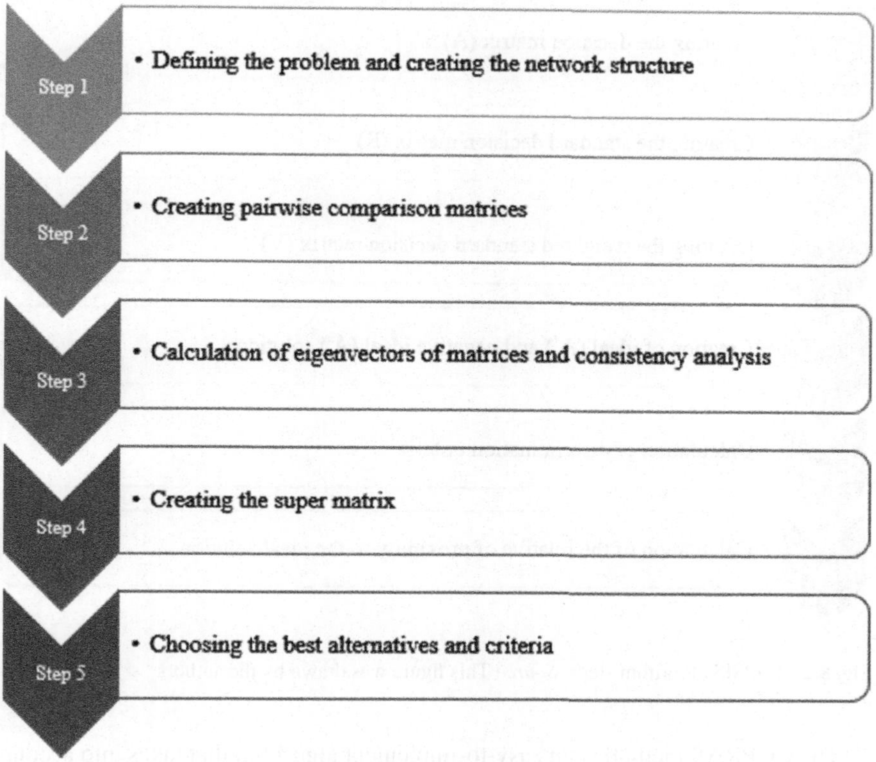

Fig. 8.1 ANP algorithm steps. *Source* This figure was drawn by the authors

8.3.2 *TOPSIS*

TOPSIS method, one of the multi-criteria decision-making methods, was developed by Yoon and Hwang (1981). In the TOPSIS method, the alternative to be selected by ranking among the alternatives should be the shortest distance to the ideal solution and the farthest distance to the negative ideal solution (Lai et al. 1994; Zhang et al. 2015). Thus, it is seen that the alternative to be chosen is the ideal choice compared to the others. Figure 8.2 shows the steps of the TOPSIS method.

8.3.3 *COPRAS*

The COPRAS (COmplex PRoportional ASsessment) method was developed in 1996 by Zavadskas and Kaklauskas. This method aims to achieve a ranking, paying attention to the importance and useful rating of the criteria considered (Ömürbek et al. 2017; Ömürbek and Hande 2019).

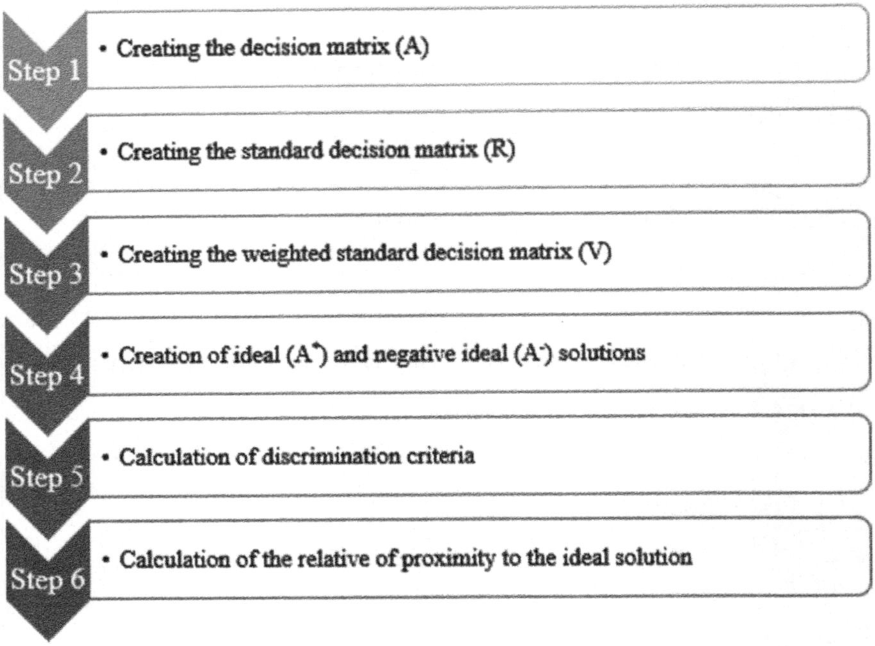

Fig. 8.2 TOPSIS algorithm steps. *Source* This figure was drawn by the authors

The COPRAS method is an easy-to-implement algorithm that takes into account the benefit values of the criteria. In addition, the ability to rank and evaluate alternatives has made this algorithm the most preferred method among multi-criteria decision-making methods (Fig. 8.3).

8.4 Case Study

The study focused on the selection of cloud service providers for middle tier companies. Companies need to use the cloud service to process, transfer, and store their data on digital platforms. For this, they need to choose the appropriate cloud service provider. The multitude of cloud service providers and their differences in their services make this choice difficult. To select the optimal cloud service provider, companies must first identify the criteria in this selection and then carry out the evaluation between companies that provide the service. This will provide consistent results with real life. The flowchart of the problem is shown in Fig. 8.4.

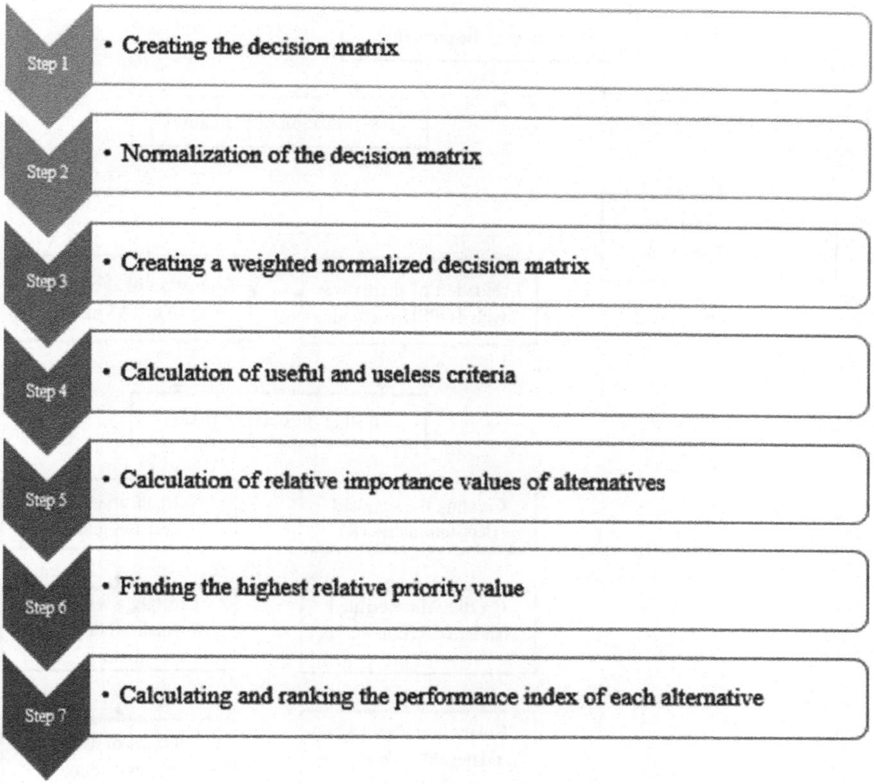

Fig. 8.3 COPRAS algorithm steps. *Source* This figure was drawn by the authors

8.4.1 Evaluation of Effective Criteria in Cloud Computing

In the literature review, the criteria that affect the choice of the cloud service provider were examined. As a result of the examination of 7 experts, 6 main criteria and 21 sub-criteria were determined. The determined criteria and explanations are presented in Table 8.1.

Internal and external dependencies of the determined criteria were determined by experts and ANP method was applied. The network structure, which is the first step of the ANP method, was created as in Fig. 8.5 so that the dependencies of the criteria between each other can be understood more clearly.

In the network structure shown in Fig. 8.5, the internal and external dependencies of the criteria were compared. Here, while performance affecting customer happiness is defined as internal dependency, processing speed affecting performance criteria is defined as external dependency.

The ANP method was applied with the help of the Super Decision package program for inter-criteria evaluation due to the dependencies between the criteria that

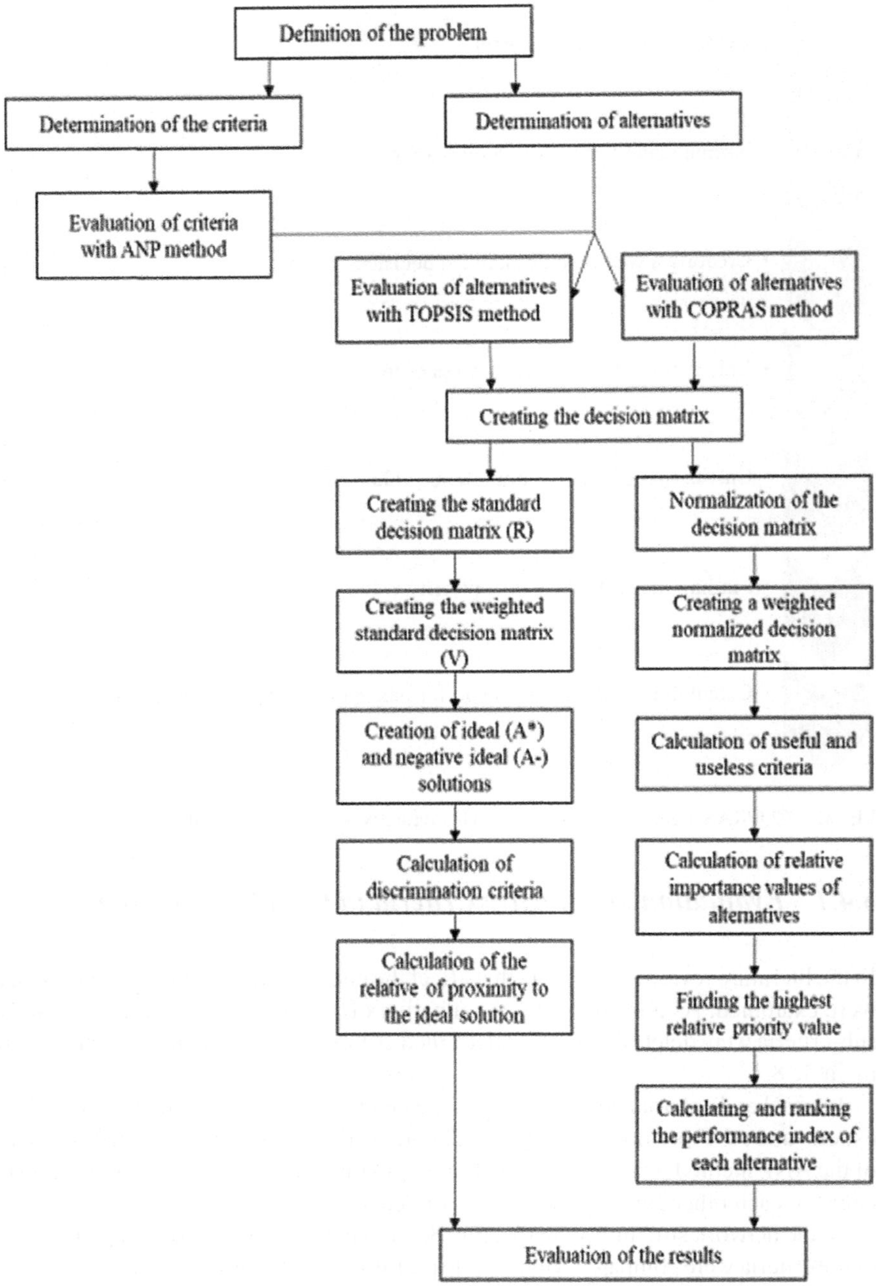

Fig. 8.4 Flowchart of the problem. *Source* This figure was drawn by the authors

Table 8.1 Explanations of criteria

Main criteria	Sub-criteria	Explanation
Memory usage (C1)	Usability (C1.1)	How well the available memory space provided by the cloud service to the company is used
	Security (C1.2)	
	Portability (C1.3)	
	Capacity (C1.4)	
	Adaptability (C1.5)	
CPU usage (C2)	Data center location (C2.1)	How well the available computer resources provided by the cloud service to the company are used
	Interoperability (C2.2)	
	Clearness (C2.3)	
	Suitability (C2.4)	
Reaction time (C3)	Data rate (C3.1)	Mean time a system needs to respond to a service request through the common portal
	Processing speed (C3.2)	
	Network*latency (C3.3*	
Service (C4)	Efficiency (C4.1)	Evaluation of the system offered by the cloud service provider to the user
	Reliability (C4.2)	
	Customer happiness (C4.3)	
	Performance (C4.4)	
Cost (C5)	Service (C5.1)	Cost of every service that can occur in the cloud, taking into account cloud service configurations and workload
	Maintenance cost (C5.2)	
	Functionality (C5.3)	
Maintenance (C6)	Virtual machine update rate (C6.1)	Updating and maintaining the data in the available space provided by the cloud service provider
	Maintenance time (C6.2)	

(*Source* This table was created by the authors)

are effective in the selection of cloud service providers. Weights of interconnected criteria applied in the ANP method are shown in Table 8.2.

When Table 8.2 is analyzed, it is seen that it follows Service criteria with 67.7%, Maintenance time criteria with 59.6%, and Performance criteria with 49.4%. Due to the effect of internal and external dependencies between the criteria, it is observed that the Service criteria provide internal dependency in the Functionality and Maintenance fee, while providing external effects to the criteria such as Performance, Processing speed, Data rate, Capacity, and Security. The performance criteria affect criteria such as Maintenance time, Data rate, Usability. It is concluded that the ranking obtained can be accepted by experts. The impact of the criteria on each other has affected their significance level.

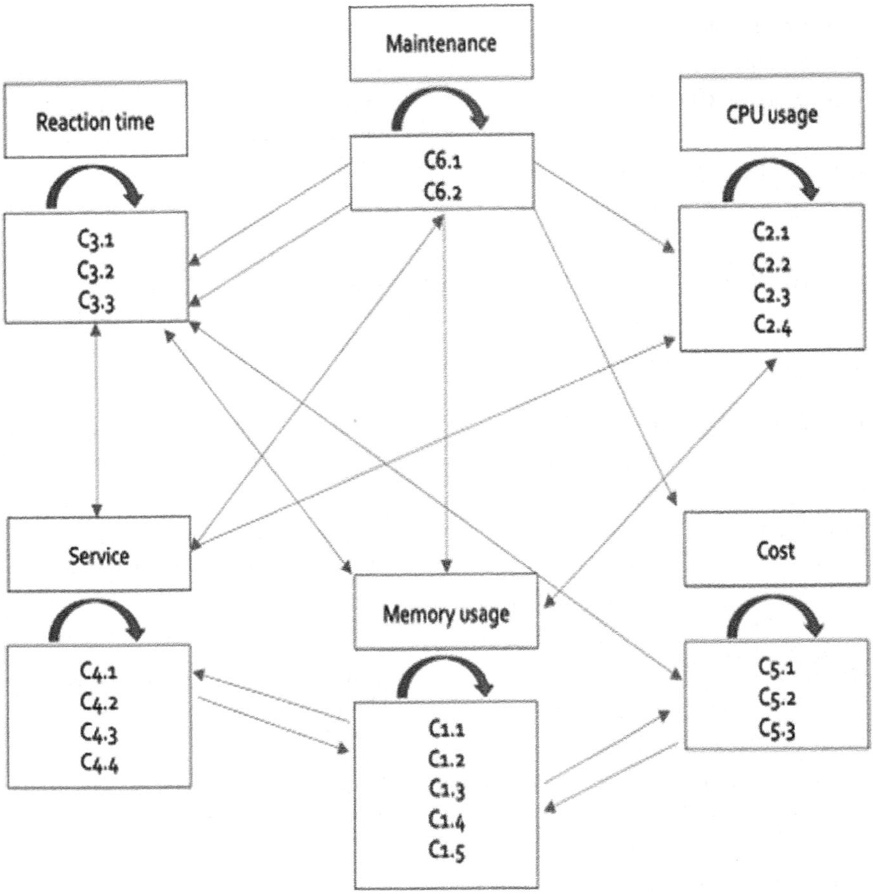

Fig. 8.5 Network structure. *Source* This figure was drawn by the authors

8.4.2 Cloud Service Provider Selection

In the literature, in the study of many cloud service providers, a certain part of the effective criteria was examined. Therefore, there is a ranking study according to certain criteria in the cloud service provider selection. In this study, a cloud service ranking will be made by considering the general criteria in the selection of the cloud service provider of middle tier companies.

The number of companies providing cloud services in the market is increasing day by day with the increase in Internet usage and the need for digital data storage. In line with the researchers and experts' opinions, Box, OneDrive, GoogleDrive, Dropbox, Amazon Drive, Mega, IDrive, IBM Bulut, and Tresoit cloud services were selected as 9 alternatives in the study. The determined alternative-criteria relationship is shown in Fig. 8.6.

Table 8.2 Evaluation of the criteria

Sub-criteria	Weights
Usability	0.2992
Security	0.2875
Portability	0.0346
Capacity	0.1979
Adaptability	0.1806
Data center location	0.2655
Interoperability	0.3487
Clearness	0.0630
Suitability	0.3254
Data rate	0.3436
Processing speed	0.2599
Network latency	0.3963
Efficiency	0.0037
Reliability	0.3385
Customer happiness	0.1634
Performance	0.4942
Service	0.6767
Maintenance fee	0.2003
Functionality	0.1228
Virtual machine update rate	0.4037
Maintenance time	0.5963

Source This table was created by the authors

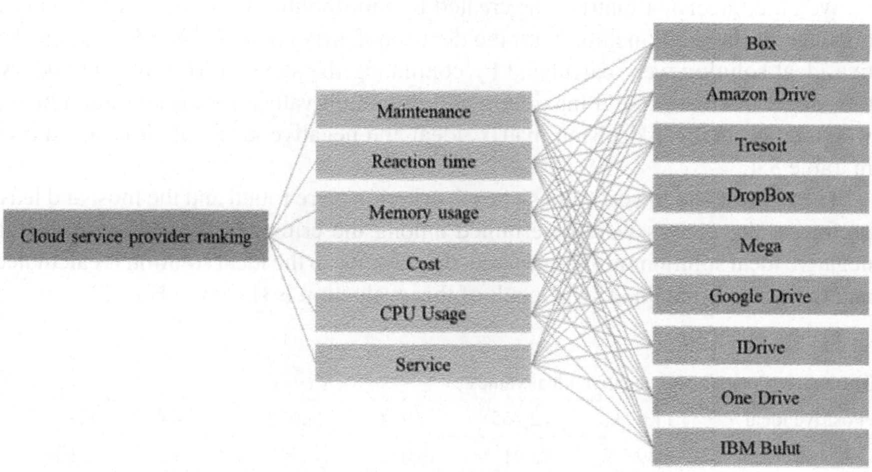

Fig. 8.6 Hierarchical structure. *Source* This figure was drawn by the authors

Table 8.3 Decision matrix

Alternative/criteria	Memory usage	CPU usage	Service	Reaction time	Cost	Maintenance
Amazon drive	14	14	16	16	18	20
Box	13	13	15	15	15	16
Dropbox	15	18	16	17	16	18
OneDrive	17	15	17	17	14	14
Google drive	18	16	18	15	13	17
Mega	10	15	12	13	14	12
Tresoit	11	11	13	13	11	13
IBM bulut	12	12	14	14	13	19
IDrive	10	12	11	12	12	15

Source This table was created by the authors

TOPSIS and COPRAS methods, which are multi-criteria decision-making methods, were applied for the ranking of the cloud service provider.

TOPSIS

Criteria affecting the cloud service provider selection problem were determined and their weights were calculated by using ANP method, which is one of the multi-criteria decision-making methods. A comparison matrix has been created between the cloud service providers and the criteria affecting cloud service selection. The decision matrix, the first step of the TOPSIS method with the assessment of 7 experts for the ranking of the cloud service provider, is shown in Table 8.3.

Weighted decision matrix was created by multiplying the weight of the criteria obtained in the ANP method with the decision matrix created. The ideal and negative ideal solution were calculated by continuing the steps of TOPSIS method. At this stage, there are found maximum and minimum values for each column of the weighted normalized decision matrix. Ideal and negative ideal solutions are shown in Table 8.4.

The highest and lowest values of each criterion were found and the most and least preferred alternatives were determined among the criteria. After the positive and negative ideal solutions are determined, the distance to the ideal solution is calculated and the result is obtained. The result of this evaluation is shown in Fig. 8.7.

Table 8.4 Positive and negative ideal values

Positive ideal	11,693	12,765	4,974	8,632	3,444	0,182
Negative ideal	6,496	7,801	3,04	6,093	2,105	0,109

Source This table was created by the authors

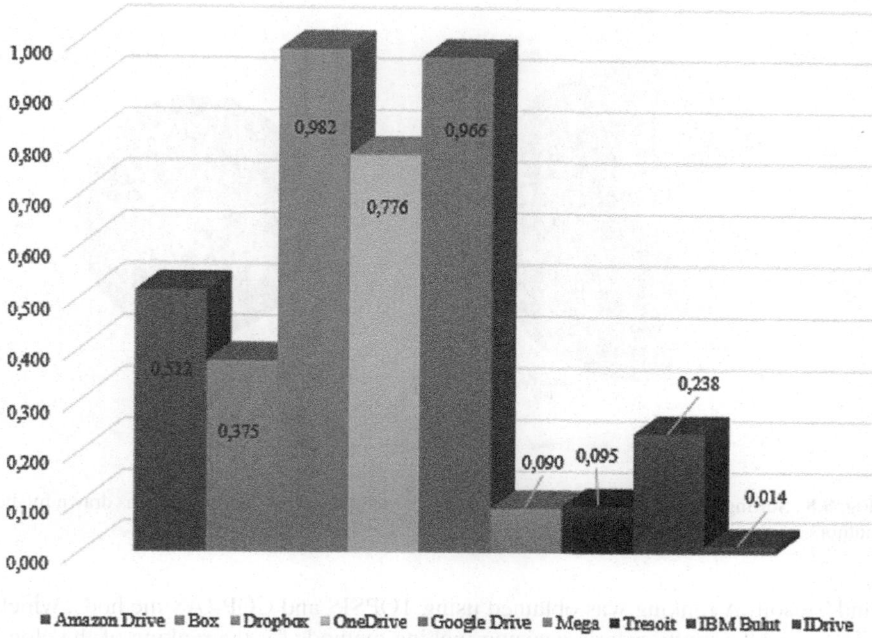

Fig. 8.7 Sorting alternatives according to TOPSIS method. *Source* This figure was drawn by the authors

COPRAS

The decision matrix among the criteria influencing cloud service providers and cloud service provider selection is shown in Table 8.3. Decision matrix values are the same as those used in the TOPSIS method. Thus, it is aimed to compare the results with the same evaluation. Following the COPRAS method steps (Fig. 8.3), the result of the calculation of the performance indices is presented in Fig. 8.8.

8.5 Results and Discussions

For the cloud service provider ranking problem, the criteria that are effective in selecting the cloud service provider should be determined first. Memory usage, CPU usage, Reaction time, Maintenance, Service, and Cost criteria were determined by literature review and 7 expert opinions. The sub-criteria of the main criteria and their interdependencies were determined by the literature and expert opinions. In the ANP method, which is one of the multi-criteria decision-making methods, the importance weights of each sub-criteria were calculated.

Then, experts identified 9 cloud service providers from the cloud service providers: Box, OneDrive, GoogleDrive, Amazon Drive, IBM Bulut, Mega, IDrive, Dropbox,

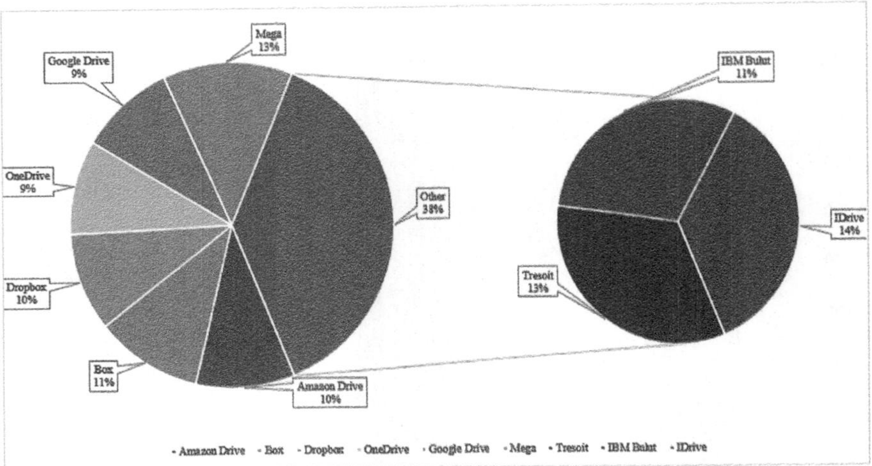

Fig. 8.8 Sorting alternatives according to COPRAS method. *Source* This figure was drawn by the authors

and Tresoit. A ranking was obtained using TOPSIS and COPRAS methods, which are among the multi-criteria decision-making methods for the ranking of the cloud service provider with the determined criteria and alternatives.

In order to use quantitative data in the problem, the TOPSIS method was used in order for the experts to obtain a ranking easily, in a short time and by paying attention to the ideal distance. When comparing the alternatives among themselves, the COPRAS method was used to make a ranking taking into account the useful effects of the criteria. Thus, by comparing the results obtained in the two methods, it was discussed in which cases which method could be used and evaluated.

In the TOPSIS method, cloud service providers have sorting as Dropbox, Google Drive, OneDrive, Amazon Drive, Box, IBM Bulut, Tresoit, Mega, IDrive. In COPRAS method, it has been sorted as IDrive, Tresoit, Mega, Google Drive, IBM Bulut, OneDrive, Box, Dropbox, Amazon Drive. The fundamental differences of the two methods addressed have also changed the ranking of cloud service providers. In the TOPSIS method, the relationship between the criteria is considered by using quantitative data. In the COPRAS method, the performance indices of the alternatives are created by taking into account the useful values of the criteria.

8.6 Conclusions

Increasing Internet usage every day also leads to the development of the technology used. Developing technology has brought along the need for data to be available, stored, and changed in a digital media. Companies are appealed to the cloud service to perform operations such as accessing, storing, and changing their data through a portal. In this way, they can access the data in a faster and shorter time by getting rid of the excess cost and time loss. In order for companies to choose the most suitable cloud service provider, they should identify and prioritize the factors that affect the cloud service provider selection. They aim to be able to get out of the minimum loss from the cost and time loss that may occur in the future by choosing the cloud service provider within the impact of the prioritized criteria.

Within the scope of the study, the problem of choosing a cloud service provider of a middle tier company is addressed. In line with the review of the literature studies and 7 expert opinions, 5 main criteria were determined. Sub-criteria affecting the determined criteria were also determined and a network structure consisting of 5 main criteria and 21 sub-criteria was created. In this network structure, the internal and external factors of the criteria have been determined and evaluated using ANP method, which is one of the multi-criteria decision-making methods.

As a result of the ANP method, the Service criteria ranks first with a rate of 67.7%, followed by the Maintenance time criteria with 59.6% and the Performance criteria with 49.4%. Considering the prioritization values of the criteria, it is concluded that the priority of the Service criteria is the influence of the cloud service providers authority on the data. Considering the following sequence, the importance rate of maintenance time and performance criteria in cloud service is an acceptable result by experts. When the interdependencies of the criteria are examined, it is seen that the maintenance time affects the performance and the performance affects the criteria such as data rate, processing speed, and network latency.

The interest of companies and individual users in cloud service causes many cloud service companies to be on the market today. The presence of many cloud service providers in the market creates many choice possibilities for users. The ability of users to choose the most suitable cloud service provider saves them from wasting time and cost for the cloud service type and model and re-experiencing the selection process. Also, since data privacy is of great importance for companies, it is known that the possibility of moving their data to another cloud service company is risky. Therefore, in this study, the cloud service providers were evaluated with their criteria for selecting cloud service providers.

In accordance with the opinions of 7 experts, 9 cloud service providers including Box, OneDrive, GoogleDrive, Dropbox, Amazon Drive, Mega, IDrive, IBM Bulut, and Tresoit were included in the scope of this study. Determined cloud service providers have evaluated using TOPSIS and COPRAS methods, which are multi-criteria decision-making methods. In the TOPSIS method, cloud service providers have sorting as Dropbox, Google Drive, OneDrive, Amazon Drive, Box, IBM Bulut, Tresoit, Mega, IDrive. In COPRAS method, it has been sorted as IDrive, Tresoit, Mega, Google Drive, IBM Bulut, OneDrive, Box, Dropbox, Amazon Drive.

The ranking is very different as a result of TOPSIS and COPRAS methods. This was because the use purposes of the methods addressed have different from each other. Although the decision matrices and criterion weights used in the two methods are taken the same; In the TOPSIS method, the proximity of the quantitative evaluations to the ideal positive and the distance to the negative ideal are examined. In the COPRAS method, the performances of the alternatives are measured by making an evaluation based on the useful of the criteria addressed. Companies that want to hire a cloud service provider should choose a method by analyzing the assessment they consider when choosing the appropriate cloud service provider. In future studies, the criteria that are effective in selecting the cloud service provider can be evaluated with the mathematical model and their impact levels can be checked. This study will be a guide in terms of criteria, alternative, and interpretation for future studies.

References

Arpaci I (2019) A hybrid modeling approach for predicting the educational use of mobile cloud computing services in higher education. Comput Hum Behav 90:181–187

Arunarani AR, Manjula D, Sugumaran V (2019) Task scheduling techniques in cloud computing: a literature survey. Futur Gener Comput Syst 91:407–415

Çakır E, Karabıyık BK (2017) Bütünleşik SWARA - COPRAS Yöntemi Kullanarak Bulut Depolama Hizmet Sağlayıcılarının Değerlendirilmesi. Bilişim Teknolojileri Dergisi. 10(4):417–434

Colajanni G, Daniele P (2019) A mathematical network model and a solution algorithm for Iaas cloud computing. Netw Spat Econ 1–21

Condori XG (2013) Community cloud computing. Revista de Informacion Teconologia Y Sociedad 70–72

Esen H, Vahit Y, Güldan S (2019) Hastanelerde AHP ve Topsis Yöntemi ile Personel Seçimi: Acil Bakim Hemşireliği Sertifikali Eğitim Programi Örneği. Avrasya Sosyal Ve Ekonomi Araştırmaları Dergisi. 6(12):112–128

Garber D, Malik J, Fazio A (2013) Windows azure hybrid cloud. John Wiley & Sons

Gireesha O, Somu N, Raman MG, Reddy MS, Kirthivasan K, Sriram VS (2020) WNN-EDAS: a wavelet neural network based multi-criteria decision-making approach for cloud service selection. In: Computational intelligence in pattern recognition, pp 853–865

Gür Ş, Uslu B, Eren T, Akca N, Yilmaz A, Sönmez S (2018) Analitik Ağ Süreci Yöntemi ile Hastanelerde Ameliyathane Performansini Etkileyen Kriterlerin Belirlenmesi. Gazi Sağlık Bilimleri Dergisi 3(3):10–25

Korkusuz AY, İnan UH, Özdemir Y, Başlıgil H (2020) Entegre Çok Kriterli Karar Verme Yöntemleriyle Sağlık Sektöründe İş Sağlığı ve Güvenliği Performansının Ölçülmesi. Journal of the Faculty of Engineering & Architecture of Gazi University 35(1)

Küçüksille EU, Özger F, Genç S (2013) Mobil Bulut Bilişim ve Geleceği. Akademik Bilişim Konferansı Bildirileri 23–25

Kumar RR, Mishra S, Kumar C (2018) A novel framework for cloud service evaluation and selection using hybrid MCDM methods. Arab J Sci Eng 43(12):7015–7030

Kumar H, Desai N (2019) Machine images and application marketplace-Cloud PaaS and SaaS for enterprise

Lai YJ, Liu TY, Hwang CL (1994) TOPSIS for MCDM. Eur J Oper Res 76(3):486–500

Nadhira A, Dachyar M (2020) Selection factor analysis for Internet of Things (IoT) implementation using DEMATEL based ANP and COPRAS method at the hospital intensive care unit (ICU)

Ömürbek N, Hande E (2019) PROMETHEE, MOORA ve COPRAS Yöntemleri Ile Oran Analizi Sonuçlarının Değerlendirilmesi: Bir Uygulama. Mehmet Akif Ersoy Üniversitesi Sosyal Bilimler Enstitüsü Dergisi 8(16):174–187

Ömürbek V, Aksoy E, Akçakanat Ö (2017) Bankalarin Sürdürülebilirlik Performanslarinin ARAS, MOOSRA ve COPRAS Yöntemleri ile Değerlendirilmesi. Süleyman Demirel Üniversitesi Vizyoner Dergisi 8(19):14–32

Özcan EC, Ünlüsoy S, Eren T (2017) A combined goal programming–AHP approach supported with TOPSIS for maintenance strategy selection in hydroelectric power plants. Renew Sustain Energy Rev 78:1410–1423
Rani D, Ranjan RK (2014) A comparative study of Saas, Paas and Iaas in cloud computing. Int J Adv Res Comput Sci Softw Eng 4(6)
Şahin C (2019) Ülkelerin Endüstri 4.0 Düzeylerinin Copras Yöntemi Ile Analizi: G-20 Ülkeleri ve Türkiye
Şengül G, Bostan A (2013) Bulut Bilişimde Bilgi Güvenliği ve Standardizasyon Çalışmaları. 6. Uluslararası Bilgi Güvenliği ve Kriptoloji Konferansı, 263–267
Sert YO, Gür Ş, Eren T (2020) Dördüncü Sanayi Devriminin Personel Seçimi Süreçlerine Etkisinin Değerlendirilmesi. Erciyes Üniversitesi Iktisadi Ve Idari Bilimler Fakültesi Dergisi 55:191–202
Shawish A, Salama M (2014) Cloud computing: paradigms and technologies. In: Inter-cooperative collective intelligence: techniques and applications
Sowmya SK, Deepika P, Naren J (2014) Layers of cloud–IaaS, PaaS and SaaS: a survey. Int J Comput Sci Inf Technol 5(3):4477–4480
Sun L, Ma J, Zhang Y, Dong H, Hussain FK (2016) Cloud-FuSeR: fuzzy ontology and MCDM based cloud service selection. Fut Gener Comput Syst 57:42–55
Ur RZ, Omar KH, Farookh KH (2012) Iaas cloud selection using MCDM methods. In: 2012 IEEE ninth international conference on e-business engineering. IEEE
Uslu B, Gür Ş, Eren T, Özcan EC (2019a) Çok Kriterli Karar Verme Yöntemleri ile Bulut Hizmet Sağlayıcı Sıralaması. Pamukkale İşletme Ve Bilişim Yönetimi Dergisi 6(1):20–34
Uslu B, Gür Ş, Eren T (2019b) Bulut Hizmet Sağlayici Seçiminde Etkili Olan Kriterlerin Çok Kriterli Karar Verme Yöntemleri İle Değerlendirilmesi. Yönetim Bilişim Sistemleri Dergisi 5(1):31–51
Uslu B, Eren T, Gür Ş, Özcan E (2019c) Evaluation of the difficulties in the Internet of Things (Iot) with multi-criteria decision-making. Processes 7(3):164
Uslu B, Gür Ş, Eren T (2019d) Endüstri 4.0 Uygulaması Için Stratejilerin AAS ve TOPSIS Yöntemleri Ile Değerlendirilmesi. Anadolu Üniversitesi Bilim ve Teknoloji Dergisi B-Teorik Bilimler 7(1):13–28
Wang Y, Wen J, Zhou W, Tao B, Wu Q, Tao Z (2019) A cloud service selection method based on trust and user preference clustering. IEEE Access 7:110279–110292
Yalçıner AY, Çaylak İ (2020) Endüstri 4.0–Dijital Dönüşüm Kapsamında TOPSIS ve AHP Yöntemi İle Sektörel Sıralama. Akademik Platform Mühendislik ve Fen Bilimleri Dergisi 8(2):258–265
Zhang J, Chang W, Zhou S (2015) An improved MCDM model with cloud TOPSIS method. In: The 27th Chinese control and decision conference, pp 873–878
Zhou Z, Zhang H, Du X, Li P, Yu X (2013) Prometheus: privacy-aware data retrieval on hybrid cloud. In: 2013 proceedings IEEE INFOCOM

Evrencan Özcan received his PhD in industrial engineering department at Gazi University in 2013, and he is an associate professor of industrial engineering at Kırıkkale University, Faculty of Engineering. Before this post he has worked as a project manager at Electricity Generation Co. which is the biggest generation organization of Turkey's energy sector about ten years. His research focuses on operations research, multi-criteria decision making, maintenance planning and electricity generation. His numerous articles and conference papers have been published on prestigious academic platforms. Dr. Özcan teaches operations research, network optimization, maintenance management, energy systems planning and energy management courses at undergraduate and graduate level in industrial engineering and computer engineering departments.

Tamer Eren is professor at department of industrial engineering at the University of Kırıkkale in Turkey. Prof Dr. Tamer Eren obtained his BS degree from Selçuk University Industrial Engineering Department in 1996. Then, he got the MS and PhD degrees from the industrial engineering department of the Gazi University in years 2000 and 2004, respectively. His research focuses on scheduling, production planning, game theory, multicriteria decision making and operation research. Prof. Eren has authored and contributed to numerous articles on health, logistics, industry 4.0, energy, transportation, waste management, recycling management, supply chain management, staff scheduling and maintenance optimization.

Chapter 9
Logistics, Supply Chains and Smart Factories

Hazar Dördüncü

Abstract The automation systems and the human factors advancing within the scope of technologies that develop under the impact of Industry 4.0 become of secondary importance in every field where human dominance exists. While labeling companies as smart factory wherein all courses of action that involves the replacement of human brain with artificial intelligence within the production processes are digitalized, and all transactions are made by digital technologies, it is revealed that this new process provides a more opportunistic system for companies that focus on supply chain management. Smart and data-based technological improvements along with the developments in barcode, sensors, internet of objects and RFID technology, and advancement in Industry 4.0 have provided great convenience to companies that use supply chain processes; logistics companies in particular. The use of artificial intelligence and new technologies ensures supervision of customers and the better management of customer relations, and companies gain a competitive advantage over their competitors. In this study, the transition of companies using traditional supply chain management to the new smart and data-based supply chain process will be explained.

Keywords Logistics · Supply chains · Smart factories

9.1 Introduction

Just as our personal devices, production equipment in production facilities are becoming smart every passing day. This progression involves machines speaking to each other, production process wherein humans are replaced by machines as well as the emergence and conceptualization of smart factories. Most recently, in this fashion, the fourth industrial revolution or Industry 4.0 has taken its place among the expressions frequently used in consort with the concept of smart factory. Automation technologies merged with Industry 4.0 collect real-time optimized data through

H. Dördüncü (✉)
Department of International Trade and Management, Nişantaşı University, Istanbul, Turkey
e-mail: hazar.dorduncu@nisantasi.edu.tr

İ. İyigün and Ö. F. Görçün (eds.), *Logistics 4.0 and Future of Supply Chains*,
Accounting, Finance, Sustainability, Governance & Fraud: Theory and Application,
https://doi.org/10.1007/978-981-16-5644-6_9

establishing key connections between the machine and the human. This process, in turn, develops the formation of "smart factory" in production by reason of cloud informatics, sensor access area and its potential on extra intelligence. In this chapter, primarily the developing and changing logistics and supply chain processes in the light of Industry 4.0 consequent to industrial revolutions, and the concept of smart factory will be explained.

9.2 From Industry 1.0 Toward 4.0

The principal factors triggering the industrial revolution by the end of eighteenth century involve steam power and steam-powered machines. The second industrial revolution on the other hand took place after the invention of electricity in company with mass production associated with the rapid development of high-speed and industrious machines. The third industrial revolution started in 1970s when electricity gave its place to electronics. Following the third industrial revolution, production process has become more efficient owing to matters of production automation, computer usage and total quality management. In a nutshell, Industry 4.0 (as shown in Fig. 9.1) accounting for computerized production processes has an impact on breakthrough changes on logistics, quality control along with subsequent progress of delivery, operational management and production process by providing ground for the automation and robotization technology as well as internet of things and communication of machines with each other.

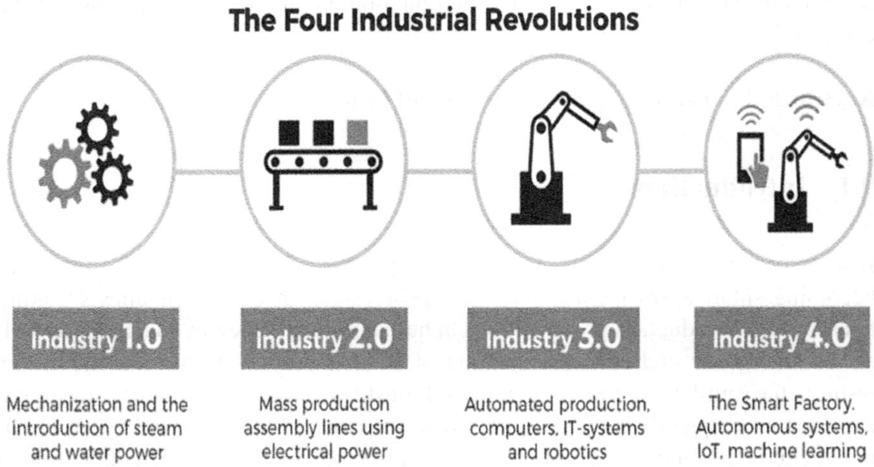

Fig. 9.1 The four industrial revolutions (Spectral Engines 2018)

- **Industry 1.0**
  - **Operation of Mechanical Production Facilities (Eighteenth Century)**

    Invention of Steam Engine (1712)
- **Industry 2.0**
  - **Transition to Mass Production based on Electricity and Division of Labor (Nineteenth Century)**

    Inventions of Telegraph (1840) and Telephone (1880)
    Taylorism (Scientific Management) (1920)
- **Industry 3.0**
  - **Automation of Production Processes (Twentieth Century)**

    First Micro Computer (Altair 8800) (1971)
    Apple I (S. Jobs and S. Wozniak) (1976)
- **Industry 4.0**
  - **Autonomous Machines and Virtual Platforms (Twenty-First Century)**

    AutoIDLab. (MIT) (1988)
    Internet of Things (2000)
    Cellular Carrying Systems (2010)
    Autonomous Interaction and Virtualization (2020) (Şekkeli and Bakan 2018)

9.3 Industry 4.0

The primary definition of Industry 4.0 characterized as the fourth generation of industrial revolution was postulated by the working team brought together by the German Federal Government at the Hannover Messe Trade Fair in 2011 (Sniderman et al. 2016). It conceptualized Industry 4.0 as "a paradigm shift reversing the logic of conventional production process that would be possible in the course of technological developments"; and described as a course of action wherein industrial production machines do not only "process" a product, but a "product" also communicates with machines and transmits them what they should do exactly (Sniderman et al. 2016). German Trade and Investment Group evaluates Industry 4.0 as "a technological revolution providing transition from embedded systems to cyber-physical systems"; and "an approach interconnecting embedded production technologies and smart production processes" (Invest 2014). Based on these explanations, it can be concluded that Industry 4.0 is "not only a condition that ensures production systems and produced objects are interconnected or a form in which physical data transmitted into digital platform; but at the same time a condition that supports transition from physical to digital, and digital to physical by communicating, analyzing and utilizing these data for sustaining smart actions in the physical world" (Sniderman et al. 2016).

Industry 4.0 is the harmonized combination of all parts of the value chain of technological integrity. For structural reasons, the existence of Industry 4.0 has given rise to the materialization and more importantly the widening of smart factories (Endüstri 4.0 Nedir? 2020).

9.4 Which Technologies Does Industry 4.0 Benefit from?

The advanced automation surfaced with Industry 4.0 and the verbal communication among machines might be considered as a frightening situation for humankind. There is an actual concern that there would be a minor need for humans in production processes. Nonetheless, experts indicate that even if Industry 4.0 will cause the erosion of certain occupations or positions that require muscle power, it will also create the need for qualified manpower capable of utilizing this new technology. Moreover, in this new period, it seems almost certain that we will adopt a new perspective comprising low-cost and high-intensity production processes. The main technological substances that Industry 4.0 will benefit from include (Endüstri 4.0 Nedir? 2020):

- Cyber-Physical Systems
- Horizontal/Vertical Integration
- Smart Robotics
- Internet of Things
- Cyber Security
- Additive Manufacturing
- Big Data.

9.5 Advantages of Industry 4.0

The innovations that will be introduced by Industry 4.0 both in individual and institutional life can be listed as:

- Simplification of system monitoring
- Having enhanced awareness
- Creation of environmentally friendly system
- Intensification of productivity and sustainability
- Significant reduction of unnecessary waste of time, energy resources and costs
- Creation of new business lines and models
- Creation of flexible production processes capable of swiftly responding to consumer demands.

9.6 Logistics Industry

The word "logistics" derives from the Ancient Greek "logistikos" referring to the wars and the struggle for survival; and has the meaning of "arrangement of finances and skillfulness in calculation"(Özdemir 2012). In this context, the concept of logistics comes into view both in the fields of military and management. In the field of military, logistics implies the preparedness of troops at times of peace and war by providing necessary equipment and supplies(Stroh 2002) concerning the transportation, maintenance and personnel procurement of military units and facilities. In the field of management, logistics denotes the arrangement of a product to be available at the right time and the right place within the framework of a business planning on the management of material, service, information and capital flow.

The roots of the history of logistics can be traced back to ancient times. In fact, the military officials known as *logistiks* were responsible for the supply and distribution of essential resources and services during the war between the Greek and the Roman Empires. It was believed that having such military officials was one of the fundamental factors that impacted the outcomes of the war. Besides, these military officials known as logistics were responsible for stocking up and protecting their own supplies while at the same time damaging the supply materials and the supply delivery routes of their enemies. This system has gradually led the way to the current logistics system in the world (Dördüncü 2018).

In general, logistics systems have developed during the Second World War period (1939–1945). While the United States of America (USA) and its allies could use the logistics systems in their favor during the war, Germany lagged behind on the same matter. In fact, the German supply warehouses and supply roads were significantly damaged and Germans failed to evenly harm their enemies. Under any circumstances, however, the American legion could manage to provide necessary support to its troops at the right time, the right place and with the right economic modus. In this context, new and advanced military logistics terms were used; and this became one of the leading factors that caused the Germans to lose the war (Dördüncü 2018).

In today's world, logistics is accepted as a field of art and science accompanied by logistics specialists who conduct their duties based on their talents, experiences and knowledge. In modern businesses, the duty of logistics managers involves providing suitable and productive logistics systems to businesses organizations. As part of their competition strategy, the managers of Logistics Systems guarantee their customers on meeting their expectations by assuring the delivery of the right goods to the right customers at the right time and place, and in the economically most efficient way (Dördüncü 2018).

While analyzing the logistics system, it would be misleading to take it as a single variable. In fact, logistics is a variable comprising of various components such as purchasing, planning, warehousing, coordination, distribution, customer service and after sale service.

The main aim and the purpose of the process of logistics involve the delivery of the right quantity and quality of goods or services to the right customers at the right time, place and price. Most of the customers directly or indirectly overlook the logistics system unless a flaw or setback transpires. The logistics process thus comes to forefront as an integrated progress connoting the transfer of goods from the origin to final the consumer involving different yet successive courses of action such as planning and coordination. Albeit, the Council of Supply Chain Management Professionals (CSCMP) specifies the scope of logistics management and logistics as "a process of fulfilling the requirements of customers as part of the supply chain that involves the providing of the effective distribution and reverse distribution service as well as storage operation under efficient controls starting from the point of departure until the point of arrival"(Professionals 2017).

9.7 Digitalization in Supply Chain and Logistics

Professor Steven Alter defines six different information systems applicable to any business function. In the context of logistics and supply chain, these can be examined in the following way (Murphy and Knemeyer 2018).

9.7.1 Office Automation Systems

Office automation systems provide effective ways and methods to process personal or institutional business data, make calculations and documentation. Software packages and office automation systems include word processors, electronic spreadsheets, presentation and database applications.

The electronic logistics spreadsheet applications that expended along with the software packages in early 1990s were used to calculate economic order amount, warehouse measurement, transportation method, carrier, production planning, center of gravity-oriented position stabilization. Today, however, the ability of these tables and calculations has advanced; and the logistic optimization models have become part of this endeavor (Murphy and Knemeyer 2018).

9.7.2 Communication Systems

Communication systems provide an opportunity for suppliers and customers to work together and interact with each other in various ways. The previous technologies such as radiopagers, fax machines and radio transmitters were superseded by the new communication methods such as electronic mail, smartphone applications and tablet computers using modern technologic infrastructures. Although presently wireless communication is considered as one of the fundamental elements of logistics and has various applications in this sector, the most important and popular of these emerge as the Global Positioning System (GPS) and the map within the navigation system (Murphy and Knemeyer 2018).

Companies utilizing advanced global positioning systems could free themselves from high fuel costs, increase employee productivity, reduce managerial costs and increase overall productivity and efficiency.

9.7.3 Transaction Processing System (TPS)

Transaction Processing System (TPS) collects, stores and operates data on product movement when required. At present, due to the development of cheap and easily accessed cloud system, the load of processing and storage of the internet of things have been transferred to the cloud system. In today's world, it is validated that over 10 billion devices are currently interconnected; and it is estimated that this figure will rise up to over 50 billion devices by 2020. While Ashton in his study, gives a ballpark figure of 26 billion devices for 2020; other studies show that although the number of interconnected devices per person in 2003 was 0.008, this number has approximately risen up to 6.48 in 2020 (Sevim et al. 2019). The internet of things on the other hand supports this process by reducing the costs of international trade and by increasing global trade efficiency. First of all, it would eliminate the problem of missing goods during transportation. Second, with the dispatch tracking system, companies would start to use dispatch routes more effectively. With this tracking system, companies could track containers they dispatched to various regions around the world; and maintain around 10–15% container efficiency by utilizing the internet of things technologies (Murphy and Knemeyer 2018).

The electronical data exchange process on account of transaction processing system provides positive advantages on logistics such as time efficiency in documentation and processing, cost reduction in inventory performance, reduction of dispatch mistakes and missing, as well as reducing time of purchase orders and associated costs.

9.7.4 Management Information System (MIS)

Management Information System is utilized for monitoring the TPS data performance and its transformation into information. At the end of this process, Logistics Information Systems (LIS) come into view as the software systems that meet the self-planning or operational needs of every company that takes place in the supply chain; and that can integrally work with associative systems within and/or outside of the company. LIS principally comprises Strategic Planning Systems and Operational Information systems in consort with other systems. Strategic Planning Systems are used for the modeling and design of activities. These systems are used either for one occasion only (Optimum Network Design) or for the purposes of the revision of the system based on new developments (Optimum Distribution Planning, Demand Planning), and used periodically for monitoring (Demand Planning, Stock Optimization) (Ertek 2012).

9.7.5 Decision Support System (DSS)

The primary objective of Decision Support Systems is to support the decision-making of managers with the tools of information, modeling and analysis. The use of these systems in the field of logistics and supply chain is mostly associated with the subjects on vehicle routing, decisions on inventory controls, development of order-picking systems and development of buyer–seller relations.

One of the most important characteristics of Decision Support Systems is its cost-efficient testing ability concerning a company's decision on making a change by using simulation technique. In essence, simulation technique is the modeling of the real world using the mathematical equations to represent the relations between system components. The more real-like a simulation modeling is its reliability increases (Murphy and Knemeyer 2018).

9.7.6 Transportation and Store Management System

Companies using transportation management system incorporate the generation of demands, preparation of goods, monitoring of distribution and the rendering of payments into automation. Store management systems on the other hand provide an opportunity for surveillance on the storage and material flows of a company. With storage management system, inventory planning, product acceptance, designation of storage location and purchase distribution processes can be planned.

Companies using Transportation and Store Management Systems can reduce fuel consumption and transportation costs, eliminate the problem of overstocking, reducing deadheading and increase inventory consistency and customer efficiency while reducing operating expenses (Murphy and Knemeyer 2018).

9.8 The Impact of Internet on Logistics and Supply Chain Processes

Internet of Things (IoT)

This term has been used for the first time in the presentation prepared by Kevin Ashton for Procter & Gambler in 1999. It is defined as "a worldwide spread network formed by the incomparably addressable objects and the interaction of the objects in this network with a specific protocol" (Brock 2001). Along with this formal definition, the IoT can be described as the system of devices that created a smart network in which more than one device can identify itself by using various communication protocols, commune with each other in real time (online) and share information among each other. The main aim of this technology is described as people's gathering of information without sitting behind a computer or without interacting with another person. Another important issue is that aside from a person's need of accessing certain daily information, the IoT is useful for organizing daily living activities. It is observed that IoT as a technological revolution can potentially change many things. The reality of establishing a connection with anyone, with anything at any place and at any time is interpreted as that IoT can be placed at the base of the new world order (Kutup 2011).

With the common platform established by the Danish shipping company Maersk and the digital technology solution partner IBM, it is recorded that in the shipment of a typical dispatch, more than 30 organization and more than 100 people interact with each other more than 200 times. The interaction among all these parties and the access to data in the correct way is considered as a success all by itself. This study is a study intending on eliminating systems that do not commune with each other including the generated paper recordings. For that matter, the late clearing of the goods from the ports due to a missing form or late approval would be eliminated; and a transparent network system that transforms the centralistic supply chain model into a digital ecosystem dispersed from end to end in which parties can follow up reliable, constant, uninterrupted and real-time shipment information will be created (Scoot 2018).

9.8.1 Blockchain

Blockchain Technology can be defined not only as virtual money that we frequently hear in Turkey lately but also as an encrypted digital accounting book wherein the values of all economic and financial transactions during the process of business deals are recorded (Kinsun 2018).

Among the characteristics of Blockchain are its nonerasable and irreversible composition that allows its transfer among concerning parties without any mediator; which in turn removes certain problems of global trade such as trust, document flow, transaction cost, mediators and the problem of speed.

9.8.2 Cloud Technology

Cloud technology in its simplest form is an online storage service that does not require any installation providing operational convenience through the web-based applications. The control of logistics sector in which various processes such as transportation, storage, distribution, vehicle tracking, and finance are managed simultaneously has become much easier due to the advantages provided by cloud technology. By means of cloud-based and advanced business intelligence tools, one can benefit from the customized access to real-time data for the daily harmonization and calibration of supply chain; have access to the situation analysis intended for the identification of weak fields, resource allocation and the optimization of decisions on distribution. The system also continuously updates itself with the regular upgrades run by the software provider (Murphy and Knemeyer 2018).

9.9 Smart Factories and Smart Supply Chain Management

Imagine a factory that is fully equipped with storerooms, furnaces, cranes, production conveyors, machines, robots and packed with people and manufactured goods; a factory where raw materials are brought in, hundreds or even thousands of components are passed through conveyors; in every hour, every minute or even in every second a new component is produced, passed through meticulous quality control processes and finally packed and prepared for delivery. Prior to Industry 4.0, all these processes that took place in paramount production chain were controlled manually by light-out factories (Prowme 2017). In these places, manpower was needed for lifting pieces up and installing them. It was a necessity for these places to employ workers as a precondition of their structures. Many factories started to alter into light-out production in between shifts with the aim of meeting the demand growth within the capacity of the technologies they have in relation to their own facilities and increasing the financial strength of the factory for the purposes of consumption sector efficiency and the prevention of the widening of the gap between supply and demand. In this day and age, Industry 4.0 has replaced the aforementioned structures with a new system wherein intercommunicating and interconnected machines are autonomously controlled. Smart factory processes have started in which all production processes are self-managed involving only a little or no human interference in any course of action from supply to storage, from production to quality control, or from distribution to service maintenance. Besides, with 4.0 procedures, numerous industries started to minimize damages with minor error margins and shorten waiting periods, make purchasing, production and distribution planning, and ultimately increase "efficiency" as a must of the production process (Prowme 2017).

The communication and interaction among things in the process of manufacturing goods have become as important as the communication among machines. As the goods become smarter and turn into conversation boxes like machines, we will soon unavoidably come face to face with objects that can be labeled as "Smart Goods". We will not only know how, where and when all these smart goods are produced in smart factories but also we will be able to follow up from which raw materials these goods are manufactured, which logistician or courier company transfers the raw materials to the factory, from which supplier the raw materials and other components are supplied, what quantity of stocks are piled in which warehouse, how these stocks are packed and all the station and routing they pass through until they reach their final customer/destination in the course of supply chain. Smart goods that interact with smart factories even after the manufacturing process can manage their maintenance processes and transmit information through analytical data on the scheduling of their maintenance requirements, when they will need new components, for how long and even for whom they are in use by using the internet of things and cloud computing (as shown in Fig. 9.2); and all these valuable information will be processed by machines to be directed by its users for maintenance requirements, which, in turn, will increase the quality perception through the widening the life cycle by offering additional services to the final users.

Smart factories clearly are not only made up of machines that can communicate through the internet of things technologies. They are also one of the concise symbols of the new object period in which every living object has mechanisms that can analyze big data, make decisions, and file a report (Duman 2016).

Smart factory and the process of smart supply chain involve the integration of machines and systems with modular devices. Subsequently, these devices allow for the learning and customization of the industrial equipment processes, which, in turn,

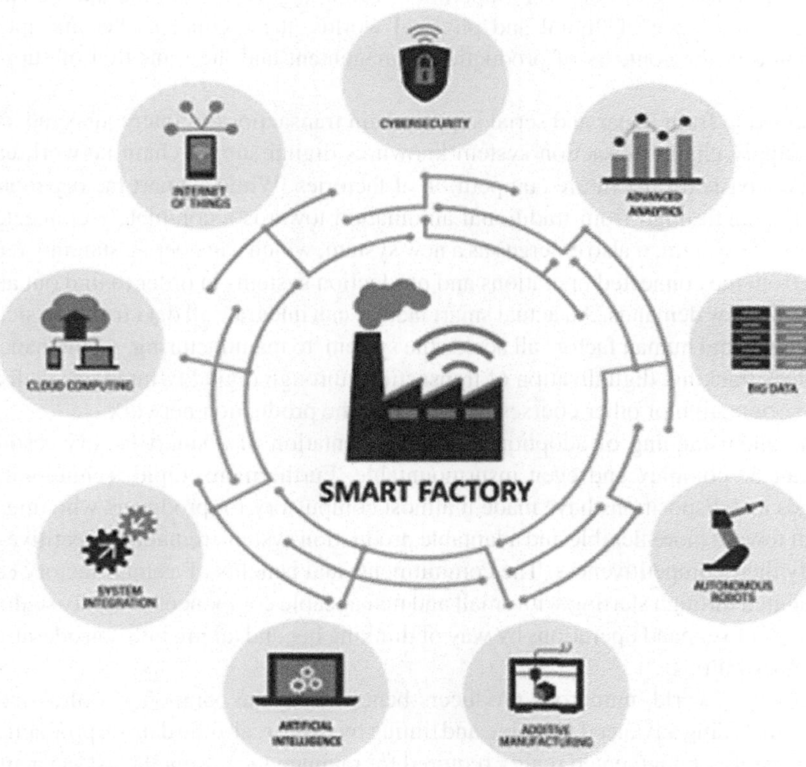

Fig. 9.2 Smart factory components (Chakraborty 2018)

establishes supply chain management and smart factories. To a certain extent, this development emerges as a result of the rise of "Big Data" comprising complex information, which cannot be managed effectively by traditional data processing applications. It is the data application that provides benefits such as predictive maintenance, flexible production and energy measurement. The key to a successful smart factory materializes through the interaction between the objects in the field and the process of supply chain. In other words, full integration in a factory can be achieved by means of total communication between objects and machines. Along with the modular structure of the facility provides flexible production/manufacturing including the communication between every machine or device. Each module generates its own data followed by the courses of storage and analysis. In order to establish a smart factory, it is necessary to transmit these data from the production phase until the end of factory management system without any problems. In smart factories, it is possible to remotely monitor and control the overall process by using smart devices (Harting 2020).

The concepts of connection and connectivity in the context of production process are not considered as new in the framework of production and supply chain management. Nonetheless, with the rise of the fourth industrial revolution including Industry 4.0, information technology (IT), operation technology (OT), and the latest trends on the convergence of digital and physical worlds, these concepts became more applicable in the contexts of production management and the transition of supply chain.

Transition from linear and serial supply chain transactions to interconnected and open supply chain transaction system known as digital supply chain network can provide a basis for the future competition of factories. While a smart factory represents a breakthrough from traditional automation towards a completely connected and flexible system, it also emerges as a new system, which can operate standing data flows from the connected operations and production systems in order to find out and customize new demands. An actual smart factory can integrate all data from physical, operational and human factors all across the system, to manufacturing, maintenance, inventory tracking, digitalization of transactions through digital twinning as well as all the orientation of other courses of actions in the production network.

The understanding of adoption and implementation of a smart factory resolution can be complex and even insurmountable. Furthermore, rapid technological changes and dispositions have made it almost compulsory for producers who intend to shift toward more flexible and adaptable production system, remain competitive or modify their competitiveness. The commitments and benefits of a smart factory can be attainted through starting with small and manageable components, rapidly scaling up in order to expand operations by way of thinking big and taking into consideration of all possibilities.

In today's world, numerous producers benefit from the components of a smart factory including advanced planning and timing by using real-time data on production and inventory or augmented reality required for maintenance. Nonetheless, an actual smart factory represents an integral effort of the business organization progressing from a standard workplace into a place impacting on a wider ecosystem. A smart factory is an inseparable part of a wider digital supply network and has multidimensional characteristics that allow producers to accommodate themselves to the changing market.

As a result, it becomes highly likely that a more efficient and flexible system will bring about minor production cut, obtain successful results in the estimation of facility and wider network changes; and hence will allow a company to better position itself in the competitive market (Proente 2019).

9.9.1 Defying Smart Factories

Automation has always been an integral part of production processes; therefore, high levels of automation are not considered as something new at present. Nonetheless, the term "automation" suggests the accomplishment of a single and separate responsibility or process. Historically, the conditions where machines make "decisions" have always been linear and automation-based such as opening of a valve or opening and/or closing a pomp based on a specific cluster of rules. Due to the increasing complexity of cyber-physical systems combining Artificial Intelligence (AI) application, physical machines as well as business processes, the automation has started to involve the complex optimization decisions typically made by the people. Finally, and may be most importantly, the term "smart factory" suggests the integration of factory decisions and estimations through the interconnected IT/OT structures with the rest of the supply chain and a wider institution. This, in turn, has essentially changed the production processes and developed relations between suppliers and customers (Sniderman et al. 2016).

In light of these assertions, it can be argued that smart factories go beyond a simple automation; and that they can self-optimize performance in a wide network, adapt themselves to real or almost real-time conditions, and learn from these; and that they constitute a flexible system independently operating all production processes. In that respect, smart factories represent a flexible learning system embarking on an open-ended evolution rather than the modernization approach of old times.

The real strength of a smart factory lies in its ability to develop and expand itself without slowing down in line with its diverging needs as well as its ability to satisfy various needs including changing customer demand, access to new markets, new product or service development. The establishment of smart factories and their ability to adapt themselves to the changes depend on having stronger information technologies, analytical abilities and infrastructure capacities along with the wider ecosystems of smarter and connected assets.

The ability to adapt to data in real time and to learn in real time provides a smart factory the capacity to be more responsive, proactive, transparent, swift and visionary; and makes it possible to prevent operational disconnection delays along with other challenges on overall efficiency of the business organization (Sniderman et al. 2016).

9.9.2 Features of Smart Factories

Smart factories principally have five common features (Şekkeli and Bakan 2018).

Connectivity: All objects in smart factories are equipped with smart sensors and integrated circuits for the purposes of providing information flow and information update. In smart factories, it is required that all the objects are in interaction with each other at the stage of information gathering needed for making decisions on certain processes. With the help of this interaction, data collected on the processes within and outside of the businesses make it possible to track the smart supply chain process right from the beginning of the operation (Şekkeli and Bakan 2018).

Optimization: In smart factories, the efficiency, service capacity and quality are increased while the costs and wastes are reduced by automatic workflows, synchronization of assets, advanced accountability, advanced-level programming and optimized energy consumption.

Transparency: Information gathered from a good through and synchronous information display are analyzed and transmitted both to humans and autonomous systems. In that respect, transparent network provides an opportunity for the organization to make the right decisions by synchronous tracking and monitoring via synchronous alarms and warnings (Şekkeli and Bakan 2018).

Proactivity: In a proactive system, employees can take actions by estimating a problem before it even occurs. Nonetheless, the estimations cannot be fully accurate at all times or the problems can be misidentified; this, in turn, causes quality faults, overstocking, security flaws etc. In smart factories, however, previously recorded data in cloud computing and constantly incoming synchronous data are compounded with the intention of making estimations more accurate and factual (Şekkeli and Bakan 2018).

Flexibility: Flexibility is the process wherein a smart factory adapts itself to a program and changes with an interference at the minimum level. Since advanced smart factories can restructure themselves, the changes on material flows or programs are managed exclusive of high levels of costs, which, in turn, results in the advancement and transformation of the smart factories (Şekkeli and Bakan 2018).

9.9.3 Conclusion

The concept of smart factory is associated with features including the interaction of the internet of things and autonomous systems with a network connection, integration of systems, technology and stock-keeping orientations, active and cross-routing, radio-frequency identification (RFID) integrated circuit and label technology, simulation systems, smart sensor and transfer of big data produced by objects into cloud computing in a broad bandwidth. In contrast with traditional factories, these features allow smart factories to befall as organizations that are competitive, proactive and suitable for flexible decision-making structures; that can achieve efficient productivity of good quality with low costs; and that have flexible working hours and conditions. In view of these features, smart factories transpire as sustainable structures that have durable economic lives.

The companies and factories that invest in resources emerged in the process of Industry 4.0 such as cyber-physical systems, artificial intelligence, augmented reality and virtual reality applications; and that have a self-improving institutional structure, which can make decisions merging emotional and industrial intelligence, will inevitably have a serious competitive advantage against their competitors since the investments on these resources will allow them to provide service-based solutions, develop skills on smart innovation and smart supply chain management as well as production of high-quality smart goods at low costs.

References

Brock D (2001) The electronic product code (EPC)—a naming scheme for physical objects. MIT AutoID Center White Paper

Chakraborty S (2018). smart factories—automating manufacturing with intelligent operations. https://www.rapidvaluesolutions.com/smart-factories-automating-manufacturing-with-intelligent-operations/

Dördüncü H (2018) Transit Merkezlerin Çok Kriterli Karar Destek Sistemi ile Perfomans Değerlemesi: İstanbul 3. Havalimanı Örneği. İstanbul: İstanbul Üniversitesi Sosyal Bilimler Enstitüsü İşletme Ana Bilim Dalı Doktora Tezi

Duman AA (2016). An emergency manufacturing age, industry 4.—smart factories. Retrieved from Innova

Endüstri 4.0 Nedir? (2020). https://www.ajansim.net/blogs/198/google-authenticator-nedir.html

Ertek G (2012). Lojistik Bilişim Sistemleri. İstanbul, Türkiye

Harting (2020) How can you create a smart factory? https://www.harting.com/IT/en-gb/topics/what-smart-factory

Invest GT (2014) National academy of science and engineering, "securing the future of german manufacturing industry: recommendations for implementing the strategic initiative industry 4.0". Germany Trade & Invest, "Smart manufacturing for the future,": http://www.gtai.de/GTAI/Content/EN/Invest/_SharedDocs/Downloads/GTAI/Brochures/Industries/industrie4.0-smart-manufacturing-for-the-future-en.pdf

Kinsun S (2018). Retrieved from Branding Türkiye: https://www.brandingturkiye.com/blokzinciri-blockchain-teknolojisi-nedir/

Kutup N (2011) Nesnelerin İnterneti: 4H, Her Yerden, Herkesle, Her Zaman, Her Nesne İle Bağlantı. *16.* Türkiye'de İnternet Konferansı inet-tr. İzmir

Murphy PJ, Knemeyer MA (2018) Contemporary logistics. Pearson

Özdemir Ş (2012) Askeri Lojistik Nedir? İstanbul

Proente (2019) Endüstri 4.0 ile Üretim ve Gelirde Artış Bekleniyor. https://proente.com/endustri-4-0-ile-uretim-artacak/

Professionals CO (2017) CSCMP supply chain management definitions and glossary. http://cscmp.org/CSCMP/Educate/SCM_Definitions_and_Glossary_of_Terms/CSCMP/Educate/SCM_Definitions_and_Glossary_of_Terms.aspx?hkey=60879588-f65f-4ab5-8c4b-6878815ef921

Prowme (2017) Akıllı ve Karanlık Fabrikalar ile Yeni Üretim Çağı. http://www.prowmes.com/blog/akilli-ve-karanlik-fabrikalar/
Scoot T (2018) Enterprise-ready blockchain brings transparency to supply chains. https://www.ibm.com/blogs/blockchain/2018/04/enterprise-ready-blockchain-brings-transparency-to-supply-chains/
Şekkeli ZH, Bakan İ (2018) Akıllı Fabrikalar. J Life Econ 203–220
Sevim MA, Kırcova İ, Çuhadar E (2019) Yerel Yönetimlerde Akıllı Şehir Vizyonu: Şehir Yönetim Araçları ve Trendleri. Strat Public Manage J 109–126
Sniderman B, Mahto M, Cotteleer M (2016) Industry 4.0 and manufacturing ecosystems exploring the world of connected enterprises. Deloitte University Press, UK
Spectral Engines (2018) Industry 4.0 and how smart sensors make the difference. https://www.spectralengines.com/articles/industry-4-0-and-how-smart-sensors-make-the-difference
Stroh MB (2002) What is logistics? Dumont, NJ

Hazar Dördüncü was born in 1985 in Izmir. He received his secondary and high school education in İzmir Tevfik Fikret High School of Science. He went to Belgium for language training. After completing his language education at Université Libre de Bruxelles, he attended "Foreign Trade" and "International Trade and Finance" education at İzmir Yaşar University, respectively. Thereafter, he completed his MBA in the field of Marketing at Dokuz Eylül University; and started his PhD in Business Administration at Istanbul University. Between 2004-2005, he worked as an information officer at the Intercollegiate Summer Olympic Games within the Image Management Unit of the Department of Marketing and Public Relations at Izmir Metropolitan Municipality. He founded his own overseas education consultancy firm in 2008. Hazar Dördüncü, currently works as a lecturer in the Department of International Trade and Management at Nişantaşı University and gives lectures on foreign trade, logistics and international marketing.

Chapter 10
Internet of Things (IoT) in Marketing Logistics

Ezgi Uzel Aydınocak

Abstract The increase in international trade around the world has expanded the markets in which businesses operate, while increasing the demand and diversity for products and services. This triggered the need for more information and visibility in managing supply chains and logistics activities. For this reason, the spread of the Internet of Things, one of the most important elements of Industry 4.0, has accelerated today. The data gathering power provided by the Internet of Things and its ability to rapidly transform this data into information is to make all activities taking place along the supply chain more visible. In this way, the control of the flow of goods and information in logistics activities has been facilitated, costs have been reduced and most importantly, it has helped enterprises increase their customer satisfaction by increasing their competitiveness. The aim of the chapter is to briefly explain the development of Internet of Things technology and demonstrate its effects over logistics and supply chain activities.

Keywords Logistics and supply chain · RFID · Internet of things · Visibility · Last mile delivery · Retailing

10.1 Introduction

Because of the globalization taking place all over the World, and with the effect of the increasing population, the needs of individuals, commercial enterprises and other institutions and organizations have made it difficult to manage the supply chains efficiently and to keep logistics costs under control in both domestic and international markets. Any logistics activity carried out throughout the supply chain should be managed with a holistic approach and visibility should be provided for all parties along the supply chain. Traceability is known to be especially important in the transport of perishable products, dangerous goods, and valuable cargo. Increasing their ability to deliver faster and more accurately to consumers due to

E. U. Aydınocak (✉)
Faculty of Business and Administrative Sciences, Beykoz University, Istanbul, Turkey
e-mail: ezgiuzel@beykoz.edu.tr

İ. İyigün and Ö. F. Görçün (eds.), *Logistics 4.0 and Future of Supply Chains*,
Accounting, Finance, Sustainability, Governance & Fraud: Theory and Application,
https://doi.org/10.1007/978-981-16-5644-6_10

the increasing demand for electronic commerce creates pressure on businesses to improve their monitoring capabilities in this regard. This requires businesses to plan their resources smarter and invest in innovative technologies while networking in supply chains, warehouses and distribution centers, transportation, and inventory management activities are carried out.

It is striking that after the introduction of Industry 4.0 into our lives, the Internet of Things (IoT) is one of the most striking elements of this latest technology. With the advent of Radio Frequency Identification technology toward the end of the 1990s, it became possible for objects to be connected and communicate with each other thanks to sensors and transmitters (Khan and Yüce 2019). Accordingly, the Internet of Things (IoT) is described as a network of electronics, software, sensors, and physical objects with network connectivity, and is referred to as a system where these objects get and exchange data using the Internet (Salazar 2016). Although the Internet of Things concept has been mentioned since the late 1990s, it is still open to development. Today, the availability and effectiveness of the Internet of Things is increasing day by day thanks to broadband Internet, Wi-Fi, and Bluetooth technologies.

Internet of Things has started to be used in many areas such as health, agriculture and animal husbandry, energy, city and environmental affairs, home automation, security, retailing, and logistics. The aim is to increase productivity and efficiency, as well as to provide higher quality services regardless of the field. This is thanks to the fact that the big data obtained through the communication network established by the objects is analyzed and enables improvements. Logistics and supply chain are some of the main areas where Internet of Things is widely used. In this sense, it is known that the Internet of Things provides significant benefits to both commercial customers and end consumers. It provides added value in many logistics activities throughout the supply chain, especially warehouse operations, transportation and last mile deliveries. In addition to increasing efficiency and effectiveness, it also contributes to customer satisfaction with safety and security, and even helps to create new business models in these areas. It is known to improve customer service levels thanks to the added value data produced by raw materials, final products, and cargo along the supply chain in addition to the possibility of real-time monitoring. It also provides benefits in managing uncertainties by providing big data to increase the predictive ability of the businesses.

The Internet of Things has emerged to be an important technology that opens the door to standardization, quality, sustainability, and continuous improvement in the field of logistics. In this part, the contributions of the Internet of Things will be explained by mentioning the usage areas in logistics activities along the supply chain. First, the Internet of Things applications in storage and inventory administration and then in transportation and distribution activities might be clarified. Finally, the contributions of the Internet of Things technology will be conveyed in retail logistics and last mile delivery concept.

10.2 Understanding Internet of Things

In our increasingly wireless world, it has become common for objects including RFID, sensors, actuators, and cell phones to communicate with one another. After the introduction of Industry 4.0, this new technology, called the "Internet of Things" (IoT), has become an important technological development that can produce many technical, social, and financial results with various applications. Internet of Things is counted as one of the heading technologies of Industry 4.0 in Europe (Ben-Daya et al. 2019). Internet of Things (IoT) can be described as an advanced technology where objects can perceive, communicate, and share information, and all these operations take place over networks, where interconnected objects regularly collect data, analyze this data, use this produced information for planning and decision making in management areas (Salazar 2016). Bhuvaneswari and Porkodi (2014) defined Internet of Things as it is a network connecting multiple objects, tools, and sensors by means of communication and data infrastructure to create value-added duties via smart data processing and administration for applications.

The Internet of Things (IoT) is a distributed Information and Communication Technology system integrating sensors, computing instruments, algorithms and physical objects famous as uniquely identifiable things (Khan and Yüce 2019). To the definition of McKinsey Global Institute, the Internet of Things is a set of tools which can monitor the environment, report the current situation, receive instructions and act according to the information it receives. Another definition for the Internet of Things is; autonomous data processing capability is achieved with systems connected with the Internet of Things without human intervention.[1] Concurrently, Internet of Things technology is based on machine-to-machine (M2M) communication systems which turns to a fully autonomous system. Also, The Internet of Things can be thought of as a different cyber-physical system than traditional internet-based systems where things can generate data that can control other things or objects in real time.

For further definitions, Table 10.1 is provided based on literature review of Ben-Daya et al. (2019).

No matter how we define it, the purpose of the Internet of Things technology is to use the processed data to connect all devices by or without wire, to collect and process data, and to perform automated tasks, and it is known that it can connect millions of heterogeneous devices over the Internet that require flexible layered architecture.

The foundations of this new technology are relied on Radio Frequency Identification (RFID) technology. With the advent of Radio Frequency Identification technology in the late 1990s, it became possible for objects to connect and communicate via sensors and transmitters. Internet of Things (IoT) as a concept was first used in 2009 by Kevin Ashton, co-founder of the Massachusetts Institute of Technology (MIT) Auto-ID Center, which creates a global standard system for RFID technology, and is it is expressed as a system in which data collects and exchanges

[1] https://www.mckinsey.com/featured-insights/internet-of-things/our-insights/the-internet-of-things-how-to-capture-the-value-of-iot. Accessed 18 June 2020.

Table 10.1 Definitions of IoT (Ben-Daya et al. 2019)

Definitions of IoT	Defined by
"Things having identities and virtual personalities operating in smart spaces using intelligent interfaces to connect and communicate within social, environmental, and user contexts."	INFSO (2008)
"The term "Internet of Things" is utilized as an umbrella keyword for covering many aspects linked to the extension of the Internet and the Web into the physical realm, with the help of the widespread deployment of spatially distributed devices with embedded identification, sensing, and/or actuation capabilities. Internet of Things envisions a future where digital and physical entities can be linked, by means of suitable information and communication technologies, in order to enable a whole new class of applications and services."	Miorandi et al. (2012)
"Interconnection of sensing and actuating tools providing the ability to share data across platforms through a unified framework, developing a common operating picture for enabling innovative applications. This is attained by seamless largescale sensing, data analytics and information representation using cutting edge ubiquitous sensing and cloud computing."	Gubbi et al. (2013)
"A loosely coupled, decentralized system of cooperating Smart Objects (SOs). An SO is an autonomous, physical digital object augmented with sensing/actuating, processing, storing, and networking capabilities. SOs are able to sense/actuate, store, and interpret information created within themselves and around the neighboring external world in which they are situated, act on their own, cooperate with one another, and replace information with other kinds of electronic tools and human users."	Fortino and Trunfio (2014)
"Dynamic global network infrastructure with self-configuring capabilities relied on standard and interoperable communication protocols in which physical and virtual 'Things' have identities, physical attributes, and virtual personalities and benefit from smart interfaces, and are seamlessly integrated into the information network. "	Xu et al. (2014)
"Group of infrastructures interconnecting connected objects and enabling their management, data mining and the access to the data they create. (Connected objects are defined as Sensor(s) and/or actuator(s) implementing a specific function and that are able to communicate with other equipment. It is part of an infrastructure allowing the transport, storage, processing, and access to the generated data by users or other systems."	Dorsemaine et al. (2015)
"The Internet of Things is a network of physical objects which are digitally linked to sense, monitor, and interact within a company and between the company and its supply chain allowing agility, visibility, tracking, and information sharing to make easier timely planning, control, and coordination of the supply chain processes."	Ben-Daya et al. (2019)

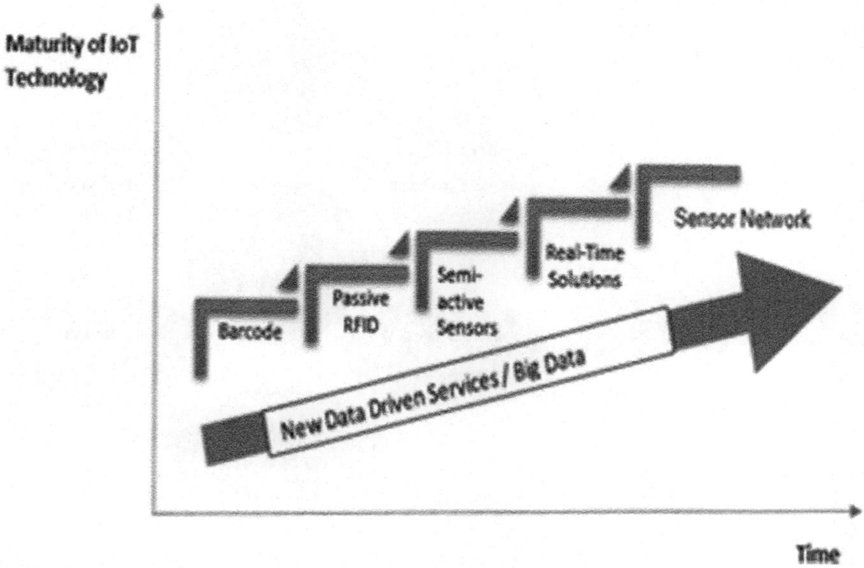

Fig. 10.1 Journey of IoT. *Source* DHL Trend Research (2015)

using the Internet of Things as a network of objects (Thiesse et al. 2009; Mengru 2017). Figure 10.1 shows a maturity of IoT through a time period as follows.

The working principle of Internet of Things technology is to create new applications and services through various objects that can connect and cooperate with one another through wireless and wired relations and unprecedented addressing schemes (Khan and Yüce 2019). The Internet of Things is definitely a revolution. With the Internet of Things, objects make themselves recognizable and provide intelligence by making context-related decisions or activating this context, as they are able to convey information regarding themselves. They can reach data collected by other objects or be aspects of complex services. This transformation has become a reality because of cloud computing technologies and the transition of the Internet to IPv6 with a virtually unlimited addressing capacity (Salazar 2016). So, we may say that the Internet of Things facilitates things to connect anytime, anywhere, with anything and anyone who ideally uses any network and any service (Salazar 2016). Figure 10.2 illustrates this statement as follows.

Wired and wireless sensor networks, 2G / 3G / 4G, GSM, GPRS, RFID, Wi-Fi, GPS, microcontroller, microprocessor, etc. are the enabling technologies of "Internet of Things" functions (Giusto et al. 2010). There are three enabling technologies for the Internet of Things; the first one is technologies that enable objects "*to get contextual information*", the second one is the technologies which allow objects "*to process contextual information*", and the last one is the technologies that allow objects "*to increase securityand privacy*" (Salazar 2016*)*. The first two of these are structures that distinguish the Internet of Things from the Internet as functional building blocks that require the transformation of "intelligence" into "things". The third one is a de

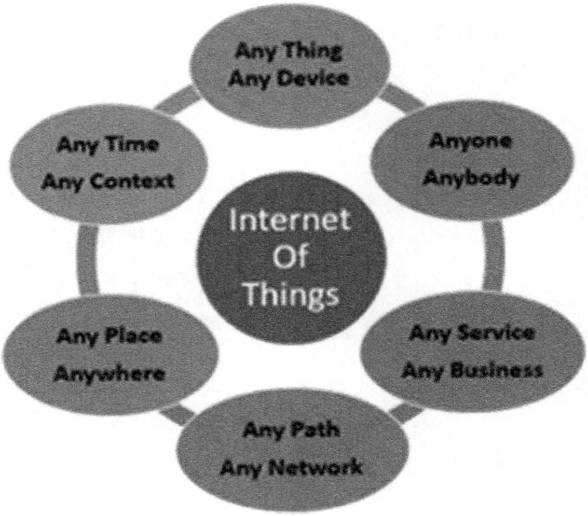

Fig. 10.2 Internet of Things. *Source* Salazar (2016)

facto necessity rather than a functional one, and without this the effect of the Internet of Things will be drastically reduced (Salazar 2016).

In fact, the Internet of Things is the combination of information technology, referring to the combination of different hardware and software technologies, involving electronic systems used for communication between individuals or groups to store, retrieve, and process data. There are five IoTTechnologies listed today: Radio Frequency Identification (RFID), Wireless sensor networks (WSN), Middleware, Cloud computing, and Io applications. The working principles of Internet of Things are based on its specific characteristics. Some of the features of Internet of Things technology can be listed as follows (Sintef 2014; Miorandi et al. 2012):

- **Connectivity**: Everything can be related with the Internet of Things and the global information and communication infrastructure. In order for objects on the network to communicate with each other, they must connect to the Internet of Things platform over the Internet. Devices in the Internet of Things network need a high-quality Internet connection to provide quality communication.
- **Heterogenity**: Tools in IoT are heterogeneous as they are dependent on various hardware platforms and networks. They can contact with other tools via different networks.
- **Dynamic changes**: The state of the tools changes dynamically basing on the conditions of the environment. Most of the devices with the Internet of Things feature have built-in sensors to detect these changes in the environment. Those sensors are stimulated when there is a change or any event.
- **Intelligence**: The Internet of Things enables objects to work smarter at a definite time or in a specific action.
- **Real-time analysis**: All instruments work simultaneously and data are transmitted without delay. Simultaneous analysis of data is important, especially in sensitive applications such as smart tools.

- **Scalability**: The number of instruments needing to be managed and communicating with one another will be at least one time greater than existing Internet-connected devices. This means IoT applications should be scalable enough to meet needs requirements in view of its future growth potential. In addition, as the data produced is so large, large devices are required to store and process the data.
- **Architecture**: IoT architectures can be different than each other as the devices to be connected made by different manufacturers. This requires a new architectural design due to new functions added to the IoT application.
- **Security**: While benefiting from IoT, security issue is important. Considering the security of our personal data and our health, the designs of this technology should be made accordingly.
- **Object-related services**: IoT can supply services connected to objects in certain frameworks. Technologies in both the physical world and the information world will shift to supply things related services in the constraints of objects.

Besides its characteristics, it is important to understand the technology of Internet of Things. Xu et al. (2014) and Peng et al. (2020) listed main essential layers of IoT as "*a sensing layer*" which helps to integrate different things like RFID tags, sensors, and actuators, "*a networking layer*" which enables information transfer through wired and wireless network, "*a service layer*" that integrates services and applications through a middleware technology, and "*an interface layer*" that displays information to the user which allows interaction between the system.

In recent years, Internet of Things (IoT) has helped the development of many new applications in the field of logistics and supply chain, and it seems to have more impacts on supply chain management in the future. What is expected to happen in Internet of Things technology is that every object has its own digital object identifier (Gershenfeld et al. 2004). It is also known that development of a global network for enabling the exchange of goods, services, and information with objects as infrastructure via the Internet of Things (Kortuem et al. 2010; Liu and Sun 2011).

The Internet of Things provides visibility, flexibility, and customer-specific services in supply chains (Ellis et al. 2015). When data from objects are collected and analyzed effectively, and then transformed into useful information, it provides visibility into each stage of activity along the supply chain, with alerts of issues that require improvement. Responding to these alerts in a timely manner can increase efficiency in the supply chain.

What is lacking in business today is not lack of the availability of information, but rather large data collection and processing technologies and the delays between data collection and action. The Internet of Things will help reducing the time between data collection and decision-making by providing the infrastructure that will enable timely response to sudden changes in supply chains (Ellis et al. 2015). In addition, the Internet of Things will enable all parties to work collaboratively throughout the supply chain, remote management of logistics operations, and obtain the quality information needed to make more effective decisions.

10.3 Internet of Things in Logistics

Rob Siegers—President of Global Technology at DHL—alerts all logistics service providers about the power of Internet of Things technology to be ready for new business models in the next 5 years (DHL Trend Research 2015). He stated that the assistance of this technology can only be possible with well-connected assets and strong collaborative and cooperative relationships between all parties in the supply chain. Indeed, the Internet of Things technology has started to be used rapidly in various logistics activities throughout the supply chain. The most important factor that triggers this is to increase the availability of products needed or desired by various customers all over the world and to target their distribution in a traceable manner. As it is known, one of the most important goals of companies is to plan and coordinate their supply chains in an efficient, effective, transparent, and smart way from the raw material supplier to the distribution of the finished products to the desired points.

Here again, we underline the facts of logistics; the issue of distributing the right products to the right customers, at the desired time and conditions, at the right price, without any damage, with the desired features draws attention. The full management of the process by meeting all these conditions is a very difficult task and all supply chains need support in this regard. Recently, developing technologies have made great contributions to supply chains in this sense, Internet of Things technology is one of the greatest that provides many advantages to supply chains.

We can list many advantages, but the first thing that comes to mind is that goods and vehicles can be followed simultaneously, and traceability capability with the Internet of Things has increased. Thanks to the other features of this technology, the risks that may occur along the supply chain has been reduced, and thus the way to manage assets more efficiently. Increasing efficiency will of course maintain serious reductions in supply chain costs. Tracking vehicles in particular contributes to fuel savings and increases timely delivery rates. Telematics sensors on trucks and multi-sensor labels on products also contribute greatly to safety and security issues, as the whereabouts of a load can be easily detected at any time. It is also supported the energy efficiency by using various sensors in some areas such as lighting and heating of the warehouses, and following the status of equipments such as forklifts and conveyors which also contributes to green logistics activities. All other benefits are provided in Fig. 10.3 as follows.

The reflections of Internet of Things technology on various logistics activities are explained below.

10.3.1 Internet of Things in Transport

As mentioned earlier, Internet of Things technology is based on identifying real-world objects that are linked through communication interfaces. The success of the Internet of Things stems from its ability to accurately and efficiently localize network

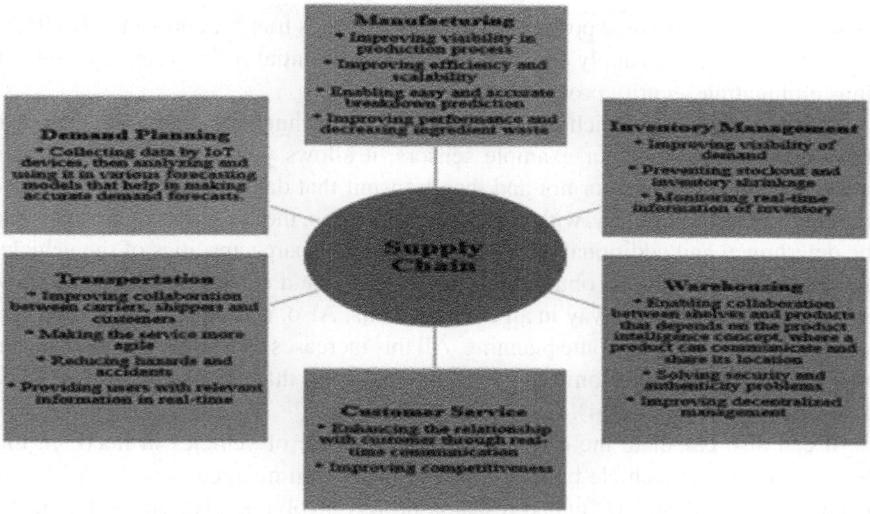

Fig. 10.3 Benefits of using IoT in SCM functions. *Source* Mostafa et al. (2019)

components, information, and processes. Transportation activities are one of the most important application areas of this technology in the field of logistics. Assets of freight transportation present great potential for Internet of Things networks. Internet of Things stands out in the following three areas in transportation activities*: location and situation tracking, fleet and resourcemanagement,* and *maintenance and repair estimate.*

Tracking and tracing of location of vehicles is only a simple offering of Internet of Things since today it is available to track and monitor containers or freighters in the middle of an ocean or in midflight. Internet of Things has better tracking and tracing capability so it offers faster, more accurate and predictive, and more secure operations (Mioransi et al. 2012). Besides geographical location information, Internet of Things can provide condition monitoring which helps related parties to get information about the situation of their cargoes. Especially in perishable, sensitive cargo and live animal transportation, this provides great help to the parties along with capability of preparing reports about the situation of the shipments. It is also a well-known fact that theft incidents cost shippers and carriers billions of dollars. Telematics sensors on trucks and multi-sensor tags on items are also a useful technology in these situations, as they can also address data on location, status, and possible theft.

Lots of features of Internet of Things provides stand out for fleet management as well. It can suggest more precise solutions to issues such as more efficient rotations in picking and deliveries, and preventing empty return loads. Internet of Things technology has a great contribution to the traceability of vehicles, especially in transit.

In this way, it has become possible to monitor all loads transparently and simultaneously throughout the supply chain and to prevent potential risks by anticipating and thus eliminating security problems easily.

Another areas in which Internet of Things technology used are fleet and resource management. For example sensors, it allows you to monitor how often a truck/container is used or not and then transmit that data for analysis to improve the performance. Similarly, with the sensors to be used, the capacities of the loads can be determined and additional information about the spare capacities of the vehicles for various routes can be obtained. It can then create a central dashboard to show spare capacity along the way in all business units. Also, it can offer suggestions for consolidation and better route planning. All this increases efficiency in fleet management, optimizes fuel economy, and reduces dead loads that account for 10% of truck kilometers (Lee et al. 2014).

It can also calculate the estimated asset life cycle of vehicles in fleets. In this way, it can predict vehicle breakdowns and helps plan maintenance checks. As an example, in 2012, an EU-funded research project involving DHL and Volvo developed a commercial truck with the latest sensor technology that decides when it needs maintenance. By using wireless network, the data collected by the sensors were transmitted to the central unit and then transferred to the maintenance platform center for analysis. After that, both drivers and maintenance-repair teams were warned of possible problems in the vehicle. In this way, both technical and life hazards that drivers will encounter on the road have been reduced by an average of 40–50% (DHL Trend Research 2015).

Like road vehicles, tracking of intermodal containers is becoming widespread day by day. In addition to containers, pallets and cargo boxes have started to be followed at affordable costs, both by sea and air.

IoT also helps to improve more secure and safe operations, prevent accidents, and alert drivers when they need to stop driving. With built-in cameras for long-distance truck drivers, it can monitor pupil size and blink frequency of the drivers to see they are tired. Caterpillar, the largest construction and mining equipment manufacturer of the wold, has implemented this system in order to prevent sleepy truck drivers from accidents. The solution activates audible alarms and seat vibrations if it detects that the driver has lost attention on the road.

All kinds of transport vehicles can communicate with other objects or any software system thanks to the sensors placed inside them. In addition to this, according to a study, it was stated that the Internet of Things can be used to measure the carbon emissions of automobiles in cities (Wang and Kexin 2013; Gökrem and Bozuklu 2016).

10.3.2 Internet of Things in Procurement and Order Management

Purchasing and procurement activities refer to the process from purchasing the raw materials required for producers to produce on time, until they enter the production facility. Efficient management of this upstream activity is essential for producers to produce and distribute the goods that their customers need on time. Moreover, the relationship with suppliers includes not only the flow of goods but also the flow of information and money (Alt and Zbornik 2002). Therefore, order processing, activation of raw materials, and timely payment is one of the critical points of the supply chain. The Internet of Things strives for the correct functioning of these different streams, providing greater visibility and control to the parties in this process. Here, the Internet of Things manages to relate both information and physical products.

Information is the most valuable resource today. Getting the right information on time is a sensitive issue for supply chains. Moreover, the costs of supply chains (especially the reduction of the bullwhip effect) are reduced thanks to the information and the creation of customer value is facilitated (Qin et al. 2017). If the main source of value creation is knowledge and the Internet of Things is the most important tool it offers, this technology can be an important part of the value proposition.

As mentioned earlier, information can be directly associated with objects and products. The usage, status, and location of the objects can be traceable. This can offer the consumer new value propositions, such as additional product information or full billing of products or services based on actual usage. This management style and flow of information have the same importance in downstream relations with customers and their orders, as in supplier-producer relations. Order management is behind the value created throughout the supply chain. Since the order is directly linked to the physical product, all details of this product should be able to be defined up-to-date in the system. In addition, it should be known that the product is stored under correct conditions. Real-time information should be accessible at any time and the necessary accounting and financial transactions should be completed at the end of the customer's order process.

The Internet of Things effectively leads to the distribution of physical products where they are needed, thanks to the information flow it provides in all these procurement and order processes.

10.3.3 Internet of Things in Warehousing and Inventory Management

Warehouses have become the backbone of today's supply chains. Increasing and diversifying customer demands have become more complex with the spread of electronic commerce. Increasing competition on top of the intense demand, accelerated the delivery speed and this brought the importance of efficient management of warehouses and stocks (Trappey et al. 2017). Of course, all these conditions increase warehouse costs, but the new business models developed have increased flexibility and personalized services have begun to be offered. At this point, warehouses can be defined as an important link that creates added value throughout the supply chain. Moreover, the optimum use of the entire area of the warehouses without honeycombing and of course the presence of a good system in the warehouse affect the efficiency of the warehouse. As a result, the Internet of Things technology offers various possibilities to relieve these stressful conditions in which the warehouses are located. Now, the warehouse itself and all the objects in it can be connected and work in communication. The concept of "dark warehouse" has entered our lives. Labeling, where many identification technologies from barcodes to RFIDs have revolutionized, gains a different dimension with the Internet of Things at the level of pallets, forklifts, shelves, boxes, products (Cui 2015). Let's imagine a smart warehouse using this technology. Thanks to wireless readers, it collects the data coming from each pallet or box and transmits it to the warehouse management system for processing. This process takes place very quickly in the warehouse, which is time consuming, for example, goods counting, volume scanning, etc. It eliminates such activities. In addition, damaged pallets can be detected by scanning them with cameras that can be attached to gateways (NEC Corporation 2017).

In the field of inventory management, thanks to labels, the visibility of the products increases, and the lack of stock can be eliminated. Sometimes there may be problems with placement. It is easier to trigger corrective actions in this regard. In addition, the issue of quality management is of great importance in warehouses. In particular, monitoring the status of products with an expiry date, monitoring the temperature and what condition of the products, warning mechanisms work if there is a danger. The Internet of Things enables companies to update their inventories in real time, encouraging efficient inventory management practices (Eldridge and Chapin 2015). These proactive actions ensure timely intervention of warehouse personnel, increase the efficiency of the warehouse manager, and thus contribute to the formation of customer loyalty as the service quality will increase.

A correct warehouse management means correct delivery. Thanks to the sensors, products can be scanned through the gateway and regularly transferred to delivery. By enabling Internet of Things, activities and information flows in the warehouse will be under control, stock levels are realized at reasonable levels and stock costs are reduced (Mostafa et al. 2019).

There is an increase in accidents due to the increasing business volume in warehouses. Using the Internet of Things technology, it becomes possible to communicate with other forklifts via sensors and actuators combined with radar or cameras attached to the forklifts, and the perimeter is scanned for hidden objects that may cause accidents. This way, forklifts can be programmed to slow down automatically when another forklift or person or object is remarked in any place of the warehouse. Thus, possible accidents can be prevented (DHL Trend Research 2015).

At the same time, a warning that the layout is wrong can be sent by using pressure sensors for risks that may arise due to improperly placed pallets or over-loaded forklifts. The same technology applies to trucks and supports safe loading with load center measurement technology. Thus, the safety of drivers and loads is guaranteed. With the Internet of Things, warehouse theft cases can also be prevented through cameras and sensors. In the near future, we will be faced with the widespread use of wearable devices, smart glasses and with this, machine-human interaction.

Finally, it is possible to benefit from the Internet of Things in subjects such as keeping the health and fatigue of employees in warehouses under control and monitoring their performance. By integrating active and passive RFID technology and mobile devices, it is possible to monitor the in-warehouse activities of people. In addition, energy management is an important issue in warehouses and it is predicted that energy consumption will decrease significantly with the addition of sensors to the warehouse infrastructure. This decrease in energy consumption has a positive reflection on the carbon footprint of the warehouse, which will serve the concept of green logistics (DHL Trend Research 2015).

10.3.4 Internet of Things in Retailing and Last Mile Delivery

It is known that the most successful in the retailing industry are those who provide their customers with greater efficiency during their shopping experience. It offers these new experiences for customers with the featured usage scenarios that Internet of Things offers in retail stores. Many conveniences such as shelf availability, stock optimization, mobile payments are becoming more common with the Internet of Things.

In the retail industry, IoT, on-the-shelf availability, inventory, and product optimization, compliance to increase efficiency through planogram, loss prevention, and mobile payments. This will change the customer experience.

Among the most well-known benefits of the Internet of Things for consumers in retailing are in-store guidance services for shoppers, such as payment optimization tools such as Qminder and Waitbot, and OSHbot robots equipped with sensors, cameras, and video analytics, offered by the American home decoration retailer Lowe's (DHL Trend Research 2015).

With these applications, not only brick and mortar stores but also multiple distribution channels are supported. For example, a record of the eggs in a refrigerator at home is kept and with the smart egg carton system called "eggminder" where the eggs are running low, the need is notified when the egg is running low.

With the increase in electronic commerce, the biggest problems in retailing services are experienced in the last part of the deliveries. With the diversification of consumer demands and the increase in delivery points, logistics service providers are facing many new challenges. It is very difficult to be both fast and prevent costs from increasing. In recent years, the issue of "last mile delivery" has come to the popular for this reason and efforts have been made to produce solutions in this field (Saatcioğlu and Uzel Aydınocak 2017).

It has been observed that the Internet of Things can be used in this area and can provide innovations for this last part where products meet the consumer. For example, it can communicate with dispatchers with sensors placed inside mailboxes and transmit information about whether the boxes are empty or full. Thanks to this information, the distribution officer can plan the daily collection routes more effectively. Likewise, by connecting the sensors placed in the mailboxes with the mobile phones of the customers who own the mailboxes, information about the package distribution can be shared simultaneously.

As in the refrigerator example, the trend for smart home products continues rapidly. Systems that follow the expiration dates of the products and automatically order online when they are low have begun to take place in our lives. Of course, the advantages of these systems for logistics service providers in the field of procurement and distribution are many. These systems perceive that retailers, as well as end consumers, have decreased stocks and reduce delivery times by automatically ordering from the nearest warehouse. In this way, the risk of being out of stock is eliminated and customer satisfaction increases. Amazon is one of the leading companies working in this field. In order to enable predictive delivery, it developed an algorithm that requires the customer to buy the cart before confirming it. In this way, it has the opportunity to bring the estimated products that can be purchased closer to the delivery addresses of the customer and when the customer approves the order, it can deliver faster from the nearest point (DHL Trend Research 2015).

Another development expected in the last mile delivery is that the delivery can be made at any time. This creates flexible delivery solutions. The expectation that electronic shoppers can receive their deliveries at the time and place they want and that this can be done at an affordable cost is possible thanks to the labeled packages developed by the Internet of Things. It also has the advantage of providing more visibility for the recipient for their delivery address requirement, and even if there is requirement for a change in address.

In delivery of perishable products, the information about temperature, humidity, etc. Can be measured thanks to the sensors used which share information simultaneously with customers, so that logistics service providers will have the opportunity to increase their service standards for customers and at the same time learn about their current situation.

All these developments show us that in the near future, logistics service providers will be able to provide predictive and flexible delivery services for consumers and other businesses by combining sensor data with customer data.

10.3.5 Internet of Things in Risk Management Through Supply Chains

Risk management in the supply chain is another area where the Internet of Things is increasingly used. As it is known, uncertainties in global supply chains show that traditional supply chain management models do not work. Natural disasters, epidemics, social and political events, and economic turmoil pose a threat to the management of supply chains. In the field of transport, for example, strikes in ports, closure of airports or highways for various reasons, or problems in warehouses due to fire or rainfall can be predicted by the analytical capabilities of the Internet of Things and through the data collected, which saves time to produce solutions to problems in a timely manner.

10.4 Conclusion

It can play an important role in improving the various functions of the Internet of Things in supply chain and logistics activities. In today's business world, procurement, order management, inventory management, and warehouses play an important role in meeting customers' expectations. Fast distribution of products with cost efficiency and flexibility is one of the main sources of competition. Stock status updates, product tracking, and order management can only reach a level that creates added value thanks to information integration. Transport is likewise a key activity that requires efficient management of resources, ensuring safety and security, and optimizing costs. In this sense, the Internet of Things is one of the most important technological developments that can provide information to managers on time and provide the opportunity to monitor all activities simultaneously and transparently throughout the supply chain.

According to all these developments, companies that want to take advantage of the Internet of Things technology in their operations must establish the correct connections throughout their supply chain if they want to gain advantage in their activities such as storage, transportation or last kilometer delivery. Increasing participation, strong cooperation, and a joint investment desire among all local and international suppliers, manufacturers, retailers, logistics service providers, and all other parties are required. As a result, the whole goal is to build a thriving Internet of Things ecosystem.

In order to successfully implement the Internet of Things technology in the logistics industry, companies working in different fields all over the world must have a standardized approach. In addition, it is essential to have an infrastructure that can work in heterogeneous environments of the information obtained through sensors. It is essential to resolve the confidentiality and trust issues that are anxious to all parties throughout the supply chain. Due attention should be paid to the architecture and network design of the Internet of Things technology and the costs it will entail. Finally, a change in business mindset is essential to understand and absorb the advantages of this technology. New business models need to be developed and adopted.

References

Alt R, Zbornik S (2002) Integrierte Geschäftsabwicklung mit electronic bill presentment and payment. In: Weinhardt C, Holtmann C (eds) E-commerce: Netze, Märkte, Technologien. Proceedings zur Teilkonferenz der Multikonferenz Wirtschaftsinformatik

Ben-Daya M, Hassini E, Bahroun Z (2019) Internet of things and supply chain management: a literature review. Int J Prod Res 57(15–16):4719–4742

Bhuvaneswari V, Porkodi R (2014) The internet of things (IoT) applications and communication enabling technology standards: an overview. In: Proceedings of 2014 international conference on intelligent computing applications, pp 324–329

Cui Y (2015) Improving supply chain resilience with employment of IoT. In: Proceedings of MISNC 2015: Multidisciplinary social networks research, pp 404–414

DHL Trend Research (2015) Internet of things in logistics, a collaborative report by DHL and cisco on implications and use cases for the logistics industry. DHL Customer Service & Solutions, Germany

Eldridge R, Chapin L (2015) Internet of things: an overview. The internet society (ISOC), pp 1–50

Gershenfeld N, Krikorian R, Cohen D (2004) The internet of things. Sci Am 291(4):76

Giusto D, Iera A, Morabito G, Atzori L (2010) The Internet of things 20th Tyrrhenian workshop on digital communications. Springer, New York, NY, US

Gökrem L, Bozuklu M (2016) Nesnelerin İnterneti: Yapılan Çalışmalar ve Ülkemizdeki Mevcut Durum. Gaziosmanpaşa Bilimsel Araştırma Dergisi 13:47–68

Khan JY, Yuce MR (2019) Internet of Things (IoT): systems and applications. Jenny Stanford Publishing Pte. Ltd., Singapore

Kortuem G, Kawsar F, Fitton D, Sundramoorthy V (2010) Smart objects as building blocks for the internet of things. IEEE Internet Comput 14(1):44–51

Lee S, Tewolde G, Kwon J (2014) Design and implementation of vehicle tracking system using GPS/GSM/GPRS technology and smartphone application. IEEE World Forum on Internet of Things

Liu X, Sun Y (2011) Information flow management of vendor-managed inventory system in automobile parts inbound logistics based on internet of things. J Softw 6(7):1374–1380

Miorandi D, Sicari SD, Pellegrini P, Chlamtac I (2012) Internet of Things: vision, applications and research challanges. Ad Hoc Netw 10(7):1497–1516

Mostafa N, Hamdy W, Alawady H (2019) Impacts of Internet of Things on supply chains: a framework for warehousing. Soc Sci 8:84

NEC Corporation 2017. Leveraging IoT For Logistics. https://in.nec.com/en_IN/pdf/Leveraging_IoT_for_Logistics-1.pdf. Accessed 12 July 2020

Peng S-L, Pal S, Huang L (2020) Principles of Internet of Things (IoT) ecosystem: insight paradigm. Springer, Switzerland

Qin W, Zhong RY, Dai H, Zhuang ZL (2017) An assessment model for RFID impacts on prevention and visibility of inventory inaccuracy presence. Adv Eng Inform 34:70–79

Salazar C (2016) Internet of Things-IOT: definition characteristics, architecture, enabling technologies, application & future challenges. Int J Eng Sci Comput 6(5):6122–6131

Sintef OV, Friess P (2014) Internet of Things–from research and innovation to market deployment. River Publishers' Series In Communications

Thiesse F, Floerkemeier C, Harrison M, Michahelles F, Roduner C (2009) Technology, standards, and real-world deployments of the EPC network. IEEE Internet Comput 13(2):36–43

Trappey AJC, Trappey CV, Fan C-Y, Hsu APT, Li XK, Lee IJY (2017) IoT patent roadmap for smart logistic service provision in the context of Industry 4.0. J Chin Inst Eng 40:593–602

Wang M, Kexin L (2013) Transportation model application for the planning of low carbon city—take Xining City in China as example. In: The 2nd international workshop on agent-based mobility, Traffic and transportation models, methodologies and applications

Xu LD, He W, Li S (2014) Internet of Things in Industries: a survey. IEEE Trans Ind Inf 10(4):2233–2243

Ezgi Uzel Aydınocak is an Associate Professor of Marketing at Beykoz University, Faculty of Business and Administrative Sciences. She is also director of Beykoz Vocational School of Logistics. She received her PhD in Business Management at the Institute of Social Sciences, Yeditepe University in 2013. Her research focuses on logistics, supply chain management, services marketing, industrial marketing, marketing research and online retailing. Associate Professor Uzel Aydınocak has authored and contributed to numerous books, book chapters, articles, and reports on logistics and marketing. Associate Professor Uzel Aydınocak has considerable experience in the development and delivery of academic courses and professional training programs in the areas of marketing, logistics, supply chain management, transportation management, warehouse and inventory management, project management, operations management and retail logistics. In addition to her academic career, she has experience in business life, worked in a marketing department of a global logistics company as an Inside Sales Consultant. She also holds several advanced certifications, including a specialty in carriage of dangerous goods on road, sea and railway based on her DGSA (Dangerous Goods Safety Advisor) License from the FTA (Freight Transport Association of UK).

Chapter 11
Integrated Systems and Utilization in Logistics

Haci Mehmet Alakaş and Tamer Eren

Abstract Logistics takes place as a support system in every phase of a product life cycle, starting from the supplier to the recycling or disposal process of the product. At these stages, logistics systems work both integrated with other systems and different systems work integrated with logistics. Logistics systems support systems such as supply chain, purchasing, production, storage, sales and marketing. Decision processes and outputs of these systems are effective in decision-making in the logistics system. In addition, systems such as geographic information systems, vehicle tracking systems and the Internet of Things also work integrated with the logistics system. With their structures that affect the system's performance, they play an essential role in the traceability of the logistics system and the establishment of effective system infrastructure. In this section, systems that are effective in the operation of logistics systems are discussed. Analytic network process which is one of the multi-criteria decision-making methods was used to determine the impact of systems on the efficiency of the logistics system. Thus, the systems that have the most effect on logistics system efficiency have been determined. Finally, information technologies used to facilitate and speed up the logistics system are mentioned. The effects of new systems developed with Industry 4.0 on logistics are emphasized.

11.1 Introduction

Logistics systems are systems that are established to manage the flow and storage of materials in the supply chain from the supplier to the end user, ensuring that the materials are at the desired place and quantity on time. With the logistics activities, it is aimed to meet the end-user demand on time and in the most economical way like other functions of the business. Three basic activities are carried out in planning,

H. M. Alakaş · T. Eren (✉)
Kirikkale University, Department of Industrial Engineering, Kirikkale, Turkey
e-mail: teren@kku.edu.tr

H. M. Alakaş
e-mail: hmalagas@kku.edu.tr

İ. İyigün and Ö. F. Görçün (eds.), *Logistics 4.0 and Future of Supply Chains*,
Accounting, Finance, Sustainability, Governance & Fraud: Theory and Application,
https://doi.org/10.1007/978-981-16-5644-6_11

execution and control in material flow management with the logistics system (Council of supply chain management professionals 2013). Departments such as production, marketing, planning, purchasing, storage, transportation take part in the material flow operationally.

Eight main functions are performed in the logistics system (Nebol et al. 2016). These functions and related departments that fulfil these functions are given in Table 11.1, respectively. In this structure, while operating functions are carried out, logistics functions are accompanied and logistics cannot be considered separately from other departments. From this point of view, logistics is an integrated system within the business functions and is not an activity carried out as a function of only one department. Therefore, it requires the integration of the departments within the system with each other in order to establish an efficient logistics system and to run the system efficiently.

Information systems (warehouse management, fleet management, internet of things, etc.) are used to perform functions within the logistics system and it is also aimed to perform functions more effectively (Gunasekaran and Ngai 2004). Table 11.2 shows which system is used as a support system to perform which function in the logistics system. These systems are especially important in facilitating the management of logistics operations and performing them quickly. They also make important contributions to the efficiency of the system.

11.2 Systems That Operate Logistic Functions and Their Effect on Logistic Utilization

Logistics system activities are carried out by many departments inside and outside the enterprise. The contribution of these departments to logistics functions and their impact on these functions will be discussed in this section. It is given how the departments contribute to logistics functions. The effects of these departments on the efficiency of the logistics system are determined by the ANP method.

11.2.1 Production System

Production is all the activities that people carry out by spending money, time and effort using knowledge and technique to provide the goods and services that people need. Production activities are important for obtaining the final product within the logistics system (Lambert et al. 1998). The most important information in deciding on the production is demand information. The planning department decides the products and their quantities to be produced by evaluating the customer demands and demand forecasts made by the marketing department. According to this information, production orders are created, and it is determined which product will be produced

Table 11.1 Logistics functions and related business departments

Logistics Function	Purchasing	Planning	Marketing	Production	Transportation	Warehouse	After-sales service
Customer order operation		*	*				
Stock management	*	*	*			*	
Purchase order operation	*	*					
Production order processes	*	*		*			
Transport	*	*	*	*	*	*	
Warehouse and warehouse operations		*		*	*	*	
Material handling and packaging				*	*	*	
After-sales operation			*				*

This table was prepared by the authors

Table 11.2 Logistics support systems and related logistics functions

Support systems	Customer order operation	Stock management	Purchase order operation	Production order processes	Transport	Warehouse and warehouse operations	Material handling and packaging	After-sales operation
Enterprise resource planning (ERP)	*	*	*	*		*	*	
Warehouse management systems		*				*	*	
Customer relations management (CRM)	*							*
Fleet management		*			*	*		
Geographic information systems (GIS)					*			
RFID	*	*	*	*	*	*	*	
Barcode systems	*	*	*	*	*	*	*	
Internet of things (IOT)	*	*	*	*	*	*	*	*
Cyber physical systems (CPS)		*	*	*	*	*	*	
Big data	*	*	*	*	*	*	*	*
Cloud computing	*	*	*	*	*	*	*	*

This table was prepared by the authors

when and in what quantity and how the production flow will be processed (Nebol et al. 2016).

In the logistics system, as a result of the production process, the final product is obtained and customer orders transmitted as information until this stage turn into physical output. The timely availability of the raw material required for the production of the product and the completion of the necessary activities within the production on time are critical for meeting customer orders on time. The products are either produced for direct customer orders or are produced in stock to meet customer demand later. The important decision to be taken in both cases is how much production will be made in which period and where will this production be made.

When evaluated in terms of logistics functions, the production department is directly related to production order processes. Making production orders determined by the production planning department in a timely manner will both increase operational efficiency and ensure fulfilment of customer orders on time. Since it is the production department that provides the most added values to the product in logistics, the efficiency of this department directly affects the efficiency of the logistics system. In addition, customers pay money for the added value of production and the activities that determine the real value of the product are production activities. Other activities are carried out to support production and indirectly contribute to the product.

Other logistics functions are carried out during the production order process. Material handling processes are carried out in the preparation of raw materials for production, and the incoming raw materials are allocated to the amounts needed for production. In addition, the materials must be transported and temporarily stored during production. How the material will move and how much, where and how long it will be stored is also carried out within the logistics functions.

11.2.2 Marketing System

Marketing is the outward-facing face of companies that connect with customers. The important function of marketing is to find a market for the products and to work to ensure the continuity of the markets found. In addition, tracking the customer orders and transferring the orders to other departments are among the duties of the marketing department (Bowersox et al. 2002). The marketing department sends information to customers about when their orders will be delivered. At the same time, it also tracks post-sales transactions and it ensures that defects are eliminated if there is any defect in the product (Nebol et al. 2016).

Although the marketing and sales channels change, the most important function of the marketing unit in terms of logistics is to perform the duty of serving the customer according to the principles called seven rights. These seven principles of logistics are the right product or service, the right amount, the right place, at the right time, under the right conditions, at the right price and with the right information. A sustainable service can be offered to customers in the most economical way by following these principles (Bowersox et al. 2002).

When the marketing department is evaluated in terms of logistics functions, the most important function it performs is customer order transactions. As mentioned in the first paragraph of the section, they are the duties of the marketing department to set up and operate the infrastructure to manage customer orders. In addition, it informs the transportation department that the necessary information is transmitted to the transport function for the orders to arrive and which product will be transported where and when. They follow the supply of necessary resources to ensure that orders are not delayed. Another function that it fulfils from logistics functions is to support inventory management. By making sales forecasts together with the planning department, it determines which product, at what time, how much stock is kept to meet the customer orders. Finally, within the scope of after-sales services, the sales and marketing department carries out the purchasing and evaluation of the customers' feedback.

11.2.3 Production Planning System

The planning department carries out the production plan and the follow-up of the plan without any interruption so that the customer orders can be delivered on the deadline. While fulfilling this purpose, the operations of following the raw material and product stocks of the enterprise, making the necessary resource plans for production, making production schedules in a way not to delay the delivery date of the products, determining the quantity and times of purchase orders, making production or purchasing decisions are tasks of planning department (Bowersox et al. 2002). The planning department creates production and purchase orders according to the customer orders coming from the marketing unit within the logistics system. Therefore, it has a one-to-one relationship with marketing, production and purchasing departments.

The planning department is the first responsible department in the creation of production orders, making the production decision and performing product inventory management functions within the logistics system (Nebol et al. 2016). The marketing department is responsible for meeting customer requests within the framework of seven principles, and the planning department is responsible for the production being performed within the framework of these principles. The marketing department is responsible for meeting customer requests within the framework of seven principles, and the planning department is responsible for the production being performed within the framework of these principles. Although the companies want to meet production orders on time with a stock-free production basically, they keep stock due to other production costs in order to respond to the customer quickly. The planning department decides on the amount of stock to be kept and when it should be kept according to the information it receives from the marketing department.

Another important task of planning is to provide necessary resources for production or to make production decisions according to production resource constraints.

The most important of these resources is to supply the necessary raw materials and materials for production. The planning department makes a need analysis taking into account the production plan and creates purchase orders. It transmits the information of the purchase orders to the purchasing department and starts the purchasing processes. Although purchase orders are primarily the responsibility of the purchasing department, the planning department also monitors these orders in order to avoid disruption of production. At the same time, the inventory amounts to be held from raw materials and materials are determined by the planning department. As a result, the purchasing department performs the purchasing of the materials according to the information provided by the planning department.

The planning department, which can be called as a key point in an entire logistics system, performs basic functions such as management of production orders, determination of stock levels, management of purchase order and resource planning depending on customer orders (Güleş et al. 2012). It also regulates the flow of information between departments, ensuring that all departments work in harmony. It plays a key role in directing forward and backward flow of information within the enterprise. It processes and transmits information from marketing to other units, and delivers information about the delivery of customer orders to marketing according to production orders.

11.2.4 Purchasing and Supply chain System

The purchasing department is the department responsible for the execution of the communication and transactions between the companies and the suppliers to supply the products, raw materials and materials needed. Although the function of the purchasing unit is mostly explained in relation to production, other enterprises such as the main dealers, dealers, retailers who take part in the transportation of the products to the customers also make purchases. A sales-purchasing process is carried out among all processes within the logistics system. The most important function of purchasing is to guarantee the uninterrupted flow of the raw materials, materials and services so that the organization can carry out its activities (Nebol et al. 2016).

Purchasing process begins with the purchase order information from the planning department. Information about the amount of raw materials and materials required for production and the dates needed are transmitted to the purchasing department. The purchasing amounts of the desired materials are decided together with the planning department in order to provide a price advantage (Brown and Coopers 1999). According to this information, the purchasing department is researching suppliers and determining the most suitable supplier. It makes a contract with the supplier and follows the process until the supply of goods or services within the scope of the contract. From the perspective of logistics functions, the process of transporting from the supplier to the enterprise is also carried out in the follow-up of the purchasing department and is responsible for the process until the product enters the warehouse.

At the same time, information about the fees of the purchased materials is transmitted to the accounting by the purchasing unit.

11.2.5 Storage System

Although storage is done many times in different stages within the logistics system, the part we will mention here is the systems established for long-term storage of products, raw materials and materials. Warehouses are defined as places where raw materials and materials are kept, protected and stored. In addition to these functions, warehouses have now become functional in which functions such as merging, sorting and grouping of products are performed in order to better serve customers. Although storage is an undesired activity in logistics systems, it cannot be eliminated (Nebol et al. 2016).

In storage systems, products, raw materials and materials are kept until they are needed. In addition, another logistics function in which storage systems are carried out is material handling and packaging. The materials that come in high quantities are divided into smaller quantities in the warehouse according to the quantities needed. At the same time, products from different production areas are combined and packaged in warehouses according to the wishes of the customers and made ready for transportation. Warehouses, which were defined as temporary waiting areas only in the past, are now more functional within the logistics system.

11.2.6 After-Sales Service

After-sales services department carries out the duties of delivering the products to the customer, assembling the products, eliminating the defects, if any, and eliminating the malfunctions that may arise with the use of the products (Brown and Coopers., 1999). It carries out post-sales operations from the logistics functions together with the marketing department. Since it is the department that is concerned with the satisfaction of the customers and the problems to be experienced, it plays an important role in the business.

11.2.7 Transportation System

Transport is the physical movement of the material forward or backward in the supply chain. Transportation is one of the important functions for logistics. Transport is carried out both in bringing raw materials into operation and delivering products to desired locations (Banks 2002). In addition, material movements within the enterprise are included in the scope of transportation. In the logistics system from the suppliers to the business, it is carried out either by the supplier or the business, depending on the terms of the contract. However, when the application is considered, it is generally carried out by the suppliers. In terms of products, it is largely the responsibility of businesses (Güleş et al. 2012).

Each of the processes such as finding the necessary resources for the transportation, choosing the transportation vehicles, determining the routes for transportation, following the vehicles, loading and unloading are encountered as a decision problem. Reducing the transportation costs, which constitute the major part of the costs within the logistics system, is a direct factor in increasing the efficiency and economy of the system. In addition, shortening the transportation times will increase customer satisfaction as it will reduce the delivery time of the products to the customer. Even when only these two aspects are evaluated, the importance of the transport function in logistics systems can be seen.

11.3 The Effect of Departments on Logistics System Efficiency

Various criteria, alternatives (departments) and the relationships between them were examined in order to increase efficiency in the logistics sector. As an example, the logistics system is considered for a company that produces durable consumer goods. ANP method was used in the solution of the study. The ANP model was created and solved by using the Superdecisions software. In this study, 17 sub-criteria depending on 6 criteria were used (Akman and Alkan 2006; Bilginer et al. 2008; Özdemir and Seçme 2009; Özçiftçi and Arsu 2013; Gergin and Baki, 2015; Arslan 2017; Alazzawi and Zak 2020; Budak et al. 2020). The criteria and their explanations are shown in Table 11.3.

In the model structure, departments are used as alternatives of decision-making problem. Alternatives are shown in Tables 11.4 and 11.5.

11.3.1 Analytic Network Process (ANP) Method

ANP is a multi-criteria decision-making method developed by Saaty (2001). It was developed for models that are not suitable for the hierarchical structure of analytical hierarchy process (AHP) method. In AHP, one question is discussed: how important is criterion A compared to criterion B with respect to the goal? While, in ANP, the model structure prioritizes criteria by asking how important they are in alternatives being considered. ANP can model complex decision problems for which hierarchical model (used in AHP) is not agreeable. Furthermore, ANP allows for feedback connections and enables clusters to form loops. ANP structure has no goal cluster. General ANP structure can be seen in Fig. 11.1.

Table 11.3 Criteria and sub-criteria

Criteria	Sub-criteria	Explanation
Cost	Production Cost (PC)	Criterion related to production costs
	Manpower Cost (MC)	Criterion for employee cost
	Transportation Cost (TC)	All costs incurred by transportation
	Warehouse Cost (WC)	Storage-related costs
Flexibility	Production Flexibility (PF)	Criterion for flexibility in manufacturing
	Manpower Flexibility (MPF)	Criterion for flexibility in manpower
	Material Flexibility (MF)	Criterion for flexibility in material
	Transportation Flexibility (TF)	Criterion for flexibility in transportation
Customer Service Level	Short delivery time (DT)	A criterion measuring the delivery time of the product
	Accurate information flow (IF)	Information on delivery process
	Quality (QU)	Criterion assessing quality of service and product
	Return rate (RT)	Ratio of returned products to delivered products
Resource utilization rate	Production resources utilization rate (PUT)	Criterion for how available production resources are used, can be called productivity
	Transportation resources utilization rate (TUT)	Criterion for usage of transportation resources
	Warehouse utilization rate (WUT)	Criterion for usage of warehouse resources
Adaptation to conditions	Changes in the market (CM)	Criterion for how much is adapted to changes in the market
	Technological innovations (TI)	Adaptation to technological innovations in the market
Stock reduction		Criterion interested in reduction in the number of products stocked

This table was prepared by the authors

The ANP can be implemented in six sequential steps.

Step 1: Problem formulation: At this stage, the problem is identified, thereafter goal, criteria and alternatives are determined in accordance with the problem.

Step 2: Pairwise comparisons: These comparisons are shown in matrix form. This matrix is created with the help of scale of relative importance. The matrix called A is defined as "$[a_{ij}]_{nxn}$" where i and j represent the row and column elements of the

Table 11.4 Alternatives of the problem

Alternatives (Departments)
Production
Planning
Marketing
Purchasing
Warehouse
After-sales service
Transportation

This table was prepared by the authors

Table 11.5 Pairwise comparison scale (Saaty and Vargas 2012)

Intensity of importance	Definition
1	Equal importance
3	Moderate importance
5	Strong importance
7	Very strong importance
9	Extreme importance
2, 4, 6, 8	Intermediate values

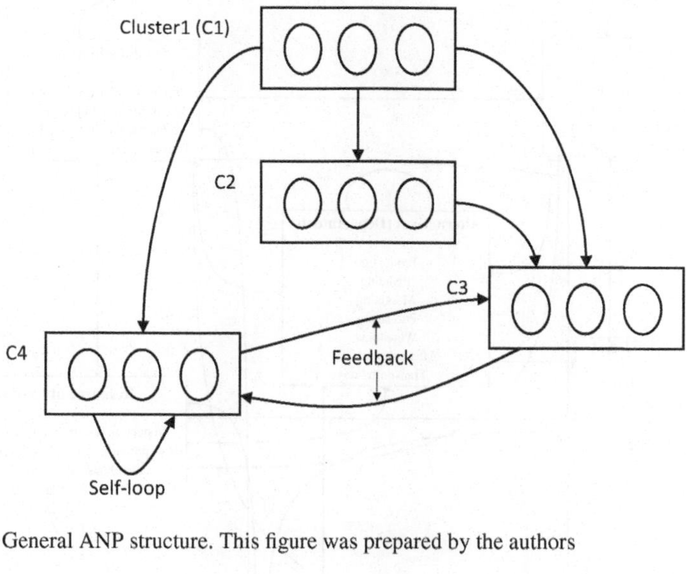

Fig. 11.1 General ANP structure. This figure was prepared by the authors

matrix, respectively. In the matrix A, a_{ij} is importance of the ith element compared to jth element. After comparing, the rest of elements of matrix is obtained by solving following equations, $a_{ij} = 1/a_{ji}$ and a_{ii}=1. If a_{ij} is equal to 1, that means elements i and j are of equal importance. Other intensity values are shown in Table 11.1, with its definitions.

Step 3: Creating unweighted supermatrix: The unweighted supermatrix is composed of column vectors that are the priorities obtained by comparing elements in pairs.

Step 4: Obtaining Weighted supermatrix: The weighted supermatrix is obtained by multiplying all the elements of the unweighted supermatrix by the corresponding cluster weight.

Step 5: Obtaining limit supermatrix: The limit supermatrix is obtained by raising the weighted supermatrix to power of itself.

Step 6: Checking stability of alternatives ranking.

The analytic network structure of the problem is shown in Fig. 11.2.

The result screen of superdecision software is shown in Fig. 11.3.

Results and network structure are shown in Fig. 11.4.

The priorities result screen of the criteria taken into account in the solution of the problem is shown in Fig. 11.5. In the Fig. 11.5, values showed in green are normalized by cluster. These values of the sub-criteria represent their importance weights. Decimal numbers showed in black state limiting values.

By means of limiting values, importance weights of each cluster (main criterion) can be obtained. The importance weights of the clusters are shown in Fig. 11.6.

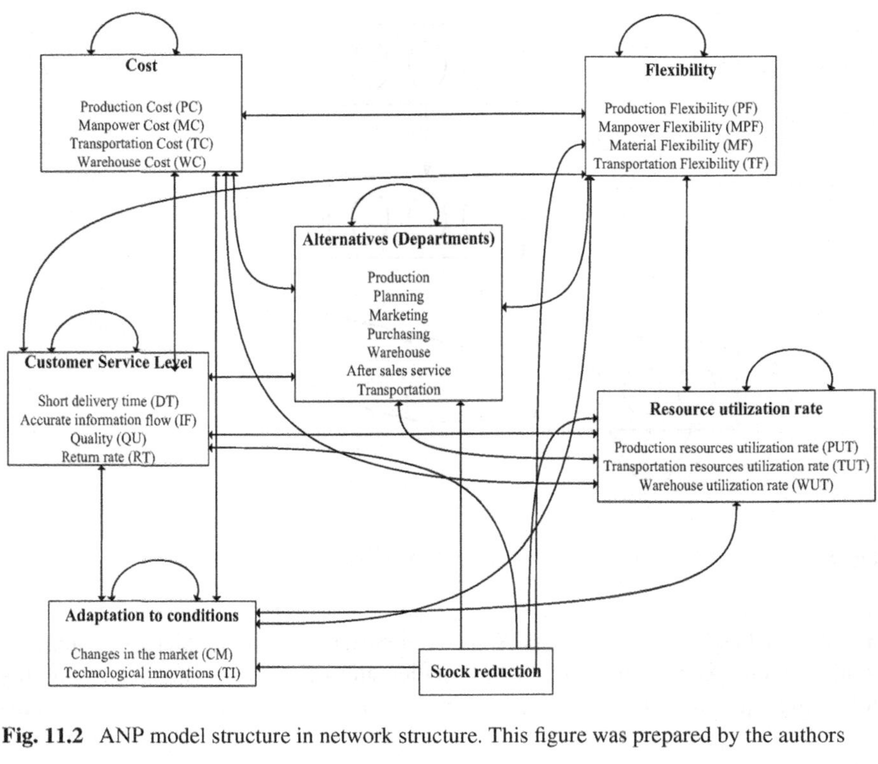

Fig. 11.2 ANP model structure in network structure. This figure was prepared by the authors

Here are the overall synthesized priorities for the alternatives. You synthesized from the network Super Decisions Main Window: Logistics.sdmod: ratings

Name	Graphic	Ideals	Normals
After sales service		0.263444	0.063361
Planning		1.000000	0.240510
Production		0.905891	0.217876
Purchasing		0.348787	0.083887
Sales and marketing		0.458791	0.110344
Transportation		0.750586	0.180523
Warehouse		0.430340	0.103501

Fig. 11.3 Weights of alternatives. This figure was prepared by the authors

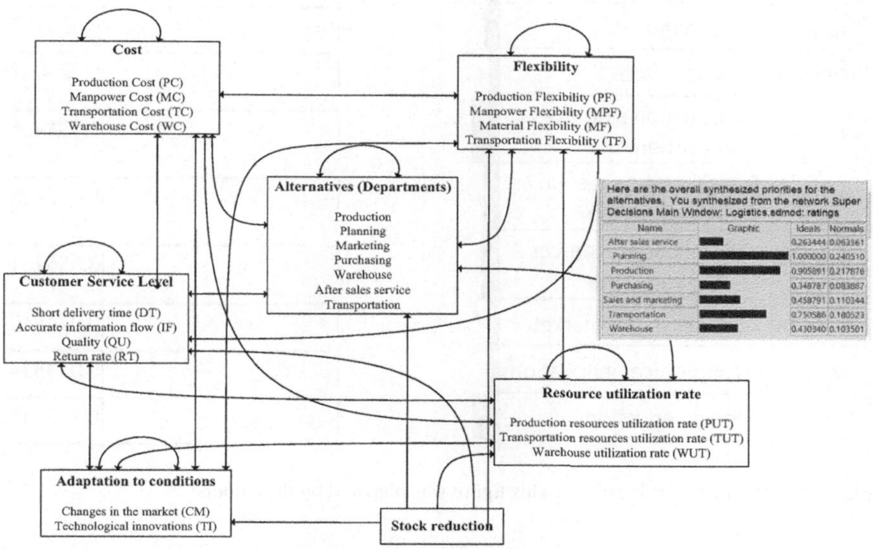

Fig. 11.4 Results and network structure. This figure was prepared by the authors

Results presented in Figs. 11.5 and 11.6 are shown in Table 11.6.

Ranks and scores of alternatives are shown in Table 11.7.

According to these results, the criteria with the highest weight are the cost and resource utilization rate. These two criteria are the most important in terms of efficiency of the logistics system. Another important criterion was found to be flexibility. When the departments are evaluated in the light of these criteria, the planning is the highest weighted department. Other important departments that affect logistics efficiency are production and transportation departments. Purchasing and after-sales service departments are the units that have the least impact on efficiency.

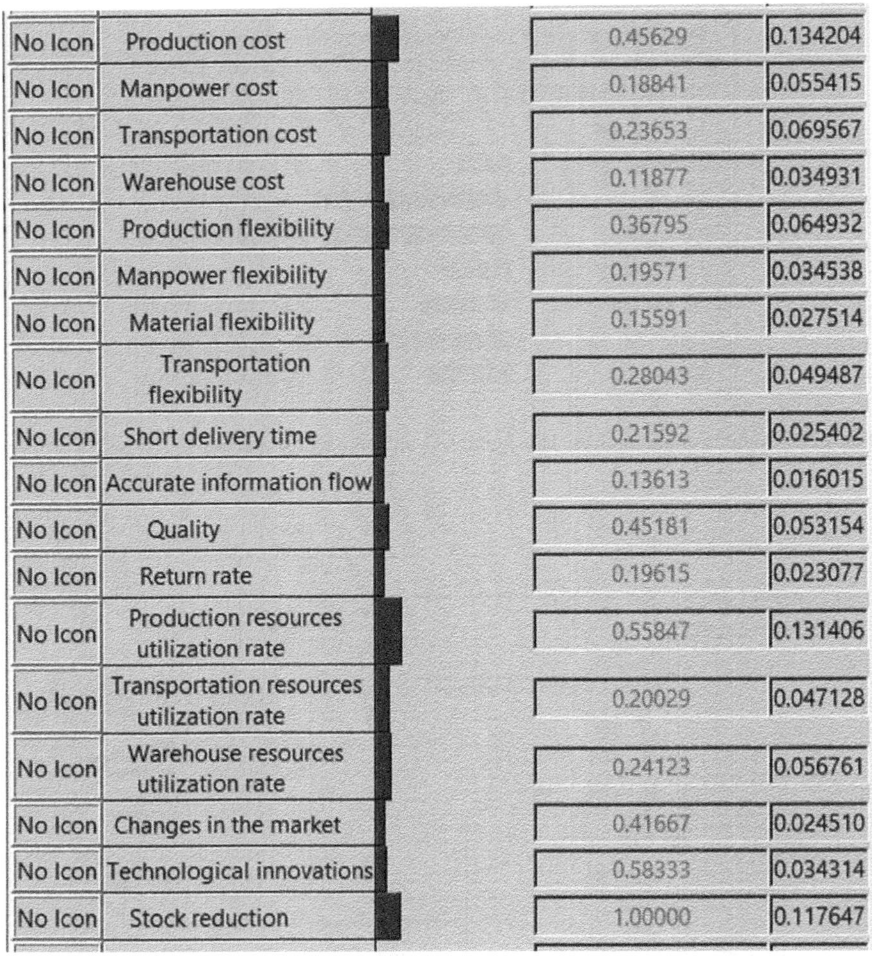

Fig. 11.5 Priorities of sub-criteria. This figure was prepared by the authors

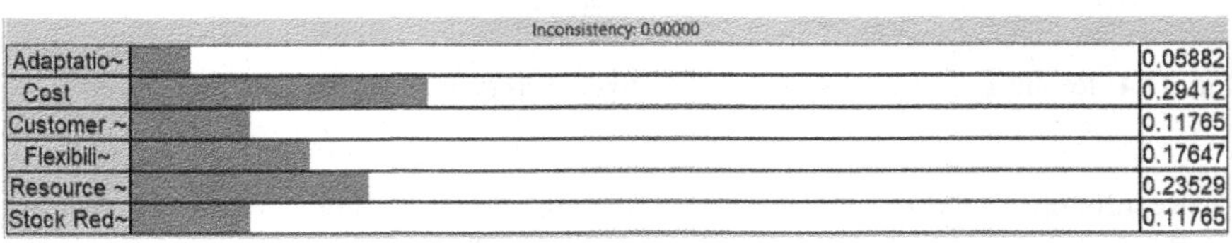

Fig. 11.6 The weights of main criteria. This figure was prepared by the authors

Table 11.6 Criteria and sub-criteria weights

Criteria	Importance Weight (%)	Sub-criteria	Importance Weight (%)
Cost	29	PC	45
		MC	19
		TC	24
		WC	12
Flexibility	17	PF	37
		MPF	20
		MF	15
		TF	28
Customer Service Level	12	DT	21
		IF	14
		QU	46
		RT	19
Resource utilization rate	24	PUT	55
		TUT	20
		WUT	25
Adaptation to conditions	6	CM	42
		TI	58
Stock Reduction	12		

This table was prepared by the authors

Table 11.7 Alternatives' ranks and importance weights

Departments	Importance Weight (%)	Rank
Planning	24	1
Production	22	2
Transportation	18	3
Marketing	11	4
Warehouse	10	5
Purchasing	9	6
After-sales service	6	7

This table was prepared by the authors

11.4 The Systems That Integrated in Logistic System

Information systems are used as support systems in the fulfilment of functions in logistics systems. In logistics, the importance of information systems is increasing day by day in order to ensure traceability, tracking and inventory accuracy of products depending on the growing market conditions. Information systems are used in logistics in a great perspective starting with the development of computer systems and extending to artificial intelligence-based systems (Talluri 2000). Information systems are among the most involved in the logistics system, and systems that have also entered logistics with Industry 4.0 are mentioned in this section. It will also be mentioned how these systems support logistic functions.

11.4.1 Traditional Systems

In logistics, systems that control the flow of material as a whole system and follow both information and material from the supplier to customer have been developed. For example: ERP, CRM, fleet management, inventory management, warehouse management systems. In this section, information systems that are widely used in logistics will be mentioned.

Enterprise Resource Planning: Among these systems, ERP systems are a holistic system for the management of business operations, which includes departments such as human resources, accounting, as well as operational departments such as production, purchasing, sales and marketing. It provides the execution of the duties of the department with modules specific to each department, as well as facilitating other departments' access to information. It is one of the most important systems that facilitates the tracking and execution of logistics operations (Nebol et al. 2016). ERP; It is a software system that contains the functions of planning, coordination and controlling the supply, production and distribution resources in different geographical regions in the most effective and efficient way in order to meet the customer demands in the most appropriate way in accordance with the strategic goals and objectives of the enterprise (Güleş et al. 2012).

When ERP software is evaluated in terms of logistics functions, it provides management of customer orders, purchase orders and production orders. At the same time, inventory management, material handling and warehouse method operations are carried out with the material management module. ERP systems are the most important software for businesses to carry out their logistics functions in a sustainable manner.

Warehouse management: In addition to the temporary storage of materials, warehouses have now taken a more important place in logistics. In addition to warehouse management and inventory control operations with ERP software, warehouse management systems are special software developed to make material movements and locations within the warehouse more effective, more efficient and easier. Warehouse management system is the system used to monitor and control the movements of the materials from the entrance to the warehouse until they leave. It is especially used for inventory management, storage and material handling operations within logistics functions. In addition, it provides the method of operations such as increasing resource utilization, warehouse optimization and transfer operations. Although warehouses are seen as an intermediate stage in the delivery of products, they have become one of the important factors affecting efficiency in logistics systems with their new functions and structure that creates added value for businesses (Nebol et al. 2016).

Customer relationship management: CRM software is an effective tool used to predict customers' behaviour, identify customer trends and set strategies to retain customers. With CRM software, it is aimed to reduce the cost of advertising, focus on customer needs, identify focused customer groups and organize specific campaigns to them, use marketing resources effectively and shorten marketing time and increase communication with customers (Brown and Coopers, 1999). In terms of logistics functions, it provides support especially in the management of customer orders and after sales transactions.

Fleet management: Operations such as determining fleet size and composition, assigning fleet, expanding/shrinking fleet, outsourcing, designing distribution networks, location of terminal and transfer points and repositioning vehicles on the logistics network are carried out with fleet management. Fleet management systems not only support logistics, but also support the management of other factors such as warehouse management in the distribution network. It also increases warehouse utilization by finding suitable distribution routes and distribution planning (Bowersox et al. 2002).

Another system used with fleet management systems is geographical information systems. In addition to tracking GIS vehicles in logistics management, it is also used as a support tool in operations such as determining efficient routes, positioning facilities, conducting analysis for customers and determining operational regions.

RFID and Barcode systems: The common feature of these two systems is that they facilitate data entry into information systems. One of the most important issues in logistics is full-time tracking of material movements. The movements of the materials are followed by wired or wireless systems with the labels and/or barcodes placed on them. By storing a high amount of information in these labels, the information about the materials can be monitored from the production to the customer. RFID tags are preferred more today because of their flexibility. Barcode systems are generally used in end products, and it forms a union with the enterprises in the supply chain with the use of the same code in different businesses (Güleş et al. 2012).

11.4.2 New Technologies in Logistics with Industry 4.0

New technologies entering our lives with Industry 4.0 are also used to support logistics functions. While more information systems come to the fore in traditional systems, new technologies such as internet of things and cyber-physical systems enable the information to be collected more easily and to use technology in decision-making by using a decision support system. Among the systems that take place in the literature as Logistics 4.0, the most frequently encountered are internet of things, cyber-physical systems and cloud computing and big data (Winkelhaus and Grosse 2020).

Internet of Things are central or decentralized systems that enable the collection of information from objects and the distribution of information to objects with technologies such as chip, sensors, RFID tags. Internet of things is the key system in Logistics 4.0 (Winkelhaus and Grosse 2020). It provides traceability of materials and products within the logistics system and provides full-time and accurate collection of information with automation (Meroni et al. 2018).

In addition, it minimizes human-related errors with automation. The Internet of Things supports all logistics functions as it allows the collection of information and tracking of products. In addition, the Internet of Things integrates with traditional systems and supports the operations of the systems (Chen and Zhao 2019). Data entry processes that are performed manually in information systems such as ERP, CRM and warehouse management are performed through automation. Data accuracy is provided and labour usage is reduced.

CPS systems are also used in Logistics 4.0. CPS provides great convenience especially in production and stock management. As in the internet of objects, data is collected through chips and sensors. However, differently, the data is processed with artificial intelligence algorithms and if there is a problem in the system, it is solved. The most important features of CPS are the use of artificial intelligence techniques and direct intervention to physical systems in the decision process (Wasesa et al. 2017). For example, if there is a machine malfunction, the system can determine the cause of the malfunction and suggest the solution method. By following the production process, it gives warnings about delayed orders and offers solutions to decision-makers. With this structure, it is expected that it will be used in wider areas in the future as an important system that supports logistics functions such as production, warehouse management and transportation (Winkelhaus and Grosse 2020).

Big data and cloud computing systems are other new systems used in logistics. The large amount of data generated by IoT and CPS brought the use of big data systems (Kim 2017). Big data and data mining are used in logistics to process data and obtain appropriate information for decision-makers. At the same time, it has brought the use of cloud computing systems to deliver the high amount of data that has emerged to users in different places around the world. At the same time, cloud computing technologies are used in collecting and recording data with IOT (Qu et al. 2016). Today, cloud computing systems are the best solution for storing data in different regions.

11.5 Conclusion

The main purpose of the logistics system is to find the products where they are needed on time from the supplier to the customer. In order to fulfil this aim throughout the whole system, business departments perform different tasks. Departments perform logistics functions while performing their own functions. Working of the departments in synchronization with each other is one of the most important factors in the efficiency of logistics systems. In order to measure the effect of departments on the efficiency of the logistics system, criteria have been determined and the effects of departments on system efficiency have been investigated according to these criteria. Departments were evaluated according to the determined criteria with the ANP method.

According to the results obtained, the planning department has emerged as the most effective department due to its critical role in the logistics system. In the second place, the production department came to the fore due to its role in obtaining the physical product. Transportation also has a high impact on productivity as it constitutes a large part of the cost in the logistics system.

Other factors addressed in terms of logistics system efficiency are information systems and new technological systems. The purpose and benefits of these systems in logistics systems are given in the second section. From these systems, it is seen that traditional systems are generally information systems and used to record the data. The systems developed with new technological developments benefit from the conveniences provided in collecting data. Another aspect is used in decision-making process as support system. In addition, easy access to data through cloud systems also benefits logistics systems.

In future studies, the impact of the functions of different departments on each other can be explored. The effect of communication between departments on the efficiency of logistics systems can be addressed. The benefits of new systems for this communication can be explored. How much of which system is effective on the functionality of departments can be found using multi-criteria decision-making methods.

References

Akman G, Alkan A (2006) Tedarik Zinciri Yönetiminde Bulanik AHP yöntemi kullanilarak tedarikçilerin performansinin ölçülmesi: Otomotiv Yan Sanayiinde bir uygulama. İstanbul Ticaret Üniversitesi Fen Bilimleri Dergisi 9(1):23–46

Alazzawi A, Żak J (2020) MCDM/A Based Design of Sustainable Logistics Corridors Combined with Suppliers Selection. The Case Study of Freight Movement to Iraq. Transportation Research Procedia 47:577–584

Arslan HM (2017) Determination of optimal vehicle election of logistics companies with AHP-ARAS hybrid method. Alphanumeric Journal 5(2):271–282

Banks JH (2002) Introduction to transportation engineering, 2nd edn. McGraw Hill, New York

Bilginer N, Kayabaşi A, Sezici E (2008) Lojistik faaliyetlerin süreçsel etkinliğine etki eden faktörlerin değerlendirilmesi üzerine ampirik bir çalişma. Dumlupinar Üniversitesi Sosyal Bilimler Dergisi 22(22):1–21

Bowersox, D. J., Closs, D. J., & Cooper, M. B. (2002). Supply Chain Logistics Management McGraw Hill. International edition.

Brown SA, Coopers PW (1999) Customer relationship management: A strategic imperative in the world of e-business. John Wiley & Sons, Inc., Canada

Budak, A., Kaya, İ., Karaşan, A., & Erdoğan, M. (2020). Real-time location systems selection by using a fuzzy MCDM approach: An application in humanitarian relief logistics. Applied Soft Computing, 92, 106322.

Chen J, Zhao W (2019) Logistics automation management based on the Internet of. Clust Comput 22(6):13627–13634

Council of supply chain management professionals (2013) “Supply chain and logistics terms-glossary”, www.cscmp.org
Gergin RE, Baki B (2015) Evaluation by integrated AHP and TOPSIS Method of Logistics Performance in Turkey’s Regions. Business and Economics Research Journal 6(4):115–135
Gunasekaran A, Ngai EW (2004) Information systems in supply chain integration and management. Eur J Oper Res 159(2):269–295
Güleş, H. K., Paksoy, T., Bülbül, H., & Özceylan, E. (2012). Tedarik Zinciri Yönetimi: Stratejik Planlama. Modelleme ve Optimizasyon, Gazi Kitabevi: Ankara.
Kim NH (2017) Design and implementation of Hadoop platform for processing big data of logistics which is based on IoT. Int J Serv Technol Manage 23(1–2):131–153
Lambert, D., Stock, J. R., & Ellram, L. M. (1998). Fundamentals of logistics management. McGraw-Hill/Irwin.
Meroni G, Baresi L, Montali M, Plebani P (2018) Multi-party business process compliance monitoring through IoT-enabled artifacts. Inf Syst 73:61–78
Nebol, E., Uslu, T., & Uzel, E. (2016). Tedarik zinciri ve lojistik yönetimi. 1. Ed. Nobel Yayinevi, Ankara.
Özçiftçi V, Arsu T (2013) Lojistik servis sağlayicisi seçiminde AHP uygulamasi. Sosyal Ve Beşeri Bilimler Dergisi 5(1):524–536
Özdemir Aİ, Seçme NY (2009) İki aşamali stratejik tedarikçi seçiminin bulanik topsis yöntemi ile analizi. Dumlupinar Üniversitesi Sosyal Bilimler Dergisi 26:71–90
Qu T, Lei SP, Wang ZZ, Nie DX, Chen X, Huang GQ (2016) IoT-based real-time production logistics synchronization system under smart cloud manufacturing. The International Journal of Advanced Manufacturing Technology 84(1–4):147–164
Saaty, T. L., & Vargas, L. G. (2012). Models, methods, concepts & applications of the analytic hierarchy process (Vol. 175). Springer Science & Business Media.
Saaty TL (2001) The Analytic Network Process, Fundamentals of Decision Making and Priority Theory, 2nd edn. RWS Publications, Pittsburgh
Talluri S (2000) An IT/IS acquisition and justification model for supply-chain management. Int J Phys Distrib Logist Manag 30(3/4):221–237
Wasesa M, Stam A, van Heck E (2017) Investigating agent-based inter-organizational systems and business network performance. J Enterp Inf Manag 30(2):226–243
Winkelhaus, S., & Grosse, E. H. (2020). Logistics 4.0: a systematic review towards a new logistics system. International Journal of Production Research, 58(1), 18–43.

Haci Mehmet Alakaş Assoc. Prof. Haci Mehmet Alakaş graduated from Yildiz Technical University as an industrial engineer in 2007. After working professionally for two years, he started to work at Kirikkale University as a research assistant. He completed his MS from Kirikkale University in Industrial Engineering in 2012. He completed his Ph.D.studies from Gazi University in Industrial Engineering in 2017. Dr. Alakaş work in Kirikkale University, as an Assistant Professor in 2017-2021 and Associated Professor from August 2021. He works on assembly line balancing, personnel scheduling, vehicle routing,constraint programming, multi-criteria decision making, and queuing systems.

Tamer Eren Tamer Eren is professor at department of industrial engineering at the University of Kirikkale in Turkey. Prof Dr. Tamer Eren obtained his BS degree from Selçuk University Industrial Engineering Department in 1996. Then, he got the MS and PhD degrees from the industrial engineering department of the Gazi University in years 2000 and 2004, respectively. His research focuses on scheduling, production planning, game theory, multicriteria decision making and operation research. Prof. Eren has authored and contributed to numerous articles on health, logistics, industry 4.0, energy, transportation, waste management, recycling management, supply chain management, staff scheduling and maintenance optimization.

Part III
Future of Industries and Applications

Chapter 12
Sustainability and Industry 4.0

Mustafa Özan

Abstract Three major industrial revolutions took place until the time we are in. Today, the fourth biggest industrial revolution, called Industry 4.0, has been entered. Industry 4.0 has a structure that includes production systems that respond to the consumer's changing needs very quickly and automation systems that facilitate coordination between business processes. With the Industry 4.0 revolution, businesses made their production processes more efficient, discovered new ways to facilitate communication with consumers, and ensured their collaboration with their partners was coordinated. It is impossible not to be affected by the logistics sector, one of the leading sectors in the world within the mentioned domain. Industry 4.0 has significantly impacted the sustainability of resources using digital technologies (Özdemir and Özgüner 2018, p. 43). They are emerging with the industrial revolution within Industry 4.0; the Internet of things, smart factories, big data analytics, cyber-physical systems, and so on. Many technologies have facilitated sustainability by significantly reducing environmental problems. This study aims to explain how to use Industry 4.0 effectively in ensuring sustainability and reveal the benefits that digital technologies can provide to the sustainability efforts of businesses. For this purpose, the literature has been studied in detail, the crucial points have been identified and focused on, and Industry 4.0 and Sustainability concepts to which it depends are explained. While working towards the stated purpose, the central premise, the complex technical concepts, and dynamics of the subject have been filed optimally, and the efforts to highlight vital points have been shown. It informs industrialists, academics, and university students about the impact of the Industry 4.0 revolution on the sustainability efforts of businesses.

Keywords Industry 4.0 · Sustainability · Cyber-physical systems · Internet of Things · Production · Employment

M. Özan (✉)
Dr. Istanbul Gelisim University Faculty of Applied Sciences Logistics Department, Instanbul, Turkey
e-mail: mozan@gelisim.edu.tr

İ. İyigün and Ö. F. Görçün (eds.), *Logistics 4.0 and Future of Supply Chains*,
Accounting, Finance, Sustainability, Governance & Fraud: Theory and Application,
https://doi.org/10.1007/978-981-16-5644-6_12

12.1 Introduction

Our world, where the only concept that does not change is the change itself, has entered into a period that is much faster, tougher, and, most notably, of different dynamics compared to the past. Our world, the only concept that has not changed, has entered into a much faster, more challenging, and most importantly, different dynamics than in the past. In this context, to understand the subject studied, first of all, the concepts connected to it must be mastered. This is one of the essential building legs of the study. The understanding of the work is made much easier by acting from particular perspectives.

For this reason, if an evaluation is made with the simplest reduction method, the starting point of the concept that gives the subject of the study is the concept of "sustainability" (Toker 2018, p. 56). Logistics 4.0 stands before us as a concept that has emerged in today's conditions to continue human needs and indirect economic order. The main point here is to ensure the "continuity" within the conditions of the day without any interruption by minimizing the damage of the world's ecological order of the logistics ecology with the technological opportunities provided by the Industry 4.0 dynamics, which is called the fourth industrial revolution (Bilgiç et al. 2020, p. 57; Strandhagen et al. 2017, p. 360).

As can be understood in the narration, although the three main titles on which the study is based seem different, it is wrong to think separately because they affect each other. Thinking of the economic order as the wheels of a machine facilitates the comprehension of the subject (Fig. 12.1).

In this study, we tried to explain the connections between these titles, which are related to each other, on a dynamical basis, and these explanations were made by examining the outputs and findings brought to the literature. Thus, in the literature,

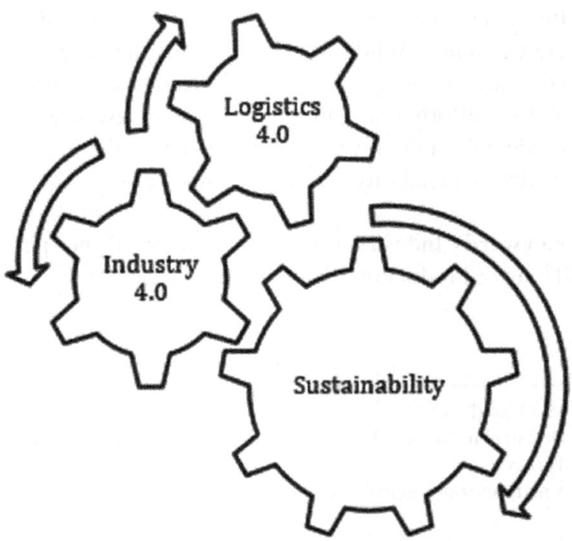

Fig. 12.1 The relationship between working concepts and each other (author)

it has been acted to gather critical information in pieces from different sources. The study was placed in a queue due to the relationships of the titles, and the narration was strengthened by adding subheadings at necessary points in this row. The order of the leading titles on which the study is based is shown below.

1. Sustainability
2. The Fourth Industrial Revolution: Industry 4.0
3. Logistics 4.0

To survive in the harsh competitive environment of today's business life is vital for all companies. Competitiveness in companies whose primary purpose is to make a profit is a concept that can be achieved in the medium and long term rather than short. This study argues that for the companies to survive, it is necessary to explain the requirements of the day's conditions and take the actions that are explained. Thus, it enables companies to obtain external support, such as credit support, from the banking sector if they are not self-sufficient. The more companies adapt to Industry 4.0, their competitiveness and market shares will increase, and their credibility will increase.

12.2 Sustainability

Today's dynamics have a highly dynamic structure in every field of life compared to the past. This dynamism brings with it many technical requirements and operations. Therefore, people have to keep up with these dynamics that are changing daily in companies at the individual level. Facing a competitive environment with many effects with Industry 4.0 possibilities can create many advantages and disadvantages. One of the sectors that will enter the restructuring process due to the effects of Industry 4.0 is the Logistics sector, which has the characteristic of being one of the leading sectors for the world economy. There are many operational and technical structuring processes to ensure the return of the economic wheels and to maintain the physical contacts of the world trade (Man ve Strandhagen2016, p. 722).

Sustainability, which is associated with all areas, is essentially a much more introverted issue with production. For this reason, while sustainability aims to transfer the natural resources of the world to the next generations, it also considers the efficiency of production to remain stable.

Limited possibilities of our planet's resources have always been known from past to present. In the past, the population of humanity, consumption habits, and other conditions were at a level that the world resources could handle. However, today, especially after the Industrial Revolution took place, humanity has become accustomed to consuming more and more than it needs with the rapid production of the produced goods and directing the entire process (Gürül 2019, p. 190). This accustomed habit has led to an increase in the amount of waste generated by production and the use of resources more with the volumetric growth of production. With all these and such negative consequences coming together, the world's capacity has started to be challenged, and many new problems have emerged in many fields. The repercussions of these problems nowadays in people's daily lives have started to be felt more clearly. The population increase was so significant that it caused the production of many utopian scenarios and even the emergence of many novel-style fictional works. While the world population is around 7 billion today, according to the current World Bank (WB) estimates, a total world population of 11 billion is predicted in 2045 (Yurdakul 2020).

It is accepted that the concept of sustainability entered the literature in its current form with the report published by the UN (United Nations) in 1983. This report is mainly focused on the possibilities obtained from nature. However, the consumption and use of natural resources within the economic wheels make it wrong to approach the issue only ecologically. To ensure the continuity of trade, keeping the health of the ecological balance at an optimal level, from the macro-level to the micro-level in the countries, some precautions should be taken, the policies are put into practice, and these policies put into practice are subject to legal supervision to give results. Therefore, ecological assessment should be related to social and economic conditions. Otherwise, there will be no economic order with the loss of a livable world. The importance of such a surveillance need was effectively recognized in the early 2000s (Oğuz 2019, p. 2).

Our world's limited resources have no more power to withstand waste than in the past (Oğuz 2019, p. 9). That is one of the main reasons for the basis of technological developments. Regardless of the area, technological developments reduce waste and become more productive and less harmful to the ecological order. Examples of these are the engines of heavy cargo transport vehicles that provide world trade logistics. The higher the Euro class, the fewer particles the engines spread and damage the environment. For the continuity of the ecological order, catching a certain technological standard has become a must for every country and company. For this reason, the realization of a sustainable production type can also be called "environmentally friendly production."

Technological developments remained in a limited area such as vehicle engines, and revealed their presence in every area in life. As in the definition of basic economics, it is possible to meet unlimited needs with limited resources, increasing the efficiency of the unit work and the more efficient use of resources (Gürül 2019, p. 1). For this reason, the most important keyword of the concept of sustainability is "efficiency" (Bähre et al. 2016, p. 380; Barreto et al. 2017, p. 1246; Oğuz 2017, p. 3).

Technological developments have taken a load on the operational areas and a specific load on the people. The use of more robotized systems in heavy-duty works has made positive contributions to protect human health. However, this change provides advantages in terms of human health and has started to change the job definition and how it works by realizing a macro-level change. The most common echo of this situation at the society level is the concept of unemployment. While the development of technology and reducing some operational processes to digital has forced companies to restructure their competitiveness, they have now forced people to develop themselves more and more individuals to make a living (Bag and Telukdaire ve Gupta 2018, p. 1411).

While the population continues to increase daily, the number of employment needs in the countries has decreased. Now, many businesses have started to be performed by robots, robotic systems, and computer systems, not people like they used to. Thus, a machine has started to perform the work done by a few people much cheaper under the economies of scale. While there is a need for humans in positions at critical locations in process-based positions, which are the background of operational affairs, people reach these positions under more challenging conditions. That is the reason why countries do not only consider the ecological system when determining their sustainability policies. Keeping the level of employment at the optimum level without disrupting the ecological order and dependence on it brings many technical and vital process designs, researches, and application studies in the background.

Industry 4.0 is the main topic that brings the most appropriate solution in the processes brought by business life to keep ecological balance healthy (Gürül 2019, p. 104). For this reason, the main reason for the emergence of Industry 4.0 is the need for a production machine that can respond rapidly to consumer demands with variable features.

There are many studies in the literature, such as the vital importance of sustainability and the evaluation of its dynamics. In a study, Oğuz (2019) has suggested that, due to the examinations he had carried out, it is necessary to turn to renewable energy sources to have a healthy commercial life without causing further damage to the ecological order. This result reveals the necessity of redesigning the production function, which has a large share in energy consumption, with the help of technological possibilities, and provides information about the elimination of the general energy need. Thus, a situation that can be achieved on both sides in both commercial and individual life can be in question. Torbacki and Kijewska (2019) analyzed the relationship between the concept of sustainability and Industry 4.0 and Logistics 4.0 in a study they conducted with the DEMATEL method, one of the MCDM methods. According to the results, it was determined that the most active concept was Industry 4.0, while the concept of Sustainability had the highest passive quality. It is seen that these findings support the expressions in this study with the power it gained from being scientific findings.

When it focuses on sustainability, the reflection of business life is named "corporate sustainability" by experts in the literature. At this point, if the word corporate is continued to be focused on, in the global order that has been globalizing since the beginning of the 2000s, companies pay attention to reflect this change to their external environment as well as going to professionalism within themselves. For this reason, a term called "corporate social responsibility" (CSR) has emerged from business life to academic literature (Özispa and Deveci 2019, p. 41). It is possible to meet this concept mostly in companies operating in the field of physical commodity production.

The meaning of this can be interpreted as follows. Apart from the studies participating in the literature, it can be said that the real sector started to realize the importance of the concept of sustainability within its dynamics (Gürül 2019, p. 186, 187).

Because, regardless of the sector, the platform to which the whole business life depends on the ecological order is nature itself. Therefore, any adverse development in the ecological system can directly affect the economic system. So the passive side here is the economic system itself. Companies are looking for ways to reduce their damage to the ecological system to protect their image, make their images more reliable in their markets, and increase their competitiveness.

Since environmental pollution, etc., may affect the operation of the companies' production facilities, making technological investments to reduce losses is directly related to Industry 4.0. However, by adding lower fuel consumption and cleaner vehicles to their fleets, companies cooperate with non-profit organizations such as associations; they also try to take actions that support sustainability in ecological life through actions such as greening forests, supporting recycling, and taking care to use maximum recyclable commodities in their production. This situation can be cited as an example of the sustainability movement realized by the real sector in itself, except for subjects that have participated in academic literature and some have been put into practice.

12.3 The Fourth Industrial Revolution: Industry 4.0

Over time, our world has started to have very different disciplines and dynamics. While every invention in the world actually creates a new era and sometimes opens the doors of a new era, the radical changes that made up today's order have started with the industrial revolution would be a true statement.

In the literature, the production-based periods of our world have been divided into four periods in total, and experts have started to discuss the future-oriented Industry 5.0 apart from Industry 4.0. The reason for this is the high speed shown by technological developments. However, in the current period, four industrial periods, their revolution, are accepted in other words. If we need to talk about all of them briefly, Industry 1.0 is considered the beginning of the world order based on production. With the industrial revolutions that took place in the eighteenth century, steam machines started to appear. Therefore, some experts remember this period with other names, such as "the era of mechanization."

Industry 2.0, which is accepted as the second industrial revolution, started to have some changes in its production activities due to the development of the machines with the electrical energy used in the early 1900s, that is, in the early twentieth century. Therefore, the central concept that this period is based on is electrical energy (Poesche and Lilja 2019, 90).

R & D facilities have expanded with the widespread use of electrical energy; R&D facilities have expanded, finding inventions, and production speeds have accelerated. The computer, which was invented towards the end of the twentieth century, caused an industrial revolution and a new era. In the literature, in the periods commonly accepted as the 1970s, computers have gradually increased globally, and computer technologies have started to be used in specific processes of companies. With its increasing involvement in production activities, this situation has changed the concept of production, and all manufacturing processes have undergone radical changes. The world, which remained in the period of Industry 3.0 until 2011, the beginning of Industry 4.0, has prepared the basis of Industry 4.0, strengthening its relations with production and computer technologies that improve its management level. Briefly, computer information technologies and manufacturing processes sometimes went beyond the physical boundaries of the production facility, removing the physical boundaries of performing company management and registration-based operational activities. Therefore, administrative activities, administrative control, and some production functions have started to be carried out remotely. That constitutes the concept of "automation," which we often hear in production activities (Şekkeli and Bakan 2018, p. 19).

12.3.1 Industrial Revolutions and Basis

A new era has started for the world after the Industrial Revolution has taken place. The concept of mass production has entered the lives of people, and the demands have become easier to meet, while the variety, volumes, and forms of demands have started to change completely. However, the concept of production has existed since the beginning of humanity. Therefore, experts have divided this issue, which is so rooted that it shapes our lives into specific periods. Continuous differentiation of production possibilities depending on the developing and changing technology has caused an unprecedented period in the past.

Industrial revolutions that have taken place throughout history have accumulated on the characteristic features of Industry 1.0 and have reached new forms and characteristics and have taken their present form. However, this change has not stopped; on the contrary, it shows changes and developments at an increasing rate compared to previous periods. The diagram below lists the industrial revolutions brought to the literature with the titles and essential characteristics.

Industry 4.0, which is defined as the last of the industrial revolutions, is also suitable to be seen as the title of today's dynamics. The development of technology, the emergence of digitality, and their adaptation to real life have led to many positive and negative changes. We are in a period in which physical movements depend on digital movements. While the preparation on the factory floor of the physical works used to include physical processes, these vital processes are now carried out entirely digitally, that is, virtually, and sometimes this process is not in the background. However, it also enables automatic physical operations in some production areas. That means that production devices and equipment can now communicate with each other. The experts who contribute to the literature call this situation "Internet of Things," mainly because of the degree of digitality in production.

This flexibility provided by the concept of digitality is limited in the sectors where production takes place and within the Logistics Sector, which ensures the physical contact of the commodities produced and the physical movement of the supply chains.

This interaction, which affects the big picture, should be seen as natural because the economic order that provides human life consists of a highly complex and variable relationship spiral.

The concept of Industry 4.0, which is active in life and continues to be active, is in the field of interest and examination of many academics in the academic sector and the real sector. Research on the concept of Industry 4.0, interviews with the designed models are understood by all segments of the business life and aim to illuminate the dark spots of the concept with the help of brainstorm. Because the companies, and in a roundabout way countries, which are not adapted to the current Industry 4.0, face losing their competitiveness entirely or almost all within the new world order. Many academics and managers define Industry 4.0 as merging the already strong structure of industrial production from past to present with other technological opportunities, especially the Internet. In short, it is possible to define it as an intelligent production period.

In a study conducted by Boston Consultancy Group (BCG) in 2015, they predicted that if Germany, which is accepted as the country where Industry 4.0 was born, expanded the use of Industry 4.0 opportunities, it would gain results with many positive economic developments at the macro-level within 15 years. This prediction seems realistic by revealing the idle capacity in the production facilities, an optimal production volume growth, and a more straightforward response to incoming demands.

The ultimate goal of Industry 4.0 can be indicated to integrate a business process, regardless of the sector, to include all dynamics and establish an optimal automation process. This automation system brings the modular structure and results in low efficiency and cost (Şekkeli and Bakan 2018, p. 24). Industry 4.0 does not have a single definition named in the literature. However, research shows that all definitions are shaped around the values and dynamics seen in the study.

Industry 4.0, which provides a radical change of all production systems and dynamics, consists of the fact that the machines with sensors and perception capability provide communication between each other through the Internet, and sometimes a large part of the production, and sometimes the entire production, without the need for human assistance in automation (Timm and Lorig 2015, p. 3119).

Industry 4.0 concept, which was brought to the literature by the world Hennig KAGERMANN at a technology fair held in Hannover, Germany in 2011, continues to expand its impact area by increasing its speed (Görçün 2017, p. 142; Stock and Seliger 2016, p. 536). It is impossible for the Logistics sector, which is considered one of the world's leading sectors, not to be affected by such growth.

If we need to sum up the dynamics that form Industry 4.0 under main titles, it is possible to rank with five main titles below. These are;

1. Intelligent Mobility
2. Smart Grids
3. Smart Products
4. Smart Buildings
5. Logistics 4.0

The merging of the stated concepts under a single frame constitutes a "smart factory" under the production perspective (Stock and Seliger 2016, p. 538).

It is possible to adapt to Industry 4.0 with technology imposing many burdens on companies in the long and short term (Özdemir and Özgüner 2018, p. 40). Companies that have reached a certain institutional maturity, have a market share, and thus have a qualified structure suitable for working with them can realize this adaptation much more quickly. However, it is an essential fact that not all companies have a large volume. For this reason, a field study on SME (Small Medium Enterprises) in the Asian region revealed that it is possible to soften the relevant adaptation process for SMEs (Haseeb et al. 2019).

It has been stated that the increasing number of experts at IT technologies in small and medium-sized firms' structure positively affects the relationship between Industry 4.0 and the firms in their internal processes. Strengthening and accelerating the adaptation of SMEs to the fourth industrial revolution has vital importance for the economy.

If the companies that are the actors of the economic order are examined, it is seen that the majority of the actors are SMEs regardless of the sector. Although the market shares of big companies are large, SMEs are accepted as the economy's driving force by some experts. If the situation is analyzed from the perspective of workflow structure, they work with small companies in some areas within the activities of large companies (Manavalan and Jayakrishna 2019, p. 929). In light of this reason, the failure of the adaptation of SMEs to Industry 4.0 shadows the success of the adaptation of big companies. The immediate solution to this problem is expressed in the fact that SMEs have an IT-based operation by performing the management within themselves through information technology systems (McAfee and Brynjolfsson 2012, p. 4).

12.3.2 Big Data for Industry 4.0 (Big Data)

The fact that the production devices communicate with each other over the internet network virtually eliminates the concept of distance. Only the machines in the same facility should not be considered from this mentioned communication statement. If it is necessary to concretize the expression with a simple example, a company centered in any region of the world can design the order received with the latest technological facilities and send the related operations to the 3D printer in the distribution facility at another location of the world, and make the products with the robots without even touching (Manavalan and Jayakrishna 2019, p. 936). In this production, while eliminating the essential items such as personnel costs, it is possible to eliminate the possible structural and design problems with the lowest level of error in the computer environment, prevent the possible structural and design problems, and increase the quality and ensure that the order is produced and distributed at the required point (inventory, enormous warehouse facility costs, etc.), it can be possible to increase the profit margin with the increase in unit work speed (Günther et al. 2017, p.191, 192).

If the production in question is a production under heavy industry, it is beneficial in health by performing specific processes that threaten human health to robotized systems.

At the same time, all the processes and related process steps are automatically recorded by the systems, allowing the formation of big data, and thus, ease of control and interpretation is obtained.

Big data means data that has gained meaning by analyzing the data obtained using appropriate metaphors, methods, and information (Kamble et al. 2018, p. 421). The ultimate goal of big data is to simplify an already complex set of information to be understood and interpreted by optimally simplifying it. Big data analytics support real-time data accumulation with many sources, help make more objective decisions, and facilitate the preparation of an action plan from a wide range of production to marketing (McAfee and Brynjolfsson 2012, p. 7). Today, the creation and processing of this data is a distinct professional specialty (McAfee and Brynjolfsson 2012, p. 7).

12.3.3 Employment in Industry 4.0

Industry 4.0 caused much flexibility in the production area by combining its mechanical infrastructure with advanced technologies. Thanks to the benefits of this flexibility in unit work, there is an increase in cost and speed per unit work. That strengthens the hand of optimization according to the logistics perspective. At the same time, with the help of technology, production can be made modular, and in some related areas, this increase in quality can occur. The situation in question may have led to the elimination of some professional fields and the emergence of a large number of new professional fields by being more software-oriented.

The concept of Digitality, on which Industry 4.0 is based, is based on cloud systems. For this reason, cybersecurity is one of the most vital points in the long term for Industry 4.0 to provide its "sustainability". "Cyber Security Experts" can be given as an example of one of the new professions that emerged with the fourth industrial revolution. This new industrial revolution based on cloud technologies constitutes the following four main benefits to the general perspective. These are (*Source* TUSİAD 2016);

1. Productivity
2. Growth
3. Investment
4. Employment

When the real sector applications regarding the effects of Industry 4.0 on employment are analyzed and the results brought to the literature, the most striking view will be the gradual increase in skilled labor and the proportion of labor positions defined as unqualified. The reason for this is that standard processes can be achieved with robotic and digital possibilities;

- Speed gain,
- Cost reduction,
- A decrease in the margin of error,
- The audit can be facilitated and reduced to an instant level,
- Facilitation in the interpretation and decision-making processes of the management level.

Such results can be obtained. Therefore, the type of employment needed is rather than the realization of standard processes;

- Sorting or designing these processes,
- Transforming the outputs obtained from the processes into findings,
- Collecting and interpreting the findings to form the big picture,
- Integration of these interpreted processes into the instant dynamics of the industry,
- Mastering the execution of all integrated concepts and processes.

An employment type based on the concept of "know-how," which has turned more abstract into concrete, has started to come to life.

Many sectors have realized such a significant change and have already entered the structuring processes on the subject. The logistics sector, which now carries out its activities professionally, is one of the sectors that adapt to these structuring processes. In their study, Ömürgönülşen et al. (2020) determined that the concepts of adaptability, level of digitalization, and flexibility are the three most important criteria when they examine the logistics sector for Industry 4.0, and they proposed a structuring model for these criteria in the structuring of logistics companies.

Digitalization, finance, human resources management, and marketing with intelligent technologies reveal a new structure and a new perspective. For this reason, especially in industrial production activities, supply chain management is among the areas where Industry 4.0 has an in-depth effect. The logistics sector is in the first ranks of the deeply affected sectors since the supply chains move along with the production support functions it contains (Özdemir and Özgüner 2018, p. 40).

12.3.4 Achievements with Industry 4.0

The development of production technologies and the rivalry of the competitive environment compelled companies to specialize in their fields of activity during this period of Industry 3.0. The reason for this is that with the help of technological infrastructure, companies combined human power with machine power; thus, with the help of more people, a much more number and variety of products were obtained from a machine. Thus, it was possible to experience expansion in the product range in some company structures. For this reason, companies that want to survive have set their expertise in their main field of activity and set the rise in quality as their primary goal, and have turned to the use of outsourcing to avoid the negative headings such as the quality being stable in other areas and time loss. That is one of the main concepts that creates flexibility in production (Görçün 2017).

Industry 4.0, on the other hand, has contributed to these achievements by providing a platform to reduce the need for human labor and reduce overall costs, reduce the margin of error, facilitate decision processes in administrative matters, and eliminate distances for administrative operations. The meaning of this statement is that a saving occurs in the general framework with the help of information technologies. Concepts such as intelligent facilities or factories are formed at this point.

The machines started to take place more actively in the production mechanisms, and that the machines started to have artificial intelligence, as examples in some parts of the world, led to the emergence of a real automation process (Öztemel and Gürsev 2018, p. 146). Thus, it has become possible to catch up with the standards and even establish the markets' standards in companies with high adaptation to Industry 4.0. The increasing use of RFID technology in the storage activities in the logistics sector has provided an instant data flow while also providing significant assistance in preventing possible accidents, and thus, internal and external dispatches have started to be automated thanks to the data flow based on big data (Öztemel and Gürsev 2018, p. 149).

The mainstays of these gains, which have been achieved with the fourth industrial revolution, are indispensable for companies. When the issue is approached correctly, the starting point of this industrial revolution is the necessity of communication between machines and people. One of the simplest feedbacks that can be given to this requirement is that each unit work should be modular by decomposing it to the smallest building block, and each part should be clearly defined within both sides.

From this point of view, it is seen that we have to separate Industry 4.0 into pieces. In this way, it will be possible to understand the building blocks and requirements of the adaptation process mentioned throughout the study. In the literature, the building blocks constituting Industry 4.0 have been collected as ten main titles. These are seen in the Fig. 12.2.

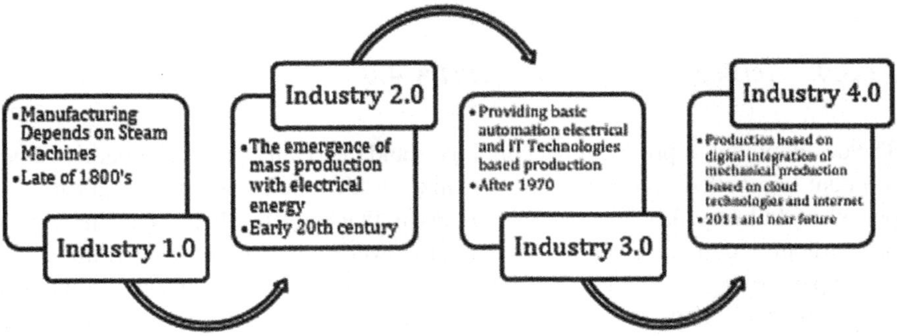

Fig. 12.2 TUSIAD (Turkish Industry and Business Association). *Source* Türkiye'nin Sanayi 4.0 Dönüşümü, Yayın No: TÜSİADT/2016-03/576. https://tusiad.org/tr/yayinlar/raporlar/item/8671-turkiyenin-sanayi-40-donusumu, (Accessed on June 15th, 2020)

Table 12.1 Sustainable manufacturing in Industry 4.0 (*Source* Machado et al. (2020), 1466)

Sustainable Production for Industry 4.0			
Technological Pillars 1. Automation Robots 2. Simulation 3. Horizontal and Vertical System Integrations 4. Internet of Things 5. Cybersecurity 6. Cloud 7. Additive Production 8. Augmented Reality 9. Big Data and Analysis	Sustainable Production Area 1. Production Technologies 2. Product Life Cycle 3. Networks Which Create Value 4. Global Production Effects	Industry 4.0 ve SM 1. Business Model 2. Networks Which Create Value 3. Equipment 4. The Human Factor 5. Smart Factory Organizations	Sustainable Dynamic Model 1. Direct Effects of Environmental Dimensions 2. Direct Effects of Social Dimensions 3. Direct Effects of Economic Dimensions 4. Indirect Effects of Environmental Dimensions 5. Indirect Effects of Social Dimensions 6. Indirect Effects of Economic Dimensions
Industry 4.0 Principles Digitalization and integration of horizontal and vertical value chains Digitalized product and service offerings Digital business models and customer Access			

As a result, within the scope of Industry 4.0, information technologies, the power of big data objects (IoT), and the flow of production dynamics changed from person to machine to machine to machine. Therefore, it gains a quality increase in production, low margin of error, flexibility, low cost, and increase in volüme (Manavalan and Jayakrishna 2019, p. 926). It also has ensured that supply chains have a more unique and modular structure out of a single level, and the control and follow-up of processes can be more healthy and up-to-date.

As a final point regarding the Industry 4.0 discipline, sustainable production is mentioned; Machado et al. (2020), in a study on the subject, gathered the dependent points of sustainable production in the context of Industry 4.0. This table is as seen below.

12.4 Logistics 4.0

The effects of Industry 4.0 constitute Logistics 4.0 (Alkiş et al. 2020, p. 375). Considering the fundamental benefit of Logistics 4.0, all the processes such as transportation, distribution, storage, and production support are part of the supply chain process instantly, and it will be possible to manage the process with the help of instant communication. In this way, the capacity increase will occur without any physical growth. In other words, idle capacity can be revealed. This development means an increase in productivity. Productivity increases directly contribute to the competitiveness of logistics companies.

With the widespread use of the Internet, advances in computer technologies have started to fade. In this way, access to information has increased. Thus, the countries' trade enabled them to move to foreign markets more quickly, starting from their domestic markets. Thus, globalization has increased globally, and there has been a definite volume increase in international trade. Each region of the world has different demographic and sociological values, and experiencing cultural differences has started to change business life like never before.

With the effects of globalized trade and developing technologies, the service sector, especially the production sector, had to get out of the general patterns and make more personal, that is, modular productions (Şekkeli and Bakan 2018,p. 18). Since each production operation has its supply chain, the backgrounds of the operations started to become more complex and detailed. In light of this reason, it has changed in many areas, such as production and marketing, sales, and human resources management.

The restructuring process, which started to take place mainly in the field of production for the renewed market conditions, was, of course, valid in the logistics sector (Büyüközkan and Güler 2019, p. 22). The logistics sector has started to try to adapt to the new world order following its activities and adapt to the renewed supply chains at the same time. For this reason, while logistics activities have been active only as transportation in the past, they have started to make more financial investments to respond to intelligent warehouses, production support operations within the same warehouses, innovations such as business models that must act simultaneously with production and demands from supply chains that operate with open dynamics (Lee and Shen 2020, p. 13). For this reason, logistics is no longer just about transportation (Lee and Shen 2020, p. 3).

With the help of developing technology, it has become easier for the management level to make decisions, and the execution of the decisions made has become healthier. These factors, which help form concepts such as speed and efficiency, bring another gain in logistics. Logistics has gained "flexibility" due to the combination of more up-to-date, reliable, and healthy information logistics and expansion of the qualifications of logistics activities. Logistics, which can move away from the stereotype, gives a tremendous competitive advantage to the companies that own it with its flexibility.

12.4.1 The Concept and Importance of Logistics

The global economic order is shaped in specific sectors, and all other sectors are dependent on them. If it is necessary to give an example to the main sectors mentioned, software, banking, insurance, health, and the logistics sector will appear at this point.

The logistics industry has a sine qua non for its physical contact for world trade. Therefore, logistics is a sector in itself, and each sector also has its logistics. In the broadest framework, if we consider each unit of trade as a supply chain, the physical movement of the commodities provide the formation of these supply chains, as long as humanity exists. In other words, the concept of logistics can be accepted as the background and factory floor of the trades seen from past to present. For this reason, it is possible to encounter expressions such as "Logistics is like a breath, understandable when it is valued" by experts in the sector and academicians specialized in the field. The COVID-19 pandemic process, which erupted in the world in early 2020, once again proved the vital importance of logistics. Except for the pandemic, a completely different crisis would be possible for people other than the pandemic if the quarantine period experienced the lack of commodities on the shelves of the markets and pharmacies that provided the daily needs to be met. Regarding this problem, which is mentioned in some regions of some countries where logistic processes are troubled, some case studies of the world press have been experienced.

Even though the concept of logistics is perceived as just a field of business today, its origins are much earlier. In a wide range of time spanning even in the early history of BC, logistics has emerged from the need for military equipment in wars and the need for soldiers to return to the battlefields or vice versa. For this reason, logistics had its strategic importance in the present times before. When the development of civilization and the quality of wars started to increase, logistics began to be stripped from the simple shell it had and started to have many different and professional disciplines (Karagöz et al. 2020).

Today, it is possible to list professional logistics in nine main titles. Logistics operations are formed as a result of the merger of the following activities:

1. Transportation
2. Prediction Management
3. Storage
4. Stock management
5. Order Processing
6. Packaging
7. Buying
8. Handling
9. Customer service

Since logistics management is a part of supply chain management, the start and endpoints of all process supply chains come to life or end with customer demand. For this reason, it has the power of speedy response time, so every unit of work is required to keep productivity at an optimal level (Bilgiç et al. 2020).

12.4.2 Logistics Periods

When logistics activities started to gain their current meaning from ancient history, they gained the professionalism of the dynamics of business life. Like the concept of Industry 4.0, in the light of the developments, the dynamics and features of logistics activities are also grouped within the historical periods. The concept of Logistics

4.0 is directly related to the concept of Industry 4.0, and specialists in the field have divided Logistics activity periods into four groups within the literature. This grouping is very close to the grouping seen in Fig. 12.3, and its difference is due to its concentration on the sector. Logistics periods and fundamental dynamics are as shown below.

Just like in the transition period from Industry 3.0 to Industry 4.0; Although Logistics 4.0 is a relatively new concept in the literature, the main difference from Logistics 3.0 is that the already used machines, tools, and devices become more digitalized and that some devices are fully adapted to computer systems thanks to electronic chips, in other words, cyber-automation with the help of building physical systems (Kamble et al. 2018, p. 408).

As shown in Industry 4.0, a transaction flow from machine to machine has started in Logistics 4.0. Mainly based on registration, basic process operations are carried out automatically with the help of technologies like sensors, IoT, etc. (Ünlü and Atik 2018, p. 439). Today, while there is a physical movement based on automation in warehouses or in-facility areas, materials are transported using workforce utilizing

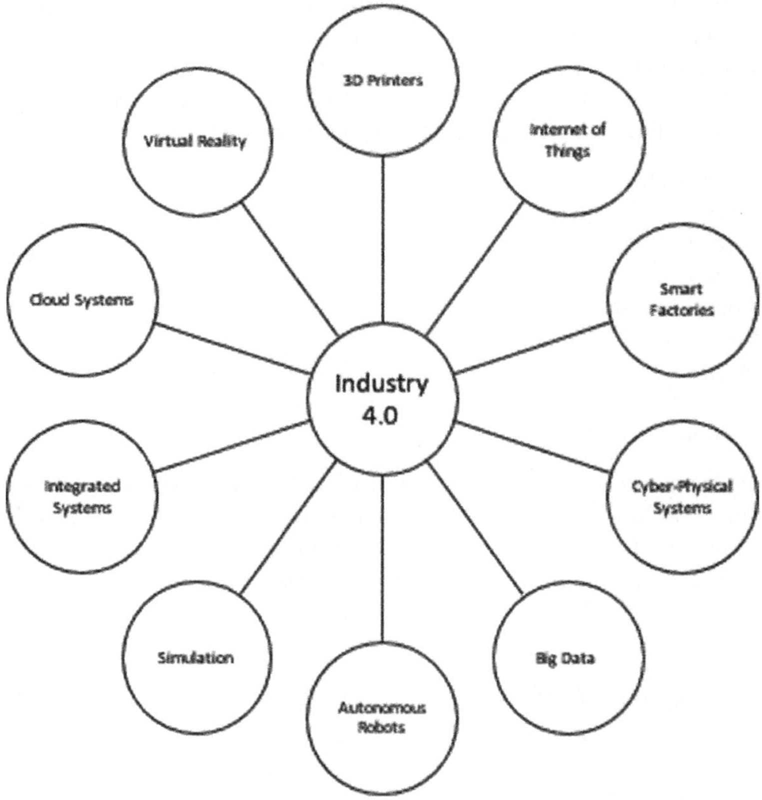

Fig. 12.3 Things Industry 4.0 Depends on (*Source* Görçün 2017, 146)

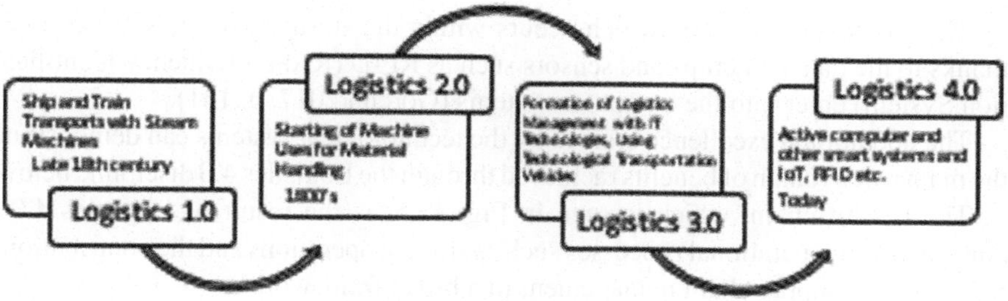

Fig. 12.4 Historical phases and basic structures of logistics activity. *Source* Karagöz (2020, 41)

freight vehicles, whose technological level is relatively high compared to the past, in larger transports between cities and countries. This transport, which is carried out with human power, takes place in the light of the decisions taken as a result of the information obtained from digital data related to the subject. That is the point to be underlined. However, the handling operations required for the loading and unloading of loads to vehicles can be carried out in some ways in automation, depending on the type of load and operation.

12.4.3 Supply Chain Management in Logistics 4.0

The main contribution of Logistics 4.0 can be collected in a sentence. A supply chain can be monitored instantaneously throughout the entire process. All the researches conducted in the literature, comments, and directions made by experts in the industry are based on this statement entirely.

Within the logistics activities, the area where the concept of automation is concentrated is in the logistics warehouses, where a high level of storage activities are performed (Lee and Shen 2020, p. 5). That is because the main transportation by land, air, and sea vehicles still need to be done with human power. However, even in these activities, there is a difference from the past. This difference can be briefly defined as "digital trace."

Automation, mainly carried out in storage activities, enables operational excellence in increased and continuously improved processes. The operational excellence here is shaped around the concept of speed. The benefits of speed increase can be summarized into several items. These are;

- Ability to instantly determine the stock level that falls below the optimal level and perform the necessary supply;
- In storage operations with instant intervention opportunity, it is possible to have an optimal stock and a decrease in inventory costs,
- Providing the opportunity to realize inventory and production plans more smoothly according to the flow performance according to the type and type of commodities, significant achievements such as can be given as an example.

The automation process, which occurs within the storage activities, is achieved thanks to the data that chips and sensors such as RFID (Radio Frequency Identification System) enter into the computer systems (Görçün 2017, p. 161).

The operational excellence created by the technological systems can demonstrate the main result (chain of benefits) achieved through the Logistics 4.0 discipline below.

The benefit chain, which is seen in Fig. 12.5, is the result of Logistics 4.0's contribution to operational processes such as storage operations and the contributions it provides to supply chain management in a broad framework.

For the achievement of operations in the logistics sector;

1. Internet of Things
2. Big Data
3. Processing images with sensors
4. Automation Technologies

are mainly used. Thanks to this discipline, it is possible to create maximum occupancy in each transport unit, which is almost the golden rule of logistics operations. This result shows us that instant information obtained from big data strengthens the basic building block of the operation every day in logistics transportation (Russom 2011, p. 7).

Another contribution of big data to logistics operations is the contribution of the operations to the planning processes that must be completed before it takes place. In other words, network and network designs now base their foundations on big data in the logistics literature (Saatçioğlu et al. 2018, p. 1683). With the change of ratios at the macro and micro levels, making plans such as order estimation and creating

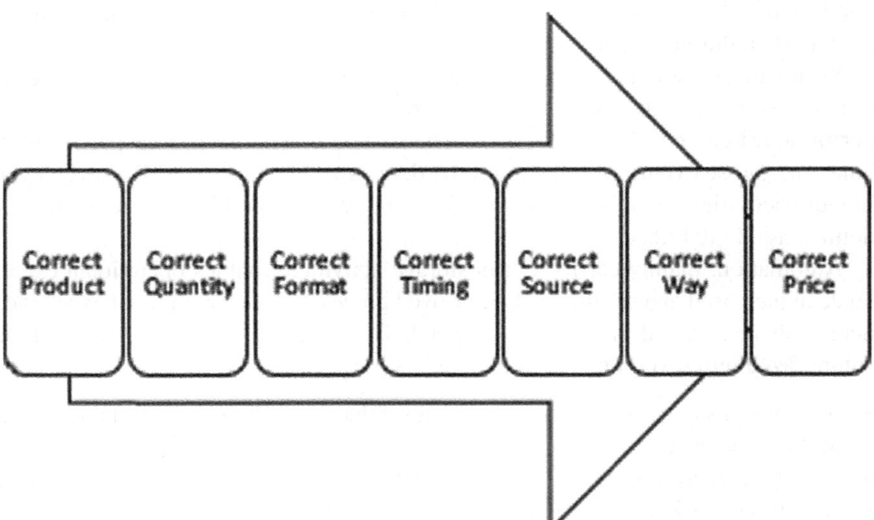

Fig. 12.5 Benefit chain with Logistics 4.0 Discipline (Yılmaz and Duman 2018, p. 190)

a moving plan is possible (Russom 2011, p. 34). Thus, optimization is possible for the related operations and logistics operations that have not been realized yet.

12.5 Conclusion

The literature shows that the key concept for the continuity of the activity in business life during the Industry 4.0 period is the concept of "know-how." Nowadays, when an automation process is dominant, it can be seen that automation administration has evolved towards artificial intelligence at some points in the future. Therefore, one of the most important results to be drawn from Industry 4.0 is that the active opponent of man is now the machine rather than himself. Because the concept of production is becoming more and more digital, the software has started to outweigh the hardware (Demir et al. 2019, p.689). Based on this statement, the employment structure in business life will change considerably in the long run as it is today.

With the help of technological developments, production and some service sectors have become more flexible as well as more efficient in the light of the disciplines brought by Industry 4.0 (Ünlü and Atik 2018, p. 443). Research emphasizes that today's productivity technology usage is becoming widespread and the technologies used are improving. Research shows us that the concept of sustainability has a very passive structure, and this passivity is based on the use of technology and technology today.

Logistics 4.0, one of the main study areas of this research, has a passive position against Industry 4.0 and an active position against sustainability. Industry 4.0 not only revolutionized production processes, but it is also one of the main sectors of the world and has caused radical changes in the logistics sector, which is expressed as the shining star of the twenty-first century. With the help of the logistics sector's activity and types, the production dynamics changing with Industry 4.0 have gained a flexible structure within the market.

Considering a technology-indexed process in logistics activities and production activities will increase activity speed, decrease essential cost items, reduce energy need, and help achieve operational excellence. This situation will also lead to increased occupational safety by taking some dangerous logistics activities over from humans.

The number of companies, which are the actors that make up the business life, is dense with SMEs (SMEs) and SME iris in practice in the real sector, and they may make the adaptation process to Industry 4.0 a little different (Manavalan and Jayakrishna 2019, p. 926). Therefore, the necessity of the change to be practical on a micro basis and effective at the macro-level is the necessity of the governments to provide incentives for companies regarding this issue as seen in the examples in the world.

With this supported change, production and production-related management activities will become flexible, and the concept of the competition will gain a different dimension with the flexibility such as smart factories, cyber-physical systems, integration of 3D printing devices into production, and production quality will increase (Ünlü and Atik 2018, p. 438; Şekkeli and Bakan 2018, p. 24, 31).

Industry 4.0 is analyzed from every angle perspective, general and sectoral, and it can create destruction on the concept on which business life is located. For this reason, any change and activity that takes place will be considered ecologically sustainable, and companies will be evaluated from the perspective of sustainability based on their performances. However, adopting a sense of sustainability in the eyes of companies alone will create the destruction of natural resources that are already challenging, making it impossible to leave a world that can be transferred to future generations, both human and commercial.

In short, today's order based on know-how comes to life with information and information flow. This situation affects logistics and production and even helps the emergence of processes called green logistics. With the increase in production opportunities, more accessible access to information, and the concept of flexibility, which began to take place in all areas of life, became active, there were no more constant demands and production, and demands and production began to turn into more personal characteristics. Governments should support the adaptation of their companies to Industry 4.0, not only in large-volume companies but also in the number of SMEs (SMEs).

This study is intended to be a helpful resource for managers in companies, regardless of volume, for academics and higher education students working on the subject, and anyone who wants to know about the subject. The findings reached in the light of the information obtained in the literature review can be further deepened with the help of the real sector with the help of empirical studies.

References

Alkiş G, Piritini S, Ertemel AV (2020) 'Lojistik Sektöründe Endüstri 4.0 Uygulamalarinin Operasyonel Verimliliğe Etkisi'. Business & Management Studies: an Internatinal Journal 8(1):371–395

Bag, S., Telukdarie, A., Pretorius, J. H. C., & Gupta, S. (2018). Industry 4.0 and supply chain sustainability: framework and future research directions. *Benchmarking: An International Journal.*28 (5) s.1411–1450

Bähre D, Swat M, Trapp K, Vielhaber M (2016) 'Supporting Manufacturing Technology Definition Through Multi-Level Resource Consumption Analysis'. Procedia CIRP 40(2016):378–383

Barreto L, Amaral A, Pereira T (2017) 'Industry 4.0 Implications in Logistics: An Overview'. Procedia Manufacturing 13(2017):1245–1252

Bilgiç E, Türkmenoğlu MA, Koçak A (2020) 'Dijitalleşmenin Lojistik Yönetimi Bağlaminda İncelenmesi'. BEÜ İİBF Aid 5(1):56–69

Büyüközkan G, Güler M (2019) 'Lojistik 4.0 Teknolojilerinin Analizi için Metodolojik Yaklaşim'. Girişimcilik Ve İnovasyon Yönetimi Dergisi 8(1):21–47

Demir KA, Döven G, Sezen B (2019) 'Industry 5.0 and Human-Robot Co-working'. Procedia Computer Science 158(2019):688–695

De Man JC, ve Strandhagen, JO, (2017) Sürdürülebilir iş modelleri için bir Endüstri 4.0 araştirma gündemi. Procedia Cirp 63:721–726

Görçün Ö.F. (2017), ''Dördüncü Endüstri Devrimi Endüstri 4.0'', *Beta Yayinlari*, Tipki 2.Baski, Temmuz 2017

Günther WA, Mehrizi MHR, Huysman M, Feldberg F (2017) 'Debating Big Data: A Literature Review on Realizing Value from Big Data'. J Strat Inf Syst 26(2017):191–209

Gürül B. (2019), ''Sürdürebilir Üretim Göstergelerinin Kurumsal Sürdürebilirlik Performansina Etkisinin Araştirilmasi'', T.C. İstanbul Gelişim Üniversitesi Sosyal Bilimler Enstitüsü İşletme Anabilim Dali, Yayinlanmiş Doktora Tezi

Haseeb M. vd. (2019), ''Industry 4.0: A Solution towards Technology Challanges of Sutainable Business Performance'', *Social Sciences*, 2019,8,154, 01–24

https://www.iienstitu.com/blog/surdurulebilirlik-nedir (Access Date: 25.06.2020)

Kamble SS, Gunasekeran A, Gawankar SA (2018) 'Sustainable Industry 4.0 Framework: A Systematic Literature Review Identifying the Current Trends and Future Perspectvies'. Process Saf Environ Prot 117(2018):408–425

Karagöz B, Bumin DH, B. (2020) 'Lojistik 4.0 Uygulamalari ve Lojistik Firmalarinin Bakiş Açisi'. İnsan & İnsan 7(23):37–51

KPMG (2018), ''Endüstri 4.0: Yanilgilar ve Gerçekler'', *KPMG Türkiye*

Lee H.L., Shen Z.J., (2020), ''Supply Chain and Logistics Innovations with the Belt and Road Initiative'', *Journal of Management Science and Engineering (Pre Proof.)*, 01–19

Machado CG, Winroth MP, Ribeirio da Silva EHD (2020) 'Sustainable Manufacturing in Industry 4.0: An Emerging Research Agenda'. Int J Prod Res 58(5):1462–1484

Manavalan E, Jayakrishna K (2019) 'A Review of Internet of Thing (IoT) Embedded Sustainable Supply Chain for Industry 4.0 Requirements'. Comput Ind Eng 127(2019):925–953

McAfee A., Brynjolfsson E. (2012), ''Big Data: The Management Revolution'' *Harvard Business Review*, October 2012 Re-print

Oğuz İH (2019) 'Politik İstikrar ve Çevresel Sürdürebilirlik'. Uluslararasi İşletme Ve Ekonomi Çalişmalari Dergisi 1(1):01–11

Ömürgönülşen M., Çekiç B., Ar İ.M. (2020), ''Lojistik Firmalarinda Endüstri 4.0 Uyum Sürecinde Dikkate Alinacak Faktörlerin Bulanik DEMATEL Yöntemi ile Değerlendirilmesi'', *International Uluslararasi İktisadi ve İdari İncelemeler Dergisi Prof. Dr. Talha Usta Süleyman Özel Sayisi*, 167–184

Özdemir A, Özgüner M (2018) 'Endüstri 4.0 ve Lojistik Sektörüne Etkileri: Lojistik 4.0. İşletme Ve İktisat Çalişmalari Dergisi 6(4):38–47

Özispa N, Deveci DA (2019) 'Literature Review on Sustainability in Multimodal Transportation'. Mersin University Journal of Maritime Faculty 1(1):40–47

Öztemel E, Gürsev S (2018) 'Türkiye'de Lojistik Yönetiminde Endüstri 4.0 Etkileri ve Yatirim İmkanlarina Bakiş Üzerine Anket Uygulamasi'. Marmara Fen Bilimleri Dergisi 30(2):145–154

Poesche, J., & Lilja, K. (2019). Biorefineries: pulp industry 2.0. *O PAPEL*, *80*(09), 90–97.

Russom P. (2011), ''Big Data Analytics'', *TDWI Research*, TDWI Best Practices Report Fourth Quarter 2011

Saatçioğlu Ö. Y., Tuğdemir Gök T., Özispa N. (2018), ''Endüstri 4.0 ve Lojistik Sektörüne Yansimalarinin Örnek Olay Kapsaminda Değerlendirilmesi'', *Süleyman Demirel Üniversitesi İktisadi ve İdari Bilimler Fakültesi Dergisi*, Endüstri 4.0 ve Örgütsel Değişim Özel Sayisi, (3), 1675–1696

Stock T, Seliger G. (2016), ''Opportunities of Sustainable Manufacturing in Industry 4.0'', *13th Global Conference on Sustainable Manufacturing - Decoupling Growth from Resource Use*, Procedia CIRP 40, (2016), 536–541

Strandhagen J.O., Vallandingham L. R., Fragapane G., Strandhagen J. W., Stangeland A. B. H., Sharma N. (2017), ''Logistics 4.0 and Emerging Sustainable Business Models'', *Adv. Manuf*, (2017),5,359–369

Şekkeli ZH, Bakan İ (2018) 'Endüstri 4.0'in Etkisiyle Lojistik 4.0'. Journal of Life Economics 5(2):17–36

The Boston Consulting Group (BCG) (2015), ''Industry 4.0: The Future of Productivity and Growth in Manufacturing Industries'' *The Boston Consulting Group*

Timm I. J., Lorig F. (2015), ''Logistics 4.0 A Challenge for Simulation'' *Proceedings of the 2015 Winter Simulation Conference,* 3118–3119

Toker K (2018) 'Endüstri 4.0 ve Sürdürebilirliğe Etkileri'. İstanbul Management Journal 29(84):51–64

Torbacki W, Kijewska K (2019) 'Identifying Key Performance Indicators to be Used in Logistics 4.0 and Industry 4.0 for the Needs of Sustainable Municipal Logistics by Means of the DEMATEL Method'. Transportation Research Procedia 39(2019):534–543

TUSİAD ''Türkiye'nin Küresel Rekabetçiliği için Bir Gereklilik Olarak Sanayi 4.0 : Gelişmekte Olan Ekonomi Perspektifi'' 2016 Araştirma Raporu

Ünlü F, Atik H (2018) 'Türkiye'deki İşletmelerin Endüstri 4.0'a Geçiş Performansi: Avrupa Birliği Ülkeleri ile Karşilaştirmali Analiz'. Ankara Avrupa Çalişmalari Dergisi 17(2):431–463

Yilmaz Ü, Duman B (2018) 'Lojistik 4.0 Kavramina Genel Bir Bakiş: Geçmişten Bugüne Gelişim ve Değişimi'. Bilecik Şeyh Edebali Üniversitesi Sosyal Bilimler Enstitüsü Dergisi 4(1):186–200

Yurdakul M (2020) 'İşletmelerde Sürdürebilirliğin Sağlanmasinda Eko İnovasyon Uygulamalari: Bir Model Önerisi'. İşletme Ekonomi Ve Yönetim Araştirmalari Dergisi 1(1):15–33

Mustafa Özan Dr. Mustafa Ozan is lecturer at Department of Logistics Management, İstanbul Gelişim University. He is also the deputy director of Applied Sciences in Istanbul Gelisim University School. He received his PhD in Business Administration from Istanbul Gelisim University Institute of Social Sciences in 2018. Mustafa Ozan has worked as general manager in a shopping mall between the years 1994 – 2018. His research focuses on management organization, strategic management, behavioral management and retail management.

Chapter 13
Finance and Cost Management in the Process of Logistics 4.0

Lokman Kantar

Abstract Industry 4.0 is a revolution of creating added value, based on the use of machinery and robot power instead of arm strength of works that do not require any qualification and specialization on jobs that require qualification. Replacing the manpower of the machines, making them coordinable; thanks to new developments in computers and internet technology, and making the production processes self-manageable, led to the emergence of the Industry 4.0 concept. Logistics 4.0 explains the implications of Industry 4.0 on transport and cross-functional coordination tasks and how digitalization and automation in logistics should be shaped and supported. In the case of applying Logistics 4.0 in the supply chain, it provides significant and potential cost savings. The financing of the supply chain in the Logistics 4.0 process, which is the subject of this study, is extremely important for logistics companies to gain profits with low resource cost and working capital, and maximize cash flows and increase firm values. The supplier companies and the focal company in the supply chain utilize the credibility of the focal company to finance their investments at a low cost. Firms in the supply chain also need to reduce their working capital and increase asset turnover (sales/assets) by lowering their stock levels in order to increase their return on equity (ROE). In order to reduce the working capital of logistics companies, they need to decrease their stock levels, storage costs, transportation costs, and increase their service levels to customers. According to the case study findings related to SCF, if the suppliers in the supply chain use the SCF instrument, a serious decrease in credit costs is observed. In order for Focal Company to take advantage of these transactions, the supplier companies in the supply chain must make an additional "pro rato" payment or sale discount.

Keywords Finance · Cost management · Logistics

L. Kantar (✉)
Gelişim University, Istanbul, Turkey
e-mail: lkantar@gelisim.edu.tr

İ. İyigün and Ö. F. Görçün (eds.), *Logistics 4.0 and Future of Supply Chains*, Accounting, Finance, Sustainability, Governance & Fraud: Theory and Application, https://doi.org/10.1007/978-981-16-5644-6_13

13.1 Introduction

Logistics companies have started to use smart tracking and tracking, control, and management technologies to manage their business processes and applications in real-time; thanks to automation and integration based on technological developments. The change in the way of doing business due to technological developments and models, and the change in the purchasing behavior of customers have made time, speed, quality, and cost issues much more important (Tekin et al. 2018). The rapid development of the logistics industry not only increases the speed of delivery of goods and the flow of capital, but also increases the speed of movement of logistics and economic growth. Countries implement a series of incentive measures and policies in order to maintain the rapid and healthy development of the logistics industry (Gu and Dong 2016). Due to a lack of resources, many manufacturing companies don't want to put too much material, manpower, and financial resources into their logistics services. Since most of the logistics companies are small, they only take into account the speed, security, and reliability of storage, transportation, and other services in logistics costs. In other words, they don't take into account the entire supply chain process, but it is very important to control logistics costs effectively. Many logistics companies can't accurately calculate the cost. Each stage in the logistics process needs to be taken into account in the cost calculation. However, logistics companies reflect the costs to the tables according to the financial statements determined by the relevant ministry, but in reality, it is not easy to control the costs (Gu and Dong 2016).

According to the latest survey data, it shows that the profit margins of international logistics companies are more than 3 times the profit margins of local logistics companies. Logistics companies control the external environment in the respective companies along with the cost control within the company through cooperation between the supply chain logistics company to complete the cost control.

13.2 Industry 4.0

The Industrial Revolution is the gradual transformation of small workshops into industrial workshops, which extend from the late eighteenth century to the mid-nineteenth century and gradually become larger scale productions. Thanks to the technological change, the industrial revolution has taken place, and it has been possible to discover machines that can produce faster and more efficiently than small tradesmen working in small workshops. (Frederick 2016).

The first industrial revolution began with the use of mechanical benches that operate on water and steam power, instead of human labor. While the use of mines and metals increased in this period, there have been developments in the field of transportation (Lu 2017).

The second industrial revolution covers the period that coincided with the beginning of the twentieth century when mass production was made on the assembly line where electricity was used and the division of labor was developing. In the second industrial revolution, steam power was completely used in industrial production, and railways were built.

The third industrial revolution has evolved since the early 1970s, with developments in electronics and information technologies. This period can also be referred to as "knowledge economy". Telecommunications technologies have become even more powerful with the third industrial revolution. In the third industrial revolution, machines based on digital technology started to replace the machines based on mechanical and electronic technology.

The transition to the fourth industrial revolution, which was expressed as Industry 4.0, was prepared by the Working Group that was formed by the German Government in 2012 after it came to the agenda at the Hannover Fair in 2011 and announced in 2013 (Kilic and Alkan 2018). Industry 4.0 is a revolution of creating added value based on the use of the machine and robot power instead of the manpower for those tasks that would not require any qualification and specialization on jobs that require quality (Sener and Elevli 2017). The fact that the production processes become self-manageable by replacing the manpower of the machines by becoming coordinated; thanks to the new developments in computers and internet technology has led to the emergence of the Industry 4.0 concept (Yilmaz and Duman 2019).

The defining feature of the fourth industrial revolution is intelligent networks based on cyber-physical systems (CPS) (Barreto, Amaral and Pereira, 2017). Cyber-physical systems are engineering and physical systems whose processes are monitored, controlled, coordinated, and integrated by a computer and communication system. The "Internet of Things", also known as the "Industrial Internet of Things" (IIoT), has also affected the interaction, monitoring, control, and management of cyber-physical systems (Barreto et al. 2017). This new paradigm, which is expressed as advanced digitalization and enables machines and people to communicate with each other in real time, is a result of increased internet usage. The concept of Industry 4.0 is shown in Fig. 13.1.

As can be seen in Fig. 13.1, the main purpose of Industry 4.0 is to make the digital production called "smart factory" sustainable, which means mobility, flexibility, and interoperability of industrial operations, integration with customers and suppliers, and adoption of innovative business models. The relevant terms used in Industry 4.0 include the Internet of Things, Internet services, industrial Internet, advanced manufacturing, and smart factory. Industry 4.0 race is on the "smart" track. Tools of this age; smartphones, smart cars, smart homes, smart factories, smart warehouses, smart markets, smart hospitals, smart schools, and smart offices (Tekin et al. 2018).

In smart factories, flexibility, cost-effectiveness, and efficiency will increase widely. Tasks will change in factories, workers will be freed from their routine duties. The need for staff will emerge for jobs that require common sense, creativity, problem-solving skills, and abilities. The human-robot collaboration will evolve as a new business process in the manufacturing industry (Kilic and Alkan 2018).

Herman et al. has identified three key components for Industry 4.0;

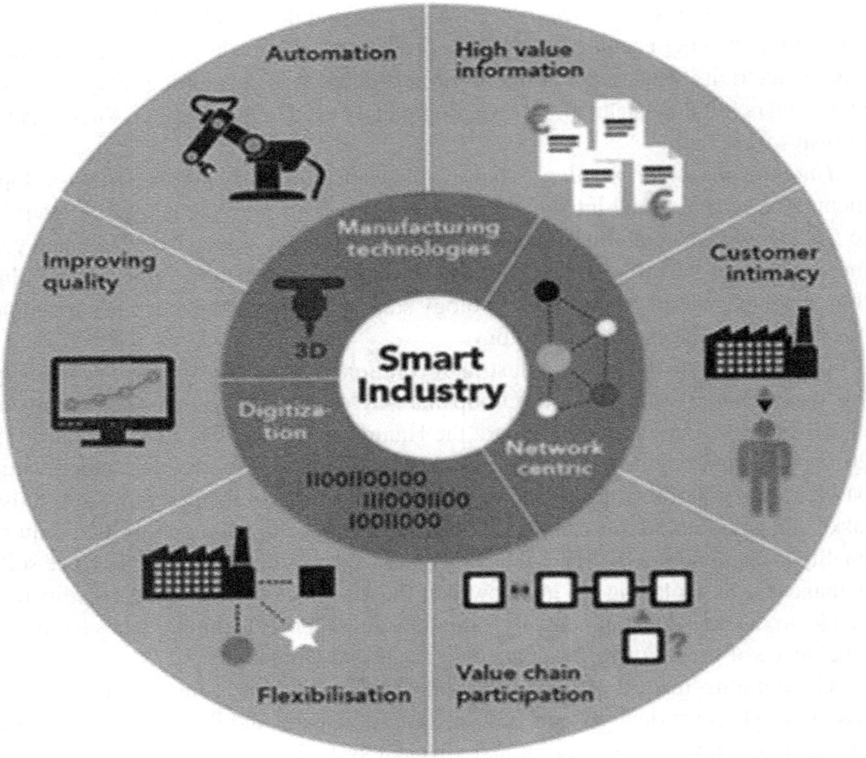

Fig. 13.1 Industry 4.0 Concept (Barreto et al. 2017)

- Smart Factory,
- Cyber-Physical Systems (CPS), and
- The Internet of Things (IoT).

Industry 4.0, which is an umbrella term covering various disciplines in industrial production, contains several different technological developments and new concepts. Some of the technological developments currently used in industrial production are as follows: (Strandhagen et al. 2017).

- Identification Technology,
- Big data,
- Industrial robots,
- Additive production technology,
- Networking technology,
- Autonomous robots and vehicles,
- Artificial intelligence, and
- Cloud manufacturing.

As of 2030, it is rumored that the digital ecosystem will replace the concept of Industry 4.0. The aforementioned concept of the digital ecosystem; is a socio-technical system inspired by the natural ecosystem, self-organizing, scalable and sustainable, separable, and open to interaction with the outside world (Ostadzadeh et al. 2015).

13.3 Logistics 4.0

In the early nineteenth century, logistics was introduced and defined by the army as the planning and movement of soldiers (Kassem and Elkader 2019). Logistics 4.0 is the development of labor savings and standardization with the evolution of the Internet of Things technology in logistics (Büyüközkan and Güler 2019). Transport, inventory management, material handling, supply chain structure, and information flow, which are key logistics activities, are affected by Logistics 4.0 (Strandhagen et al. 2017). Logistics 4.0 and sustainable business models were launched to secure the future of the German manufacturing industry. Recently, many governments have launched strategic initiatives to strengthen industrial production, and the fourth industrial revolution has become global. Examples of such initiatives are the US national production innovation network, Japan Robot Technology, and China production 2025 (Strandhagen et al. 2017). Due to the automation and digitization of the production processes of European companies using industrial robots, labor costs will decrease and therefore production abroad will not be preferred because of the low labor force (Strandhagen et al. 2017).

In the case of applying Logistics 4.0 in the supply chain, it provides potential cost savings of 20% in quality and maintenance and 30% in inventory costs (Wang 2016). The purpose of Logistics 4.0 is to improve efficiency, productivity, and supply chain partners, and to increase decentralized decision-making structures and information technologies (Yilmaz and Duman 2019).

13.3.1 Logistics Trends

With the rapid development of information communication technologies and the spread of the Internet, the importance of time, quality, speed, and cost has increased more. With globalization, the logistics sector started to be expressed with the concepts of speed, flexibility, and technology. These technological developments have also affected the way businesses do business and models, and their purchasing behavior. Customers want the products and services they will receive as soon as possible. It is possible with technology to present the right product at the right place and time with the least cost in an intensely competitive environment. Many changes such as technological developments, competition, urbanization, demographic structure have led logistics companies to research and develop different models and technologies.

Industry 4.0 and the logistics sector, which has been moved to a smart environment, will be able to gain a competitive advantage by adopting new logistics trends and technological applications in many areas such as resource use, transportation, storage, tracking, monitoring, and labor use (Tekin et al. 2018).

New approaches started to be used in the logistics sector with Industry 4.0 can be listed as follows; Multi-purpose networks, Integrated channel logistics, On-demand delivery, Sharing economy feature, Smart energy logistics, Super network-net logistics, Tube logistics (see Tekin et al. 2018).

13.3.2 Logistics Evaluation Processes

Logistics has undergone three revolutionary changes in the past. The first innovation (Logistics 1.0) is caused by the "mechanism of transport" since the late nineteenth and early twentieth centuries. The second innovation (Logistics 2.0) has been guided by the "automation of the transport system" since the 1960s. The third innovation (Logistics 3.0) has been represented by the "logistics management system" since the 1980s. We are at the beginning of the fourth logistics innovation, now called Logistics 4.0. Its main driving force is IoT & S (Internet of Things and Service). The evolution process starting from Logistics 1.0 is as shown in Fig. 13.2.

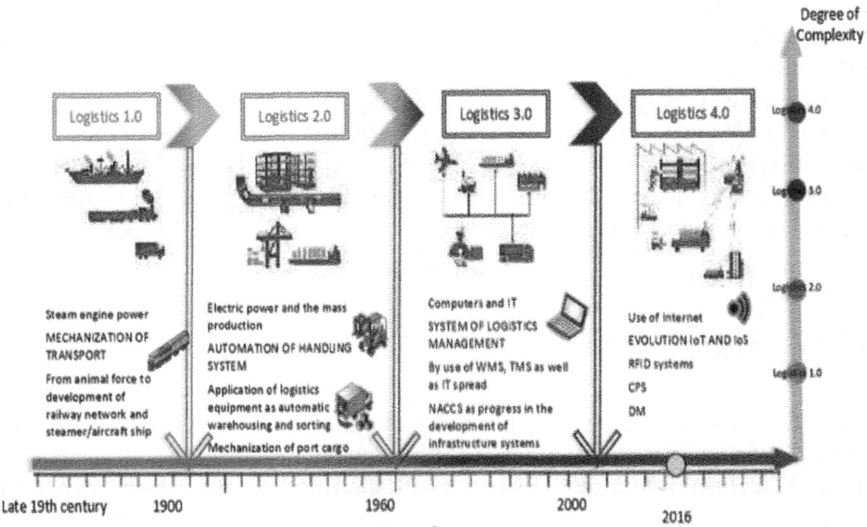

Fig. 13.2 Evolution Process of Logistics (Wang 2016)

13.4 Supply Chain Management

Supply chain management has seven important components. These components are; customer relationship management, customer service management, demand management, order fulfillment, manufacturing flow management, procurement, product development and commercialization (Lampert et al. 1998). Businesses operate to effectively build these seven important components, which form supply chain management.

The main objectives of supply chain management can be listed as follows (Kagnicioglu 2007):

- To decrease the activity costs,
- To maintain quality in production,
- Minimizing inventory costs and losses,
- Having reliable suppliers and protecting activities,
- To ensure the continuity in the goods and services with the information flow between the chains in the supply chain,
- To ensure continuity in raw materials, semi-products, parts, and services provided,
- To increase competition in enterprises,
- To establish good relations with other groups in enterprises, and
- To produce raw materials, semi-finished products, parts, and services required for the lowest cost production.

Factors that prevent supply chains from performing effectively (Fawcett et al. 2008):

- Top management not providing adequate support to the supply chain,
- Operational plans and uncertain strategic,
- Reluctant or inability to share information,
- Trust problems among chain members,
- The reluctance of chain members to share risks and rewards,
- Inflexible processes and organizational systems
- Being reluctant to change,
- Job descriptions are not explained sufficiently,
- Resistance to innovative ideas, and
- Lack of education and knowledge of the required subjects can be counted.

13.4.1 Big Data in Supply Chain

In the new century, information systems work as smart agents. There is much more information in this period than in the past. The business world calls this situation, big data (Brzozowska 2016). Big data; is called the gold of the twenty-first century. Big data refers to complex and unstructured data that are difficult to analyze and use with traditional applications and analysis. It comes from a variety of sources related to monitoring and sensor devices, mobile devices, the Internet of Things, and radio frequency identification (RFID) technology, which are widely used in big data, logistics, and supply chain management. Today, 85% of the data volume is produced by companies, but most of this volume is produced during logistics activities. Every movement in logistics activities creates data that must be analyzed efficiently afterward. Every correct analysis means optimization of logistics processes (Yilmaz and Duman 2019).

According to big data, information comes from every transaction, every person, and hardware. In other words, information is obtained not only from computers but also by smartphones, watches, cars, TVs, household items, and even kitchen items (Brzozowska 2016). Big data can be considered as a source of information in different areas. Big data for business development is often used for consumer behavior analysis

Big data collection and interpretation are not easy, mainly due to its quantity. In terms of supply chain and logistics processes, big data can be used as a key to developing new management methods. Researchers are working to use big data for asset productivity and business growth. Asset efficiency is one of the most vital measures for each supply chain, it shows the development possibilities and also points to poorly managed sections of the processes (Brzozowska 2016).

13.4.2 Blockchain Technology in Supply Chain

Technological developments in recent years have caused changes in habitual rules in business life. After new technology-related developments such as autonomous vehicles, the Internet of Things, and Industry 4.0, blockchain technology is a more prominent subject, especially with cryptocurrencies. Blockchain technology is an encrypted filing system that allows the data block to be tracked at the same time by all users who have access to the network in a network environment, or it can be specifically distributed to all users, and therefore it is permanently stored in a decentralized distribution database. Blockchain-based technologies are distinguished from other technologies; thanks to their algorithms working without being connected to any center. By reducing the factors such as cost and time that occur in the supply chain, it provides important advantages to businesses such as the transparency of the transactions performed (Kaya and Turgut 2019). The first application of blockchain technology is cryptocurrencies. Due to the development and safety of technology, this technology is expected to be used in many fields in the future. The main of these areas (Kaya and Turgut 2019) include:

- Supply Chain Management,
- Digital Identity,
- Proxy Voting,
- Copyright Registry Systems,
- Notary Applications,
- Global Payment Systems,

- Syndication Credit,
- Internet of Things,
- Customer Recognition System, etc.

The supply chain is a network that starts with the raw material and ends with the delivery of the product to the end consumer. All links in the supply chain are interrelated. According to another definition, the supply chain is the process of providing information, materials, and money flow between retailers, wholesalers, textiles, distributors, logistics service providers, and suppliers. The main purpose of the supply chain is to deliver the goods or services demanded by the consumer to the end consumer at the lowest cost. Flexibility strategies in mass production and transportation were adopted in the 1950s and 1960s to reduce unit costs. The importance of resource planning was understood in the 1970s and related strategies were developed. Until the 1980s, businesses and managers focused on demand planning, requirement planning, storage, and purchasing. After 1980, as a result of the increasing competition in international trade, businesses started to take measures such as cost reduction, quality increase, and reliability. In this period, enterprises started to struggle with some criteria such as timely production, zero inventory, productivity in production (Kaya and Turgut 2019).

13.4.3 Payment Methods in Supply Chain

Although e-commerce is not considered very safe, more and more financial transactions are carried out through e-banking services. The e-banking services used in e-commerce are as shown in Fig. 13.3. There are many ways of payment on the Internet: (Brzozowska 2016, 91).

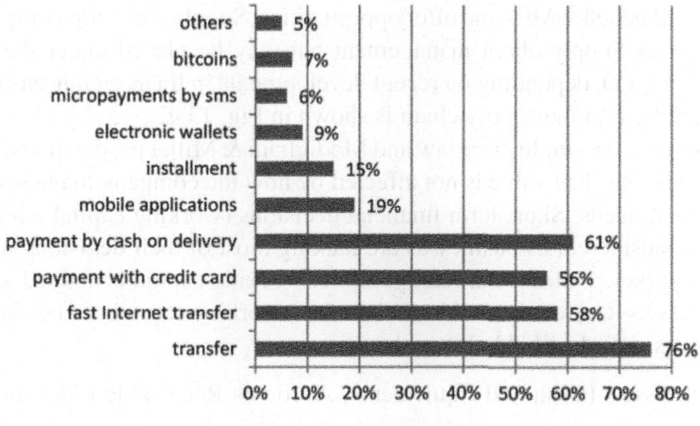

Fig. 13.3 Available payment methods (Peczak 2016; Brzozowska 2016)

As seen in Fig. 13.3, when it comes to money, customers are still very conservative. It can still be seen that the safest payment method is used—transfer from the account. However, it can be understood that fast transfers and credit card payments are also very popular. Given these facts, it can be assumed that customers do not rely on new payment methods, such as electronic wallets or bitcoins, which will revolutionize the world of e-commerce (Brzozowska, 2016). The Industry 4.0 is a definite turning point for logistics processes. The development of these inventions can revolutionize supply chains and save costs (Brzozowska 2016).

13.5 Supply Chain Finance (SCF)

Supply chain finance is a relatively new subject in logistics. Recent studies show that the focal company can reduce working capital and capital costs by 40%; thanks to higher credit ratings. The financing roots of the supply chain are based on the reverse factor. Factoring has traditionally been used to finance the receivables of the company by selling its trade receivables to a factoring institution through a bank. Here, the factor puts the receivables discounted. Then he collects his receivables from debtors and the difference is the factor's return. Supply chain management depends on the design and optimization of goods flows. Information flows are also included in supply chain management. Financial flows are not included in the process of supply chain management. However, according to the Supply Chain Finance Cube model, financial flows can be included in the model to reduce working capital usage and capital costs. The factors affecting the costs of capital in the supply chain are the volume of the working capital, the duration of the capital, and the capital cost (WACC).

Capital Costs = Volume of Working Capital*Duration*Cost of Capital (WACC)

Although supply chain financing is more widely used in multinational enterprises (MEB), it is not known much by SMEs. However, it is assumed that supply chain financing will benefit SMEs and offer opportunities. Supply chain financing is considered as part of Supply chain management but may be placed under the Business Controller or CFO, depending on recent developments in finance (Jansen 2016). The finance paradigm in the supply chain is shown in Fig. 13.4.

According to the single price law and Modigliani & Miller propositions, in perfect market conditions firm value is not affected by how the company manages its short-term financial needs. Short-term financial decisions (working capital management) need to be considered, as businesses are making most of their decisions short-term, as perfect market conditions are not possible. Working capital is defined as follows: "Current Assets–Current Liabilities", therefore, working capital can be expressed as follows by using a detailed balance sheet:

- Current Assets (Cash and Equivalents, Accounts Receivable (AR), Inventories, Other);

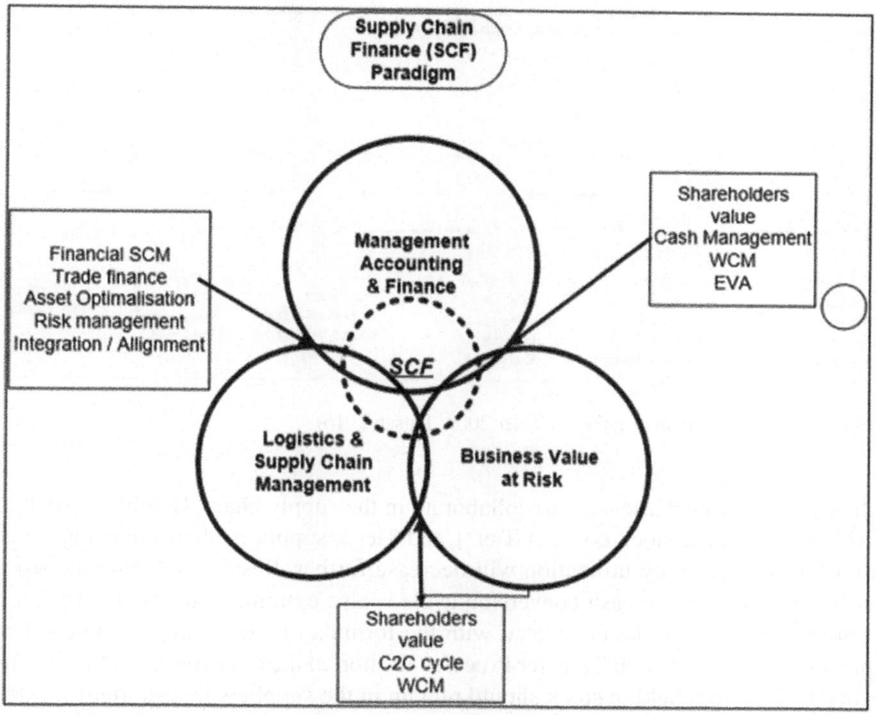

Fig. 13.4 Supply Chain Finance Paradigm (Cosse 2011; Jansen 2016)

- Current Liabilities (Notes Payable, Accounts Payable (AP), Tax Obligations, Accrued Expenses).

Since the purpose of this study is to decrease the Net Working Capital costs in the supply chain financing process, the supply chain process from the production of a bicycle to the end consumer is discussed. In this process,

- The flow of goods from left to right (dotted line);
- Information flow between companies in the supply chain (bold line with two arrows with bilateral arrows);
- Cash flow (gray line) is expressed from right to left.

The main idea of supply chain financing is to share sales forecasts from the focal company with Tier 1 and Tier 2 suppliers so that inventories in each company are fulfilled with economic order benefits following the JIT (Just in Time) / Lean concepts. In this case, not only the stock level per company is low, but also the amount and duration of trade receivables and commercial debts are shortened. Costs of inventories will also be reduced thanks to economic.

Optimum inventory costs in the supply chain are accomplished using a tool such as Advanced Planning and Scheduling (APS) techniques. This tool is especially used to manage stock levels in the supply chain of multinational companies and

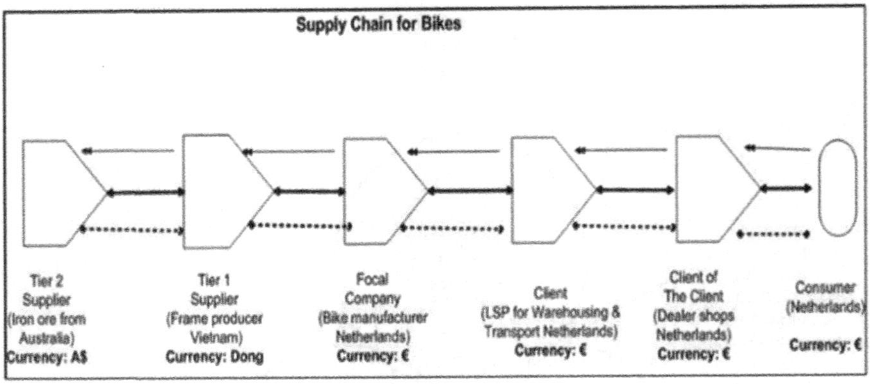

Fig. 13.5 Supply Chain of Bikes (Wisner 2009; Jansen 2016)

their foreign subsidiaries and to collaborate in the supply chain. If vehicles such as APS are used, the stock costs of Tier 1 and Tier 2 suppliers, their financing costs, and financial capacity utilization will decrease further. Besides reducing the stock costs of suppliers, the cash conversion cycle is also extremely important. The cash conversion cycle can be calculated with the formula of stock turnover time - debt payment period + Credit Transfer Process. To shorten the cash conversion cycle, the goods or services held in stock should remain in the suppliers in a shorter time and the Receivables should be collected from the debtors in a short time.

DuPont shows how activity rates and profitability ratios affect the profitability of company investments. Return on equity (ROE) is the most important factor for shareholders, and it is calculated by dividing net profit by equity.

Accordingly, Return on Equity (ROE):

$$ROE = \frac{\text{Profit}}{Equity} = \frac{\text{Profit}}{Sales} * \frac{Sales}{Assets} * \frac{Assets}{Equity}$$

Equation 1, DuPont Identity

The DuPont approach is often used in Finance and Logistics to understand the root causes of profitability. In logistics, there is a tendency to increase asset turnover (sales/assets) to reduce stock capital, thereby reducing the working capital and making the best use of assets. In supply chain financing, basic financial statements are used to calculate the net working capital cost and weighted average capital cost of firms in the supply chain.

According to the notation in the balance sheet shown in Table 13.1, to show the lowest working capital effect:

- Cash = $A_{C\text{-}C} = 0$;
- Debt / Total Asset = (γ); Equity / Total Asset = (δ), i.e., ($\gamma + \delta = 1$);

Table 13.1 Balance sheet (Jansen 2016) (developed by the author)

Current assets		A_C	Debt		$D = \gamma * A$
Cash	$A_{C\text{-}C}$		Short Term Debt Overdraft Creditors (AP)	$D_S = \gamma_S * A$	
Account Receivables (AR)	$A_{C\text{-}D}$		Long Term Liability	$D_L = \gamma_L * A$	
Inventory	$A_{C\text{-}I}$				
Fixed assets		A_F	Equity		$E = \delta * A$
Total assets		$A = A_C + A_F$	Total liabilities		$D + E = A$

- Fixed Assets are financed by Long-Term Liabilities or Equity and this is referred to as the "Golden Rule of Finance". Since current assets will be financed by short-term debts, $A_C = D_S$ and Net Working capital will be 0.

The interest rate of working capital:

Expressed in $D_S * T * R_{D\text{-}S}$ or $A_C * T * R_{D\text{-}S}$. Here, T denotes the duration of credit usage, $R_{D\text{-}S}$ represents the interest cost of short-term borrowing. Here, if the duration is accepted as 1 year for display; Weighted average cost of capital (WACC);

$$WACC = \delta * R_E + (\gamma_S * R_{D-S} + \gamma_L * R_{D-L}) * (1 - t) \tag{13.1}$$

In Eq. (1), $R_{D\text{-}S}$ represents the short-term debt cost and $R_{D\text{-}L}$ long-term debt cost. R_E expresses equity cost and can be estimated according to CAPM (Capital Asset Price Model) and Gordon models:

$$R_E = R_F + \beta(R_M - R_F)^1 \tag{13.2}$$

$$\text{According to GordonModel, } R_E = (d1/P_0) + g \tag{13.3}$$

According to the Gordon model in Eq. (3), the expected return of equity consists of dividend yield (d_1 / P_0) and capital gain ($g = (P_1\text{-}P_0) / P_0$). Thus, the expected return of the equity for the future will be estimated in the WACC equation and the weighted average funding cost of the funds that the company will use for the future will be calculated. Economic Value Added is achieved by subtracting the net operating profits (NOPAT) from firms after the tax, by subtracting the sum of assets multiplied by WACC.

$$EVA = NOPAT - WACC * A \tag{13.4}$$

To express the Eq. (4) more clearly,

NOPAT = Revenues − Cost of Goods Sold − Overhead Expenses

Table 13.2 Imaginary data for three Companies (developed by the author)

Company	DSO	Amount
Sales Z Supplier to Y Supplier	60 days	500.000 $
Sales Y Supplier to X Focal Company	45 days	1.000.000 $
Sales X Focal Company to T Clients	30 days	4.000.000 $

$$-\text{FinancialExpenses} - \text{Taxes}.$$
$$\text{WACC} = \delta * R_E + (\gamma_S * R_{D-S} + \gamma_L * R_{D-L}) * (1 - t).$$
$$A = A_C(\text{Cash Policy, CollectionPolicy of Debtors/Accounts Payable}) + A_F(\text{Fixed Asset}). \quad (5)$$

As can be understood from Eq. (4), EVA is also affected by the working capital management of the enterprise. For this reason, for companies to obtain high EVA, it is necessary to increase the turnover rate of stocks and collect their receivables in a short time. Apart from that, they should decrease the costs (WACC) of the funds they use for their activities as much as possible. The effect of working capital on EVA is as shown in Fig. 13.7.

Hofmann et al. observed that firms in different sectors developed an empirical model related to C2C (Cash Cycle), achieving the best EVA and SVA (Shareholder Value Added) value with an optimal C2C cycle. SVA value in the supply chain is obtained by multiplying each company's WACC value. However, the share of each firm in the supply chain from the earnings remains the subject of discussion. Trust and Power in the supply chain can be considered as a pioneering principle for fair distribution of earnings in the long run (Jansen 2016).

Coyle (2003) interpreted Supply Chain Finance as the effect of Supply chain strategies on operating income and EVA. Management of working capital financing in the supply chain focuses on the following:

- Decreasing the stock level,
- Reduce storage costs,
- Reduce shipping costs, and
- Increasing the level of service to customers.

In supply chain finance, the focal company focuses primarily on financing its suppliers' working capital needs. Focal Company can allocate funds that it can obtain with low-interest rates to its suppliers, since it is generally located in a business environment with better financial stability. It is an important advantage for suppliers that are in a high-inflation and high-interest environment to take part in the supply chain and benefit from the credibility of the Focal Company. In the supply chain financing process, the bank deals with the Focal Company, which has a higher credit rating. Because the important thing for the bank in this process is that the loans it has given can be collected. The Role of Banks and the Focal Company in Supply Chain Finance are as shown in Fig. 13.8.

Financial contracts and financial arrangements in the form of debt and equity, facilitated by the Focal Company and used in collaboration by at least two supply

Balance sheet Tier 1 Supplier			
Inventory			
Debtors	$ 50,00	Creditors	

Balance sheet Tier 2 Supplier			
Inventory			
Debtors	$ 80,00	Creditors	$ 50,00

Balance sheet Focal company			
Inventory			
Debtors		Creditors	$ 80,00

Fig. 13.6 Balance sheets of Tier 1, Tier 2, and Focal Company in USD-Area (Jansen 2016)

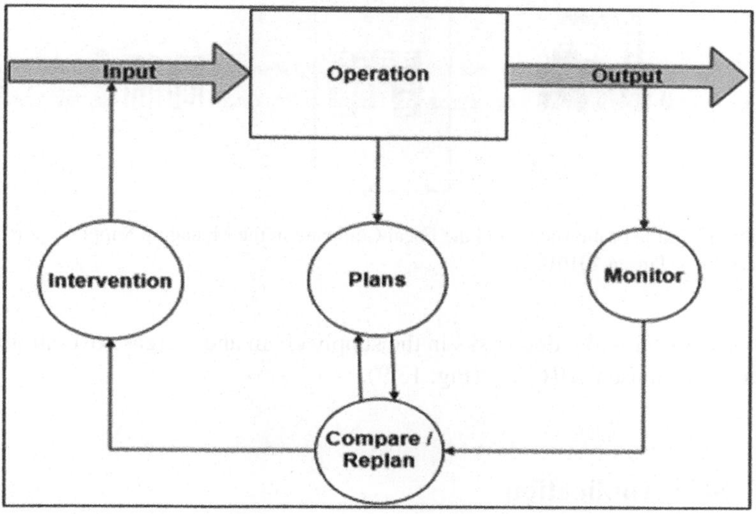

Fig. 13.7 A model of control (Slack 2011; Jansen 2016)

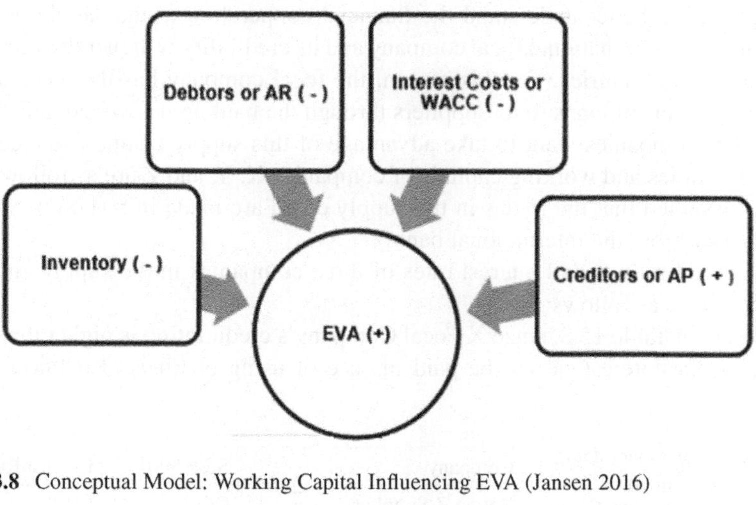

Fig. 13.8 Conceptual Model: Working Capital Influencing EVA (Jansen 2016)

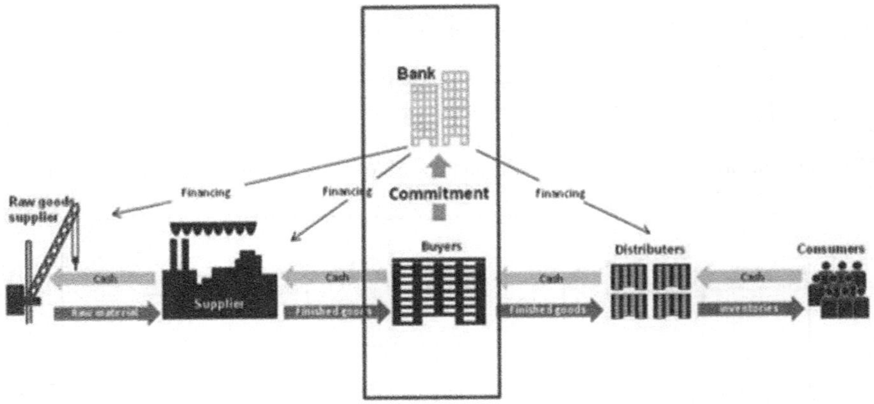

Fig. 13.9 The Role of the Banks and the Focal Company in the Managing Supply Chain Finance (Steeman 2013; Jansen 2016)

chain partners, aim to reduce risks in the supply chain and increase overall financial performance (Jansen 2016, 21) (Fig. 13.9).

13.6 SCF Application

It is assumed that there is a virtual X company (Focal Company) and two (Y and Z) suppliers to better understand the financial cooperation in the supply chain. X company is an international focal company and its credibility is higher than suppliers from different countries. For this reason, the focal company has the advantage of providing financial loans to its suppliers through the bank in its own country. Y and Z supplier companies want to take advantage of this supply chain. The receivable collection times and working capitals of companies X, Y, and Z are as follows:

It is assumed that the trades in the supply chain are made in $ (USD). $ Credit will be used from the international bank.

The credit ratings and interest rates of three companies in the supply chain are assumed to be as follows:

As seen in Table 13.3, since X Focal Company's credit rating is higher than other suppliers, the interest rate to be paid in case of using credit is also lower. Focal

Table 13.3 Imaginary data for the three companies (developed by the author)

Company	S&P 500	Interest Rate
Rating Z Supplier	CCC-	%12
Rating Y Supplier	BBB +	%7
Rating X Focal Company	B +	%4
Mark-up for SCF Facility		%0,5

Table 13.4 Interest Costs without an SCF Instrument (developed by the author)

	Z Supplier	Y Supplier	X Focal Company
Debtors (AR)	$500.000	$ 1.000.000	$4.000.000
DSO	60 days	45 days	30 days
Interest%	%12	%7	%4
Interest Cost	$10.042,63	$8.666,65	$13.171.60

While calculating the interest cost, a year is accepted as 365 days and compound interest was used.

Company can take advantage of its credit convenience to other suppliers. In such a case, the bank demands a mark-up of 0.5% as the risk will increase. Interest costs of X, Y, and Z companies in the supply chain without using SCF instrument are as shown in Table 13.4 and Fig. 13.10.

Interest costs in case X, Y, and Z companies in the supply chain use SCF instrument are as shown in Table 13.5.

As shown in Table 13.5, when the SCF instrument is used in the supply chain, Z and Y supplier companies gain significant advantages, but Focal Company cannot benefit from this situation. To compensate for this, Z and Y must pay the company X pro (focal company) with a total of $ 1,649.40 "pro rato" or the Y supplier company will have a higher sales discount to X (Jansen 2016, 24).

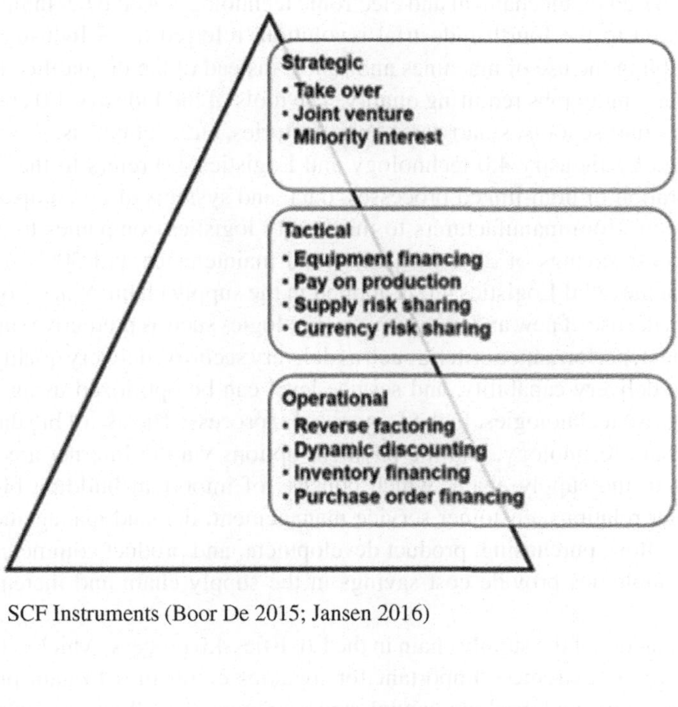

Fig. 13.10 SCF Instruments (Boor De 2015; Jansen 2016)

Table 13.5 Interest Costs with an SCF Instrument (developed by the author)

	Z Supplier	Y Supplier	X Focal Company
Debtors (AR)	$500.000	$ 1.000.000	$4.000.000
DSO	60 days	45 days	30 days
Interest%	%4,50	%4,50	%4,50
Interest Costs with SCF	$3.712,11	$5.563,02	$14.821,00
Interest Costs no SCF	$10.042,63	$8.666,65	$13.171.60
ADVANTAGE	$6.330,52	$3.103,63	-$1649,40

Interest = Focal Company Interest Rate + Mark-up for SCF Facility (%4 + %0,5 = %4,50).

13.7 Discussion

With the industrial revolution that started in the late eighteenth century, human labor has been replaced by machines powered by water and steam. From the beginning of the twentieth century, mass production started on the assembly line where electrical energy was used. As of the 1970s, with the developments in electronic and information technology, machines based on digital technology started to replace machines based on mechanical and electronic technology. At the beginning of 2011, the transition to the fourth industrial revolution, referred to as Industry 4.0, took place, enabling the use of machines and robots instead of the unqualified manpower and specializing in jobs requiring quality. The tools of the Industry 4.0 era are smart hospitals, smart schools, smart cars, smart factories, etc., defined as.

Affected by Industry 4.0 technology and Logistics 4.0 refers to the integration and integration of high-linked processes, data, and systems of all companies in the supply chain, from manufacturers to third-party logistics companies to customers. Potential cost savings of 20% in quality and maintenance and 30% in inventory costs are achieved if Logistics 4.0 is applied in the supply chain. Since Logistics 4.0 focuses on the use of new and innovative technologies such as predictive supply chain management, performance criteria such as delivery security, delivery quality, delivery flexibility, delivery capability, and service level can be optimized using these new and innovative technologies. In the Logistics 4.0 process; The use of big data, the use of blockchain technology, and the payment options via the Internet are extremely important in the supply chain, which consists of important building blocks such as customer relations, customer service management, demand management, order, production flow, purchasing, product development, and product commercialization. These technologies provide cost savings in the supply chain and increase service quality.

The financing of the supply chain in the Logistics 4.0 process, which is the subject of this study, is extremely important for logistics companies to gain profits with low resource cost and working capital, and maximize cash flows and increase firm values. The supplier companies and the focal company in the supply chain utilize the credibility of the focal company to finance their investments at a low cost. Firms in the supply chain also need to reduce their working capital and increase asset turnover (sales/assets) by lowering their stock levels to increase their return on equity (ROE). To reduce the working capital of logistics companies, they need to decrease their stock levels, storage costs, transportation costs, and increase their service levels to customers.

According to the exemplary findings of SCF, if the supplier companies in the supply chain use the SCF instrument, a serious decrease in credit costs is observed. For Focal Company to take advantage of these transactions, the supplier companies in the supply chain must make an additional "pro rato" payment or sale discount.

References

Barreto, L., Amaral, A., & Pereira, T. (2017). Industry 4.0 Implications in Logistics: An Overview, Manufacturing Engineering Society International Conferance, Procedia Manufacturing 13 (2017)1245–1252

Brzozowska, M. (2016)., Industry 4.0- Impact on Logistics Process Management, Ed. Sulkowski, L.,&Spychalska, K., D., Effects of Social Media Networking on the Market Position of Light Industry Enterprises in Lodz Region, Wydawnictwo ss-87–100

Büyüközkan, G. & Güler, M. (2019). A Medhodological Approach for Logistics 4.0 Tecnology Analysis, Journal of Enterpreneurship and Innovation Management-JEIM, 8(1), 21–47

Cosse, M. (2011). Case Studies in Supply Chain Finance.

Fawcett SE, Magnan GM, McCarter MW (2008) Benefits, barriers, and bridges to effective supply chain management. Supply Chain Management: an International Journal 13(1):35–48

Frederick DE (2016) Libraries, Data and The Fourth Industrial Revolution (Data Deluge Column). Library Hi Tech News 33(5):9–12

Gu Y, Dong S (2016) (2016). Logistics Cost Management from the Supply Chain Perspective. J Serv Sci Manag 9:229–232

Hofmann, E. e. (2010). A supply chain-oreinted approach of working capital management. Journal of Business Logistics, 305- 330.

Jansen, H., J. (2016). Supply Chain Finance "Is SCF ready to be applied in SMEs?", International Business Engineering Conference (IBEC), Malasia,

Kagnicioglu, C. H. (2007). Tedarik Zinciri Yönetiminde Tedarikçi Seçimi. Eskişehir: Anadolu Üniversitesi Yayınları.

Kassem, S., S. & Elkader, M., A., A. (2019). Logistics 4.0: Definition and Historical Background. Conferance paper, October, 2019

Kaya S, Turgut M (2019) Blockchain Technology in Supply Chain. The Journal of International Scientific Research 4(2):121–134

Kilic, S. & Alkan, R. M. (2018). Fourth Industrial Revolution Industry 4.0: World and Turkey Reviews, Journal of Research in Entrepreneurship Innovation and Marketing, 2(3):29–49

Lampert, D.M, Stock, J. R. and Ellram, L.M (1998). Fundamentals of Logistics Management, McGraw Hill Company

Lu Y (2017) Industry 4.0: A Survey on Technologies, Applications and Open Research Issues". J Ind Inf Integr 6:1–10. https://doi.org/10.1016/j.jii.2017.04.005

Ostadzadeh, S. Shervin, F. S. and Kambiz B. (2015) An Architectural Model Framework to Improve Digital Ecosystems Interoperability, Elleity and Sobh (eds.) New Trends in Networking,

Computing, Elearning, Systems Sciences, and Engineering, Springer International Publishing, Cham, 513–520. doi:https://doi.org/10.1007/978-3-319-06764-3_65

Sener, S. and Elevli, B. (2017). Endüstri 4.0'da Yeni İş Kolları ve Yüksek Öğrenim (New Businesses and Higher Education in Industry 4.0), Journal of Engineer Brains 1 (2), 25–37.

Steeman, M. (2013a). Supply Chain Finance. In EVO, Logistics Yearbook 2013. Zoetermeer (NL): EVO.

Steeman, M. (2013b). Supply Chain Finance. Train the trainer: Supply Chain Finance (p. NA). Breda: Dinalog.

Strandhagen, O., J., Vallandingham, R., L., Fragapane, G., Strandhagen, W., J., Stangeland, A., B., H.& Sharma, N. (2017). Logistics 4.0 and Emerging Sustainable Business Models, Adv. Manuf. (2017) 5:359–369, Springer

Tekin, M., Etlioglu, M., Tekin, E. (2018). Industry 4.0 and Logistics Trends, International Symposium on Social Human and Administrative Sciences, Asoscongress Conference Proceedings

Wang, K. (2016), Logistic 4.0 Solution: New Challenges and Opportunities, International Workshop of Advanced Manufacturing and Automation (IWAMA, 2016)

Wisner, J. e. (2009). Principles of Supply Chain Management. Mason (USA): South-Western Cengage Learning.

Yilmaz, U. & Duman, B. (2019). An Overview On The Concept of Logistics 4.0 Concept: Development And Change From Past to Present, Bilecik Şeyh Edebali University Journal of Social Sciences Institute, Volume 4: Issue 1, pp.186–200

Chapter 14
The Effects of Industry 4.0 Components on the Tourism Sector

Gülüm Burcu Dalkiran

Abstract The industrial revolution that emerged in Europe in the 1800s has affected all sectors by spreading to different continents with the use of different technologies over time. Developments in technology have enabled the tourism sector to develop by affecting both production and consumption dimensions. In addition to the factors that improve the tourism supply, such as transportation and communication technologies, the increase in commercial activities with the ease of production and supply provided by mass production, and the expansion of the tourism market with the increase of travel, technologies that can be closely followed digitalized tourism market, allowing customized marketing, have revealed the demand for tourism. In this section, industrial revolutions and their effects on the tourism sector are evaluated under the titles of Industry 4.0 components such as big data analytics, augmented reality applications, internet of things and artificial intelligence.

Keywords Tourism technologies · Industry 4.0 components in tourism · Augmented reality in tourism · Big data in tourism · Internet of things in tourism · Artificial intelligence

14.1 Introduction

The concept of industrialization can be defined as the functioning of machines as a partial or complete alternative to human power (Outman and Outman 2003). The distinctions of the industry that emerged as a result of periodic developments also progressed from Industry 1.0 to the Industry 4.0 period. It was not immediately possible for human beings to come to a period when they were engaged in hunting and gathering activities, where economic units could perform their basic activities with robots; each innovation led to other innovations. Another sector that is affected

G. B. Dalkiran (✉)
Department of Tourism and Hotel Management, Vocational School of Social Sciences, Trakya University, Edirne, Turkey
e-mail: burcudalkiran@trakya.edu.tr

İ. İyigün and Ö. F. Görçün (eds.), *Logistics 4.0 and Future of Supply Chains*, Accounting, Finance, Sustainability, Governance & Fraud: Theory and Application,
https://doi.org/10.1007/978-981-16-5644-6_14

by all changes, especially sociological, psychological and economic effects, is the tourism sector.

In the tourism-related literature, the framework of the concept is presented in different ways according to different disciplines. The most important reason for this is that the tourism sector is not only composed of the relationships between the tourist and the tourism business. In addition to food and beverage, accommodation, travel, entertainment and transportation enterprises within the tourism sector, it is involved in input–output relationships with many sectors such as agriculture, textile, health, banking, construction and education, it affects many sectors and is affected by them.

The prominent features in the definitions for tourism in the literature; it is related to the economic dimension of people traveling outside their places where they live for various reasons (travel, entertainment, health, business, sports, etc.) and also spending during travel and accommodation activities. In addition to its economic impact, the sociological dimension covering the interaction of local people and tourism, the psychological dimension on all parties is affected by the behavior of tourists and tourism activities, the public dimension of the state's investment and expenditure policies, and the legal dimension regulating the relations between the parties also stand out in the definitions made by different disciplines. According to the current definition of the World Tourism Organization, tourists are defined as "people who travel in or outside their region for various reasons, participate in social, economic and cultural activities and spend tourism activities by participating in tourism activities" (UNWTO 2014). Developments that highlight the economic dimension of the tourism movement in the historical process are considered as industrial revolutions. The industrial revolution has prepared the necessary infrastructure for the emergence and development of modern tourism and has become an important factor that accelerates tourism movements over time (Bahar 2016).

Technological developments that emerged with the use of steam power in machines also affected the development of tourism, and developments in transportation technology and communication technology also supported this. In addition, the increase in the welfare level created by globalization, the increase in the education level and the legal regulations for holidays, urbanization and population increase have developed the awareness of tourism in people (Akdağ et al. 2019).

In this section, the effects of industrial revolutions on tourism are examined, the changes in the tourism sector in today's technology are exemplified and the concept of Industry 4.0 is evaluated over big data, virtual/augmented reality, internet of things and artificial intelligence. Based on the effects of technological developments on the tourism sector, the affected business processes of different tourism operations and the purchasing processes of tourists were examined under different dimensions, and the tourism of the future was tried to be viewed from the technology framework.

14.2 Reflections of the Industrial Revolutions on the Tourism Sector

Although the main target of technological developments and changes is the industrial sector, which emphasizes the strategic objectives such as gaining competitive power, producing at lower costs, being a pioneer and using resources efficiently, tourism sector is among the sectors most affected by the change. Considering that people have traveled for various reasons since the early ages of history, it is obvious that every change that occurs can affect the travel motivations for the consumer, the service style for the producer and the sectoral investments for the state.

The industrialization process, which started in the 1800s in Europe and is accepted as the beginning of the globalization process, has evolved three times over the years and has affected the whole world with its latest version. Developed countries, positively affected by these processes, have experienced the advantageous effects of globalization as architects, users and close followers of changing technology.

The beginning of the Industry 1.0 era was with the advent of machines powered by water and steam in 1787. In this period when mechanization started, the phenomenon of urbanization also emerged and along with it, radical reforms in all areas of economic, cultural, social and commercial emerged periodically. The use of steam-powered machines for transportation, which started the first revolution in England, was an important step that started tourism movements. As a matter of fact, that period witnessed the first tour organization in the history of tourism. Thomas Cook, a gardener in 1841, became the first person to organize trips for members of the Teetotalers Club between Loughborough and Leicester, England, and over time became a world-renowned travel agency (Enzensberger 1996).

First Industrial Revolution (FIR) or the Industrial Revolution, affecting the period between 1760 and 1830; it has also accelerated the development of the railway by using steam, coal and iron as energy source and raw material. In this way, it became easier to dispatch the increased production with both the movement of people and mechanization to other places. In this period, the socio-economic structure that changed with the mechanization of machine work in England resulted in steam power technology facilitating printing works and communication technology brought cultural change with it. The most important result of all these developments reveals the most basic definition of globalization. According to this definition, globalization can be explained as the free movement of the means of production all over the world without or with the least restriction, in all economic, cultural and political divisions. Thus, industrial revolutions are accepted as the beginning and the most important trigger of this process (Görçün 2016).

The main source of every innovation in the Second Industrial Revolution (SER-Industry 2.0) is the studies of the previous period. The effects of the industrial revolution have manifested themselves in different ways in many countries in different geographies and the implementation part of the innovations coincided with different times (Görçün 2016). It is no coincidence that the First Industrial Revolution is European-centered. During the ancient Roman period, the city of Rome was frequently visited by non-Romans, the use of Roman baths for healing, watching the fights in the Coliseum, and the slave market functioning as a shopping center were the firsts in the history of tourism (Beckerson and Walton 2005). The fact that Rome is an important center of attraction in tourism can be explained by the development of tourism in parallel with industrial civilizations. We can give the best example of this situation with UK tourism. The beginning of the industrial revolution in England and the rapid adaptation of England to the Second Industrial Revolution (SIR) was crowned with the dominant positions of British tourists in the nineteenth century (Enzensberger 1996).

In the emergence of the Second Industrial Revolution, the use of steel, electricity, oil and chemicals along with steam, coal and iron played an important role. While the starting place of the first industrial revolution was England, and the expansion region was primarily Europe, countries such as the USA and Japan were also rapidly affected by the second industrial revolution. When we list some of the most effective inventions of this period and the changes they have created;

- Henry Ford's start of mass production in the automotive industry—Fordist Production (Eraydın 1992).
- Rapid growth of cities.
- Acceleration of industrialization with the use of electrical energy in production by factories.
- Starting the use of tools such as telephones, radios and typewriters.
- With the discovery of internal combustion engines by German engineer Rudolf Diesel, it can be listed as the development of the transportation sector.

The effects of the Second Industrial Revolution innovations on the tourism sector can also be evaluated separately for different types of businesses and activities. The world tour organization of Thomas Cook with the ship "Oceanic" in 1872, the use of steam trains and large ships for tourism purposes and the increase in the welfare of the population formed important markets (Çelik and Topsakal 2019). Not only sea travels but also airline voyages came into play in this period. Regular lines have been established with passenger services on the Paris-Brussels and London-Paris and Berlin-Leipzig routes (Gierczak 2011).

The most fundamental feature of the Third Industry period (Industry 3.0) can be considered as the digitalization of production. With the flexible production systems that emerged with the use of the developments in communication technologies for production (Yüksel and Genç 2018), administrative differences and the contributions of previous period revolutions for serial production, production costs also decreased. In this period, developments in information technologies and the use of the internet affected the way of doing business in all sectors. The use of computers in the 1960s, the use of personal computers in the 1980s, the emergence of the Internet in the 1990s, the emergence of Google's Android operating system phone "HTC Dream" in 2008 and Apple's filling the gap between smartphones and laptops in 2010 by producing tablet computers are the most important technological developments of the Third Industrial Revolution (Topsakal et al. 2018: 1629). The developments of this period resulted in an increase in demand and supply in tourism.

14.3 Industry 4.0. Revolution and Digitized Tourism

Industrial revolutions are not processes that develop suddenly but are the result of the emergence of many innovations that trigger each other over time. The concept of Industry 4.0 is used for the first time in 2011 in Germany's Hannover fair, in the report prepared by technology working groups to the German Government. The core of this concept, which is thought to have started its work in the past by developed countries, lies in the integration of industry and technology. This strategy, which aims to reduce costs, produces more in a shorter time, creates innovation or follows innovations, which are the necessity of competition in the globalizing world economy, with the advancement in technology, has turned into a new version in many sectors. In addition to sectoral designations such as Logistics 4.0, Tourism 4.0, Agriculture 4.0, concepts such as Human 2.0 are also used for people in parallel with business functions and even the use of technology.

Digitalization and technological change are considered as interrelated but distinct concepts that affect the ways of doing business in the new century. Digital transformation is a digitization process and it is accepted as a planning process that enables the transformation of classical codings and processes in business structures, especially efficiency and speed. Digitalization provided by technological advances is a business intelligence revolution and it took place after software and hardware revolutions (Yanık et al. 2019). When digitalization is compared with the concept of industrialization, it is seen as the biggest revolution after the industrial revolution (Bunz 2014), and software in which digitalization comes to life is considered as the main actor of the modern world (Matthewman 2011).

In order for digital transformations to be accepted by the consumer, it is important that the new technology responds to the need or creates the desired changes and provides user satisfaction from different angles. Technology literacy rate is also a concept that affects this situation. As in all industries, the forms of production and consumption that emerged with digitalization and digitalization have been used in tourism. Starting from the market research that the tourist will make before the destination choice, mobile devices, kiosks, social media platforms and artificial intelligence can be used in the booking and accommodation process, travel transactions and even post-travel feedback and benefiting from different personal services. Giving a few examples of the applications used within the scope of Industry 4.0 or Technology 4.0 in the tourism sector;

- Through social media accounts and websites, businesses can produce content and potential tourists can follow this content while making a holiday decision and do market research on social media.
- Tracking the digital footprints of tourists with mobile devices and mobile applications and providing them with appropriate, personalized content.
- Robots can be listed as an element of attraction in the tourism industry, as well as taking part in the business.

Considering that the developments in technology affect not only the tourism sector but also other sectors related to tourism, it is obvious that the expected benefit will increase even more. For example, in the "agriculture" sector, which is the main field of activity of eco-tourism and rural tourism types that supply food and beverage to the tourism sector, basic activities in crop and animal production can be carried out with machines and productivity can be increased with different calculations for the sustainability of natural resources. At the point where technology has reached today, different applications can be seen among the agriculture tourism sectors in terms of stock and inventory management. For example, the decreasing materials, fruits and vegetables in the refrigerators or in the equipment cabinets in the hotel kitchens or food and beverage establishments will be detected (internet of things), these can be collected by robots used in agricultural areas and transferred to the hotel. Hotel customers will also be able to view the images of harvesting from the field simultaneously with the application on their phones.

Although the concept of "authenticity" in tourism or "preserved cultural structure" is among the attractiveness factors that often direct tourism demand, the concepts of authenticity and technology, which seem to be contradictory concepts, have become complementary and helpful in the digital platform. It is possible to reinforce this situation with different examples.

- Digitalization of documents and creation of digital archives are important in terms of protecting cultural heritage and creating tourism demand. In this way, it is possible to draw attention to the cultural heritage of the young population, who is at the top of the list in terms of technology usage level and their interest in digital (EC 2018; Gürsoy 2019).
- It becomes easier to send destination promotional materials to appropriate target audiences in the digital environment and to constantly update texts and images.
- In the field of architecture, with three-dimensional imaging technology, creating reliefs and facades, determining historical and archaeological sites, detecting deformation in buildings, and documenting the preservation and maintenance of cultural assets are among the simplified procedures for the protection of cultural heritage (Yakar et al. 2005).
- With NFC (Near Field Communication) technology, GPS (Geolocation Systems) and Bluetooth technologies, the option of providing special service to the tourist before, after and during the trip (Pamukçu and Tanrısever 2019).
- It is important to preserve, share or use in commercial activities of Visual and Audio works on the digital platform, both in terms of creating different markets and protecting the works.
- Despite the protection measures applied during the exhibition process of the portable historical artifacts, it is possible to digitize them with different techniques and to investigate them in depth, in response to abrasion or destruction in different ways. The project of copying Donatello's sculpture "Magdalena" by using three-dimensional scanning and photogrammetry method together for the first time in Italy is one of the first examples that can be given to this situation (Guidi et al. 2003).

With Industry 4.0, the increase in automation in labor-intensive touristic production brings some risks. One of them is that in the future, the need for human labor will decrease and perhaps some jobs may disappear altogether. The possibility that augmented reality applications such as mobile travel guides, smart menus and smart tables may reduce the need for service personnel and tourist guides primarily show this (Özgüneş and Bozok 2017). In addition, there is the difficulty of solving the problems that arise with technical innovation with traditional methods (Topsakal et al. 2018). In addition, the adaptation process of tourism personnel to new technologies and the economic aspect of this, the security of the systems used and the cost of data protection can be considered among the important risks.

In parallel with the developments in technology, concepts with "smart" prefix have started to be used in the tourism industry. The "smartness" feature, which is attributed to new concepts that can be developed as "smart tourism", "smart destination", "smart hotel", undoubtedly means that the usage rate of new technologies increases customer satisfaction; it addresses the protection of social, cultural and natural inventory and its contribution to sustainability. The use of technology is important in all operational stages, from determining the carrying capacity to the use of digital data obtained from tourists in product planning, and from event planning to input procurement processes of tourism enterprises.

14.3.1 Big Data in Tourism

Innovation is considered to be one of the most value-added ways to gain competitive advantage on a global, national or local scale. The basis of innovation is producing information or having the information produced (Eraslan et al. 2008). The amount of information created in daily life has increased enormously compared to previous periods. Every data obtained from different sources such as social media shares, videos, photo archives, every transaction made on the internet and phones, log files that we constantly record, is very valuable for businesses. However, information gains economic value by processing these data and making them meaningful.

In the simplest way, big data is defined as "large data sets that cannot be managed by traditional tools" (Ohlhorst 2014). In another definition, big data appears as "large data sets that are insufficient in analyzing relational databases, but can be analyzed with advanced statistical methods or interpreted with visualization programs" (Jacobs 2009). The scale of the data highlighted in the definitions is very different from what we use in daily life; petabytes (one terabyte capacity external disk we use today), exabytes (million times), zetabyte (billion times) and yotabyte (trillion times) (Arıkan 2016). The concept, which was first thought to be used by the science of genetics and astronomy (Mayer-Schöberger and Cukier, 2014), has become important for many different fields for similar reasons.

Big data is used in the tourism sector for various purposes such as improving service quality, revealing the tourist profile, regulating city traffic and demand forecasting. While the d-Lab named data analysis system in Barcelona, which is among the most visited cities in the world, aims to reveal the density of tourists in their travel routes by revealing the tourist profiles (Esen and Türkay 2017), Pan and Yang (2014) used the website visit traffic, search engine reports and weather reports made demand forecasting to determine hotel occupancy rates in the short term. In many studies, social media interpretations and customer feedback have been analyzed, and they have made predictions to reveal destination visitor profiles and the intention to visit again (Gong et al. 2016).

From the point of view of the tourism sector, differentiated data to be obtained with big data can be savior in cases where the data obtained from consumers by traditional methods are insufficient to see the whole and to detect changes. Mobile phone data, data on business resources, special data such as radio frequency tags (RFID) in customer transactions, and digital traces of data obtained from other units with which all users interact, reveal a complex and large-capacity data structure (Esen and Türkay 2017). According to calculations, the amount of recorded data produced by all humanity from the beginning of civilization until 2003 is produced in 2 days in time (İHS Telekom 2015). How this situation, which is also called as information explosion, came to be seen in Fig. 14.1 (Arıkan 2016). The emergence of data with the invention of writing and printing, creation of data with information technologies, archiving of data in various ways on different platforms through internet and digitalization are among the most important stages. The fact that not only humans but also machines enter the data generation process (internet of things) has enabled us to reach big data.

By integrating big data with the data sets of all private, public and autonomous organizations, it is possible to reveal hidden information and new correlations by using data mining and statistical methods. Big data in tourism consists mostly of unstructured data obtained from internet searches, location determinations, website traffic, social media shares, call center records and sales reservation transactions made by tourists from the devices they use (Xiang and Fesenmaier 2016: 18).

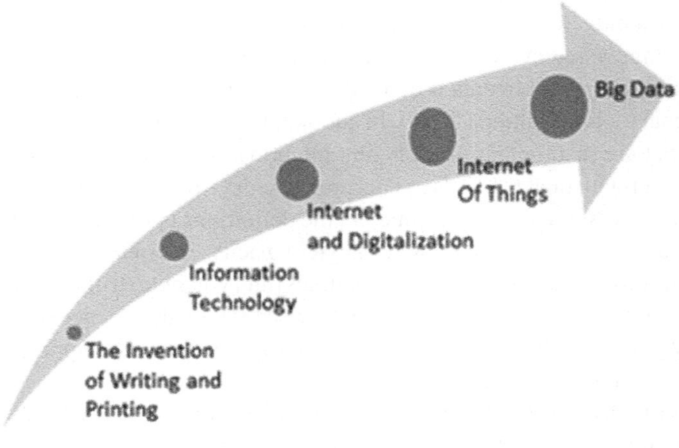

Fig. 14.1 Big Data (Information Explosion) Process (Arıkan 2016)

Big Data Analytics provides the most efficient use of informatics and marketing information from the perspective of tourism marketing. In addition to detecting the current situation in the tourism market, it can detect unknown relationships and orientations in the situation. In addition, it can detect algorithms that will predict situations that do not yet exist but may arise in the future. Big data analytics can have two effects on the tourism market as a marketing tool: Pattern recognition and Trend Analysis (Arıkan 2016). In this way, it can be said that concepts such as smart tourism and smart destinations can deserve their "smart" qualities in terms of clarifying today and planning tomorrow. Considering in terms of foreign markets, with the data to be used for the tourism demands of each target market, tourism marketing on the basis of destinations will be more successful, and even with the niche markets created, foreign active tourism income can be increased. However, it is very important to know what data and how to access them in order to be able to analyze the data and activate the analysis results in decision-making mechanisms in order to reveal the benefits of big data analytics. Bearing the costs of all these activities is also an issue to consider.

The main features of the concept of big data, which are named as dimensions, include five features called "5 V Model" (Arıkan 2016);

- *Volume*: The size of the data
- *Variety*: The structural features of the data (structured, semi-structured and unstructured data).
- *Speed(Velocity)*: The speed at which data can be generated and processed in real time.
- *Variability*: Whether the data can be used functionally or not.
- *Reality (Veracity)*: The accuracy and consistency of the data.

14.3.2 Virtual Reality—Augmented Reality Applications in Tourism

In parallel with the developments in information technology in recent years, one of the remarkable applications in many areas is the Augmented Reality (AR) applications. The concept of reality can be examined in two separate ways as "virtual reality" and "augmented reality". İçten and Bal (2017; 111) in their studies used the concept of virtual reality "the environments where the person's relationship with the outside world is broken when he enters the computer environment" and they defined the concept of augmented reality as "environments where the connection with the real world is not broken, where data and images can be added on real world images, where virtual and reality are together".

AR application, which was first used in the military field, has been used in many different areas over time. It is aimed at its first use that the application, which is placed on a helmet to be used by fighter pilots designed as transparent screens at eye level during the flight, they could see the data screen and the real image simultaneously (Livingston, Rosenblum, Brown, Schmidt, Julier, Baillot, Swan, Maassel, 2011).

In virtual reality applications, Trindade et al. (2002) mention three components: Immersion, Interaction and participation. Immersion is the creation of the perception of being physically present in a non-physical environment, and interaction is the adaptation of the limbs to the environment to increase the credibility in the virtual environment (Carrozzino and Bergamasco 2010). HMD; while it is defined as helmets that allow the perception of computer-generated images and sounds, Wired Glove is glove that provides the sense of touch (Orhan 2019). In the participation component of virtual reality, it is aimed to enrich the experience by focusing on the participant's learning.

The development of virtual reality technology is used by the private sector and the state as a promotional tool in tourism. In this way, it is possible to provide a preliminary experience to potential tourists in the domestic and foreign markets, to promote to larger masses at lower costs, and to promote destinations without the political negative directions that may be caused by international relations. Apart from this, it can provide the opportunity to become a "virtual tourist" in the disadvantaged segments of the domestic and foreign markets with travel barriers. They can do this application within the scope of social responsibility by considering social benefit or they can support it for a fee. In response to the criticism that virtual reality applications cannot connect with the real world, augmented reality applications come into play. In the near future, it is thought that tourists will be able to experience new reality applications such as stroking animal skin, contacting a fish or hugging a tiger (Dubey 2016). It is even possible to meet many extinct creatures today. Dinosaur simulations in the Disneyland entertainment center (Newsome and Hughes 2016) are one of these applications. The areas in which augmented reality applications can be used in the tourism sector are listed below.

- AR applications for different destinations that can be used within the scope of accessible tourism.
- Museum, playground or beach-sea themed applications for disadvantaged segments (children in government care or low socio-economic areas) within the scope of social tourism.
- AR applications to be made with contents prepared for different sports branches within the scope of sports tourism.
- AR applications that can support the treatment that can be prepared for psychiatric patients within the scope of health tourism.
- AR applications are prepared with content produced from destinations or events suitable for the target market in promotional activities to be carried out in distant markets in foreign tourism.
- Visiting museums with AR applications within the scope of cultural tourism list the usage areas of virtual reality technologies in the tourism industry as follows.
- Observing the situation with virtual simulations prepared during the preparation of tourism plans and creating tourism policies and to serve the purpose of sustainability by accurately determining the carrying capacity.

- Increasing sales by providing preliminary experience with simulations based on virtual reality applications of tourist attractions to be marketed to potential tourists.
- High-cost theme parks, which need to be built on large areas, can be established in smaller areas and at lower costs with virtual reality technology.
- The ability to reveal virtual/artificial tourism in cases where travel cannot be made due to adverse weather conditions or extraordinary situations (war, pandemic, etc.).

In accommodation businesses, especially chain hotels benefit from virtual reality to increase their brand value, and offer new services to their customers through websites and advanced technological equipment. The first hotel to use these technologies is Marriott Hotels. This international hotel chain cooperated with Samsung Electronics America, allowing their customers to take a virtual tour with the program called "VRoom Service" before they sell rooms. The hotel that provides personnel training with virtual reality applications is the Best Western Hotels group (Orhan 2019). In addition to hotel businesses, travel businesses also aim to promote the destinations in their portfolios by using new technologies and to increase the level of consumer satisfaction after pre-experience.

14.3.3 The Concept of the Internet of Things and Its Effect on Tourism

The Internet of Things (IoT) means that everything is connected and interacted with each other via the Internet, without the limitation of time, space and existence. Objects that interact with each other are Radio Frequency Identification Tags (RFID), sensors, processors, chips, mobile phones and other mobile devices (Çelik and Topsakal 2019; 25). The Internet of Things application, which plays an important role in the formation of big data, is accepted as the basic network in touristic production and is seen as an indispensable part of "smart tourism" (Wu 2017).

In terms of tourism enterprises, internet of things technology can be used for purposes such as analyzing consumer behavior, creating suitable products for the customer and providing automation and control. For example, it is possible that chips embedded in wristbands that a hotel will give to its customers can detect the number of entrances to the customer's room, the number of coffee they drink, and even the times when they sleep and wake up. Similarly, in congress organizations, the data obtained can be used to access the location information of the participants with their name badges or to ensure the coordination and security of tourists in major tours within the scope of faith tourism. All these applications are obtained with RFID (Radio Frequency Identification) system. It is the name of the technology that enables the electromagnetic waves emitted from RFID devices to be converted into digital data (Weis 2007). Among the tools that collect data in this system are barcode tapes, optical readers and smart cards with memory cards and processors, sticks (Esen and Türkay 2017).

Table 14.1 Examples of Applying Internet of Things in Hotels (Buhalis and Leung 2018: 48)

Area	Technology/Sensor	Function
Inside the hotel		
Rooms	Motion sensor Voice sensor Temperature sensor Door lock Wearable sensor	Providing ambience and energy saving by illuminating the guest's location in the room Sound control of tools such as curtains, lighting and room temperature Controlling the temperature in providing guest comfort Opening and closing doors with mobile applications Determining the health status of the guests (in areas such as sauna and fitness room)
Restaurant/lobby	Position sensor Position beacon	Determining the location of the tourist Sending a welcome message to the determined tourist
Hotel amenities	Availability Beacon	Send guests an availability message
Store	Inventory tag	Tracking the location of the inventory, determining the expiration date and stock level
Out of hotel		
Building	Temperature sensor Light sensor	Ensuring energy efficiency by controlling building temperature Lighting according to daylight
Road	Traffic sensor	Parking and traffic control
Social network	Content sensor	Social network system that provides feedback to managers by controlling the hotel's social networks

Examples of applications that take the Internet of Things in hotel businesses can be examined in two parts as applications inside and outside the hotel (Table 14.1). Applications inside the hotel are mostly for optimal use of heat, light and sound systems through sensors. In addition, with Beacon technologies compatible with mobile devices, sending greetings to customers and messages for facility use also provides customer-specific service. Providing customer feedback by providing energy efficiency of hotels, monitoring inventory and reporting on content in social networks are also important examples (Buhalis and Leung 2018).

14.3.4 Artificial Intelligence Applications in Tourism Enterprises

Artificial intelligence applications in tourism enterprises can also be effective in the employment structure in the sector in recent years. The concept of artificial intelligence, which supports human activities and can also be an alternative to human workforce in the future, means the imitation or revitalization of human intelligence by machines or information systems (Yıldız 2019). The artificial intelligence technology market, which has grown by 44% in 2019 compared to the previous year and reached 35.8 billion dollars, is one of the most important products of the rapidly developing technology (Sedefçi 2018). Artificial intelligence and robots are used in many different areas in the tourism sector. In the food and beverage sector, it is possible for bar robots to prepare the beverage that the customer wants by placing an order via the phone (Imbardelli 2019) or to protect the privacy of the customer room by performing room service in hotels with robots (AL-Enterprise 2018). It is possible for hotels to make customer registration (check-in/check-out) without human beings, to use multilingual voice recognition systems for security purposes, to reduce waiting times in different services in hotels with facial recognition systems (Akgül 2019; Ivanov and Webster 2017; Tung and Law 2017). It can be said that artificial intelligence applications in the tourism sector are mostly Asian countries, which are considered to be the origin of high value-added technology products. Some of the examples that can be given are listed below.

- "Flyzoo Hotel" owned by the e-commerce giant Alibaba Group in China is one of the important examples that integrates technology with tourism. The robots that were established for trial purposes and their technology licenses are planned to be sold to other businesses in time, work at the hotel, especially using the face recognition system. Starting from meeting the customers; it is thought that robots, which will be responsible for providing many services such as room lighting, air conditioning, food and beverage service, will be effective in providing customer-specific service concept (Brennan 2019; Yıldız 2019).
- Operating in the airline industry, Amsterdam-based KLM Airlines operate with the artificial intelligence robot "Spencer" for the purposes of answering customer questions and increasing the quality of travel (Akgül 2019). Similarly, the robot "Sanbot" working in Chinese airports guides tourists by providing information on issues such as gate, platform, flight time (Block 2017).
- The robot "Sara", operating in Singapore, works like a digital tour guide, helping tourists with directions, showing the city's historical and cultural attractions, and restaurant recommendations (Kim and Banchs, 2014).
- All employees in the hotel chain called Henn-na operating in Japan, except the security personnel, consist of robots. The most basic feature of the robots used in the entrance and exit operations of the hotel, such as guidance, cleaning, food and beverage service, is that they look like real people (Dursun 2020; Henn-na 2020).

14.4 Conclusion

The industrial revolution that started with the use of steam power has revealed the phenomenon of globalization in the world and increased the level of competition over time. With the provision of mass production, the input procurement process in tourism enterprises has become easier, and the concept of mass tourism has emerged with the developments in transportation and communication technology. The concept of "smart tourism" has emerged with "smart destinations" using new technologies to create sustainable tourism policies by keeping up with technological revolutions. With the third industrial revolution, the use of the internet and mobile phones created the consumer model that can be defined as "smart tourist". All technological revolutions trigger change in businesses, which are the main components of the tourism sector, and all other sectors related to tourism.

The "digital revolution" that emerged with the new generation technology tools called Industry 4.0 also significantly affects the tourism sector. With its components such as augmented reality, internet of things, artificial intelligence applications and big data, it is possible to create a service difference in tourism as in other sectors, to reduce costs in certain areas or to activate the control mechanism. With big data, it is easy to recognize target markets; it is possible to create new markets in different tourism types and in this way spread the tourism activities to all four seasons. In the digitalized tourism market, with the facilitation of introducing tourist attractions to the global market, reaching potential tourists at a lower cost and determining the factors affecting the demand in a short time, the tourism market is growing.

References

Akdağ G, Güler O, ve Çakıcı C (2019) Turizme Genel Bakış, (ed) C Çakıcı, *Genel Turizm* içinde,Detay Yay.,2019, 1. Baskı, Ankara

Akgül-Özdemir S (2019) Turizm 4.0 ve Yapay Zeka Uygulamaları, (ed) Akgül-Özdemir S, ve Atar A, *Turizm 4.0.* içinde, 1. Baskı, Detay Yayıncılık, Ankara

AL-Enterprise (2018) The internet of things fort the hospitality industry. https://www.al-enterprise.com/-/media/assets/internet/documents/iot-for-hospitality-solutionbrief-en.pdf

Arıkan A (2016) Pazarlamada Yeni Bir Evren: Büyük Veri, (ed) AC Kırgız, *Ortaya Karışık Pazarlama Tarifleri* içinde, 1. Baskı, Beta Yay., İstanbul

Bahar O (2016) Turizm ve Ekonomi, (ed) Çakır, P.G. ve Esen, E., *Turizm Ekonomisi* içinde, Anadolu Üniversitesi Açıköğretim Fakültesi Yayını, No: 2267

Beckerson J, Walton JK (2005) Selling air: marketing the intangible at British resorts. In: Walton JK (ed) Histories of tourism represantation, identity and conflict, Channel View Publications, Clevedon

Block D (2017) Yakında Sevimli Robotlar Hizmet Sektörüne Giriyor. https://www.amerikaninsesi.com/a/yakinda-sevimli-robotlar-hizmet-sektorune-giriyor/3727747.html

Brennan T (2019) Introducing Alibaba's flyzoo future Hotel Alizila. https://www.alizila.com/introducing-alibabas-flyzoo-future-hotel/

Buhalis D, Leung R (2018) Smart hospitality interconnectivity and interoperability towards an ecosystem. Int J Hosp Manag 71:41–50

Bunz M (2014) The silent revolution: how digitalization transforms knowledge, work, journalism and politics without making too much noise. Palgrave Macmillan, New York

Carrozzino M, Bergamasco M (2010) Beyond virtual museums: experiencing immersive virtual reality in real museums. J Cult Herit 11:452–458

Cheung C, Tsang IT, Wong KH (2017) Robot avatar: a virtual tourism robot for people with disabilities. Int J Comput Theor Eng Singap 9(3):229–234

Çelik P, Topsakal Y (2019) *Endüstri 4.0 ve Akıllı Turizm*, Deyay Yayınları, 1. Baskı, Ankara

Dubey AK (2016) Future technology and service industry: a case study of travel and tourism industry. Glob J Enterp Inf Syst 8(3):65–69

Dursun MT (2020) Endüstri 4.0 ve Turizmde Yapay Zeka Uygulamaları, (ed.) Sert, A.N., Karacaoğlu, S. ve Karacaer, S.S., *Turizmde Değişim, Dönüşüm ve Güncel Eğilimler* içinde, Detay Yayıncılık, 1. Baskı, Ankara

Enzensberger HM (1996) A theory of Tourism New German Critique. Spec Iss Lit **68** 117–145

Eraslan H, Bulu M, Bakan İ (2008) Kümelenmeler ve İnovasyona Etkisi: Türk Turizm Sektöründe Uygulamalar, *Seyahat ve Otel İşletmeciliği Dergisi*, Cilt 5. Sayı 3:15–29

Eraydın A (1992) *Post-Fordizm ve Değişen Mekânsal Öncelikler*, Ortadoğu Teknik Üniversitesi Mimarlık Fakültesi Yayınları

Esen MF, Türkay B (2017) Turizm Endüstrilerinde Büyük Veri Kullanımı. J Tourism and Gastronomy Stud 5(4):92–115

European Comission (2018) Digital solutions in the field of cultural heritage. https://www.interregeurope.eu/fileadmin/user_upload/plp_uploads/policy_briefs/2018-08-06_Policy_brief__on_digital_technologies.pdf

Gierczak B (2011) The history of tourist transport after the modern industrial revolution. Pol J Sport Tourism 18(4):275–281

Gong L, Liu X, Wu L, Liu Y (2016) Inferring trip purposes and uncovering travel patterns from taxi trajectory data. Cartogr Geogr Inf Sci 43(2):103–114

Görçün ÖF (2016) *Dördüncü Endüstri Devrimi Endüstri 4.0*, Beta Yayıncılık, 1. Baskı, İstanbul

Guidi G, Beraldin JA, Ciofi A, Atzeni C (2003) 3D acquisition of Donatello's Maddalena: protocols, good practives and benchmarking. Electron Imaging Visial Arts Conf EVA 174–178

Henn-Na Otels Group (2020) https://www.glamping-holiday.com/hotel-japan-all-hennna.html?gclid=Cj0KCQjwwuD7BRDBARIsAK_5YhXKPgkIcB_nDlNetYiwf0WTTssoJk9r5PSkgfIKu-FMYhJ4DD5fkGAaAmTmEALw_wcB

Imbardelli AP (2019) How smart rooms can transform hotel brands. Hotel Management. https://www.hotelmanagement.net/tech/how-smart-rooms-can-transform-hotel-brands

Ivanov S, Webster C (2017) Adoption of robots, artificial intelligence and servis automation by travel, tourism and hospitality companies – a cost- benefit analysis. International Scientific Conference Contemporary Tourism-Traditions and Innovations, Sofia University

İçten T, Bal G (2017) Arttırılmış Gerçeklik Üzerine Son Gelişmelerin ve Uygulamaların İncelenmesi, *Gazi Üniversitesi Fen Bilimleri Dergisi* Part C: Tasarım ve Teknoloji, Cilt 5, Sayı 2, ss. 111–140

İHS Telekom (2015). Büyük veri hakkında 10 ilginç bilgi, URL: https://www.ihs.com.tr/blog/buyuk-veri-hakkinda-10-ilginc-bilgi/ Erişim: 15.09.2020.

Jacobs A (2009) The Pathologies of big data. Commun ACM 52(8):38–44

Kim S, Banchs RE (2014) R-cube: a dialogue agent for restaurant recommendation and reservation, Annual Summit and Conference (APSİPA), Asia-Pasific Signal and Information Processing Association, Siem Reap, pp 1–6

Livingston MA, Rosenblum LJ, Brown DG, Schmidt GS, Julier S, Baillot Y, Swan JE, Ai Z, Maassel P (2011) User interface design for military ar applications. Virtual Reality (UK)

Matthewman S (2011) Technology and social theory. PalgraveMacmillan, New York

Mayer Schöberger V, Cukier K (2014) Big data: a revolution that will transform how we live, work and think. First Mariner, New York

Newsome D, Hughes M (2016) Jurassic world as a contemporary wildlife park allegory. J Curr Issues Tourism 1–9

Ohlhorst F (2014) Big data analytics: turning big data into big Money. John Wiley&Sons, New Jersey

Orhan BK (2019) Turizmde Yeni Bir Akım: Sanal Gerçeklik, (ed) Sezgin, M., Özdemir Akgül, S. ve Atar, A., *Turizm 4.0 Dijital Dönüşüm* içinde, Deyay Yayıncılık, 1. Baskı, Ankara

Outman JL, Outman EM (2003) Industrial revolution: almanac, Thomson Learning, Farming Hills

Özgüneş RE, Bozok, D (2017) Turizm Sektörünün Sanal Rakibi (Mi?): Arttırılmış Gerçeklik, *Uluslararası Türk Dünyası Turizm Araştırmaları Dergisi*, Cilt 2. Sayı 2(146):160

Pamukçu H, Tanrısever C (2019) Turizm Endüstrisinde Dijital Dönüşüm. Sezgin, M. Özdemir Akgül, S. ve Atar, A. (eds), *Turizm 4.0 Dijital Dönüşüm* içinde, Detay Yayıncılık, Ankara

Pan B, Yang Y (2014) Forecasting destination weekly hotel occupancy with big data, J Travel Res
Sedefçi K (2018) Endüstri 4.0 bakış açısıyla nesnelerin interneti ve müşteri deneyimi açısından incelenmesi, Marmara Üniversitesi, Sosyal Bilimler Enstitüsü İşletme Anabilim Dalı Yüksek Lisans Tezi
Topsakal Y, Yüzbaşıoğlu N, Çuhadar M (2018) Endüstri Devrimleri ve Turizm: Türkiye Turizm 4.0. Swot Analizi ve Geçiş Süreci Önerileri, *Süleyman Demirel Üniversitesi İktisadi ve İdari Bilimler Fakültesi Dergisi*, C.23, Endüstri 4.0 ve Örgütsel Değişim Özel Sayısı, s. 1623–1638
Trindade J, Fiollas C, Almeida L (2002) Science Learning in virtual environments: a descriptive study. Br J Edu Technol 33(4):1–18
Tung VW, Law R (2017) The potential for tourism and hospitality experience research in human-robot interactions. Int J Contemp Hosp Manag 29(10):2498–2514
UNWTO (2014) Understanding tourism: basic glossary. https://www.unwto.org/glossary-tourism-terms
Weis S (2007) Rfid (radio frequency identification): principles and applications. System 2(3):1–23
Wu X (2017) Smart tourism based on internet of things. Revista de la Facultad de Ingeniaria U.C.V. 32(10) 166–170
Xiang Z, Fesenmaier DR (2016) Analytics in smart tourism design: concepts and methods. Springer
Yakar M, Yıldız F, Yılmaz HM (2005) Tarihi ve kültürel mirasların belgelenmesinde jeodezi fotogrametri mühendislerinin rolü, TMMOB 10. Türkiye Harita Bilimsel ve Teknik Kurultayı, 28 Mart- 1Nisan, Ankara. https://www.hkmo.org.tr/resimler/ekler/54KT_148_ek.pdf
Yanık A, Atas U, Batu M (2019) Dijital Turizm: Turizm ve Pazarlamada Dijital Dönüşüm, International Congress on Digital Transformation in Tourism I-DIGIT-19, Düzce. https://www.researchgate.net/publication/338403520_DIJITAL_TURIZM_TURIZM_VE_PAZARLAMADA_DIJITAL_DONUSUM
Yıldız Ö (2019) Nesnelerin İnterneti, Yapay Zeka ve Turizm, (ed) Kurgun, H ve Kurgun, O.A., *Turizm 4.0 Kavramlar ve Uygulamalar* içinde, 1. Baskı, Detay Yayıncılık, Ankara
Yüksel M, Genç KY (2018) Endüstri 4.0 ve Liderlik, II. International Symposium on Innovative Approaches in Scientific Studies, November 30- December 2, Samsun, Turkey. SETSCI Conference Indexing System

Gülüm Burcu Dalkiran Asst. Prof. Dr. Gülüm Burcu Dalkiran completed her undergraduate studies in Trakya University Agricultural Economics, Anadolu University Economics and Hospitality Management departments, respectively. She completed her master's degree in Trakya University Department of Economics and her doctorate in Agricultural Economics department. Dalkiran, who has been doing academic studies on sustainability in tourism, women's employment in the tourism sector, women's cooperatives and tourism economy in her graduate theses and academic studies, started to work as a lecturer at Trakya University Edirne Social Sciences Vocational School in 2006. Since 2010, she has been working as the Head of the Hotel, Restaurant and Catering Services Department.

Chapter 15
The Problem of Employment and Growth in the Fourth Industrial Revolution

Duygu Yücel

Abstract Having been the products of arising needs, industrial revolutions led to many innovations in economic and social life during their respective periods. The Fourth Industrial Revolution has started the digital age we are in, thanks to the development of the Internet of Things, the Internet of Services, Cyber-Physical Systems, Robots, Big Data, and Cloud Computing Systems. The technological development of automation and information systems in production paves the way for certain changes in macro dimensions such as economic growth and employment. Digital advances in the Logistics sector, which cannot be considered independent from production, on the one hand provided increased productivity in the sector, yet, on the other hand, gave a rise to an unemployment problem for the labor sector. The unfavorable landscape that the Fourth Industrial Revolution initially created regarding employment can be overcome by lifelong learning, higher education, and the supply of a qualified and competent workforce. First of all, it has been the human capital itself that triggered technological developments. In the present study, the first three industrial revolutions are briefly discussed in terms of their economic effects and the Fourth Industrial Revolution is examined theoretically. The effects of digitalization on economic growth and employment, and especially the changes induced thereby in the logistics industry are mentioned. The fact that technological developments are inversely associated with employment at the beginning, is examined as a disadvantage along with the many facilities it introduces.

Keywords Industrial revolution · Technological development · Employment · Economic growth · Logistics

---

D. Yücel (✉)
Department of Accounting, and Taxation, Vocational School of Social Sciences, Trakya University, Edirne, Turkey
e-mail: duyguyucel@trakya.edu.tr

İ. İyigün and Ö. F. Görçün (eds.), *Logistics 4.0 and Future of Supply Chains*, Accounting, Finance, Sustainability, Governance & Fraud: Theory and Application,
https://doi.org/10.1007/978-981-16-5644-6_15

15.1 Introduction

Industrial revolutions from past to present have been a product of arising human needs. The recent Fourth Industrial Revolution was also formed with the incorporation of the innovations required by the age into economic and social life. The Fourth Industrial Revolution, which will radically change the production process, employment structure, personal qualifications, and competencies today and in the future, is based on such advanced technological components as Cyber-Physical Systems, Internet of Things, Internet of Services, Big Data, Smart Factories, Robotics, Artificial Intelligence, and Cloud. Intelligence, communication, and information network come to the fore in this period, also known as the age of digitalization.

The Fourth Industrial Revolution, which enables low cost, flexible, and customized production, also has many effects on macroeconomic events such as employment and economic growth. As in every industrial revolution, the Fourth Industrial Revolution will initially lead to a shrinkage in employment and lost professions. However, the system will continue to create lines of business tailored to its needs. In this new process, the most important element for the labor life is that the demand for the workforce will turn towards a skilled workforce with high education. Therefore, the importance of education and lifelong learning will continue to increase each day. The Fourth Industrial Revolution, which enables lower cost and faster production of goods and services thanks to its smart production systems, will also create economic growth for countries.

With the digital age affecting many economic sectors, it has been possible to minimize failures and increase productivity in the "just in time" logistics service by minimizing costs; thanks to the development of new fuel and energy resources, robotics and artificial intelligence technology, and advances in communication techniques. Cost reductions in the logistics sector have also increased investments in the sector. From an employment point of view, the logistics sector is expected to suffer the same fate. The workforce providing logistics services will have to face the problem of unemployment in case of a lack of required qualifications. However, the new process will create new professions such as design, coordination, machine advocacy, and will continue to provide new job opportunities to the qualified, competent workforce.

15.2 Evolution of the Industrial Revolution from the Past to the Present

Human beings so far have led to many different transformations in economic, social, and political order as a result of their inventions and revolutions. These transformations have had an impact on economic functioning in particular and led to a series of industrial revolutions. The industry is defined as the entirety of the collective production activities, methods, and tools applied in the form of the processing of

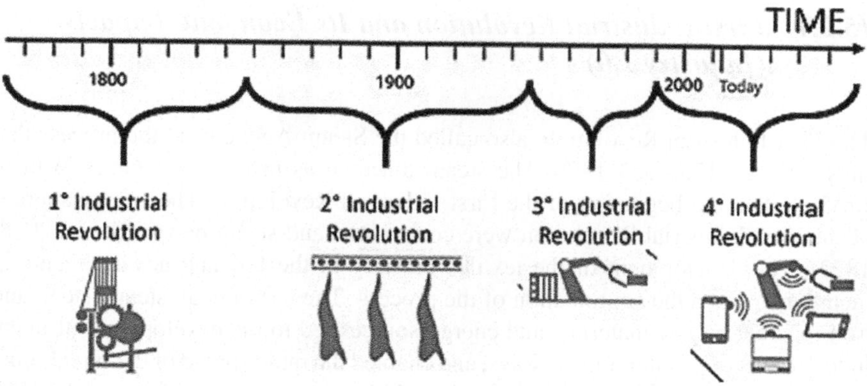

Fig. 15.1 Evolution of the industrial revolution from industry 1.0 to industry 4.0. *Source* Zambon et al. (2019: 4)

raw materials and manufactured and semi-finished products using machinery, and the creation of energy resources. Rapid, radical, and qualified changes in a particular field are called revolutions (TDK). The industrial revolution is expressed as a radical and rapid change in the production activities, methods, and presentations used to process raw materials, manufactured and semi-finished goods into the final commodity (Sözen and Mescioğlu 2019: 291). The concept of the industrial revolution was first used in a letter written by Louis Guillaume Otto of France in 1799 (Sözen and Mescioğlu 2019: 291). This term refers to a change in technological, economic, and social systems within the industry. The industrial revolution focuses on changes in working and living conditions and economic wealth (Fig. 15.1). The first industrial movement in the UK began in the middle of the eighteenth century, and the US and European countries started to transform their respective agricultural societies into an industrial society (Dombrowski and Wagner 2014: 100).

As regards Industry 1.0, the invention of machines powered by water and steam power started the first industrial revolution. The second industrial revolution began with the start of mass production with machines powered by electric and combustion engines. During this period, the first examples of assembly lines were also seen. In Industry 3.0, automation was introduced to manufacturing processes with developments in the fields of electronics, information technologies, and industrial robots, which paved the way for the beginning of the information age. Industry 4.0 started with the introduction of the digital supply chain, intelligent manufacturing, digital products, services, and business models, along with data analysis activities as a basic competence. Flexible and interconnected value chain networks, virtual processes, virtual customer interfaces, and industrial cooperation activities as the key value factor will be established in 2030 and beyond, a period called as a digital ecosystem (Yilmaz and Duman 2019: 188). The economic effects of industrial revolutions are examined in a more detailed approach below.

15.2.1 First Industrial Revolution and Its Economic Impacts (Industry 1.0)

The First Industrial Revolution, also called the Steam Age, covers the process that runs between 1750 and 1890. The steam machine as invented by James Watt in 1765 marked the beginning of the First Industrial Revolution. The energy sources of the First Industrial Revolution were coal, water, and steam power (Stone, 2018: 1822). The development of the textiles industry in the UK and advancements in metallurgy drove the continuation of the process. The use of coal, steam, steel, and iron together as raw materials and energy sources led to the development of many industrial sectors, including railways, and enabled the rapid spread of the First Industrial Revolution throughout Europe (Gürün 2019: 70; Bulut and Akcacı 2007: 52). The Bessemer Method in use made steel production cheaper, making railway tracks cheaper. The newly established rail networks increased trade by providing convenience in transportation and shipments. Steamships and trains increased the volume of foreign trade by allowing goods to be transported to more distant destinations. Another development during this period was the provision of communication facilities such as telegraphs and telephones (MÜSİAD 2017: 32). The most important economic result of the First Industrial Revolution was the suppression and beginning of the extinction of old crafts and production methods. Unemployed artisans began to turn to factories, and as a result, a capital-intensive production mode replaced the labor-intensive production mode. The new production style challenged the aristocratic structure that upon the emergence of new industrialists and new traders (Sözen and Mescioğlu 2019: 293).

15.2.2 Second Industrial Revolution and Its Economic Impacts (Industry 2.0)

The Second Industrial Revolution covers a period that spanned from the 1870s, the last quarter of the nineteenth century to 1914s, the first quarter of the twentieth century. The most basic characteristic of this period was the use of electricity and the transition to mass production based on the division of labor. Therefore, electricity and oil came to the fore as energy sources in this period. Electricity use became widespread in factories and workshops, the automotive industry was developed, some durable consumer appliances were invented, and utilization of oil, oil derivatives, and chemicals started in the industry. In addition, the increase in the efficiency of durable steel production accelerated rail transport and trade (Taş 2018. 1821–1822; MÜSİAD 2017: 32).

The first assembly production line was carried out in Cincinnati Slaughterhouses in 1870 (Sözen and Mescioğlu 2019: 293). Yet, Fordism has been considered the start of the Second Industrial Revolution. The production line technology, called the Fordist Production Style, was applied in 1913 by Henry Ford at the automobile factory, forming the beginning of the mass production and division of labor (Bulut and Akçacı 2017: 52). With the invention of the oil-based internal combustion engine in the last quarter of the nineteenth century, the textile industry also entered a different process (MÜSİAD 2017: 32).

One of the most important economic consequences of the Second Industrial Revolution was the rapid growth in many sectors such as chemistry, iron and steel, aviation, and telecommunications. New inventions, machinery and production modes, and improved factories increased trade and thus revenue sources. Factory owners and main producers achieved great wealth, while the workers started to negotiate for wages and working hours. The biggest economic growth in world history was experienced (Sözen and Mescioğlu 2019: 294).

15.2.3 *Third Industrial Revolution and Its Economic Impacts (Industry 3.0)*

The Third Industrial Revolution began with the introduction of electronics and information technologies. The process that started with the invention of the calculator called Z1, which was operated with mechanical electricity in the 1950s (MÜSİAD 2017: 35), continued until the end of the twentieth century. The characteristic of this period was the introduction of computers, digital technologies, and the Internet into human life. The rapid development of the Internet, mainframe and personal computers, the introduction of programmable industrial robots into the manufacturing process, the emergence of digital technology and information technology enabled this revolution to be called also as the Digital Revolution (Bulut and Akçacı 2017: 52; Sözen and Mescioğlu 2019: 294).

Small computers, mobile phones, hybrid cars, nano-technology products, sending a shuttle to space have been the milestones of the Third Industrial Revolution. Rapid development in electronic, information, and communication technologies allowed automation in production, and this automation was improved to a very advanced level with the development of Programmable Logic Controllers (PLCs) (Taş 2018: 1822).

The economic result of the Third Industrial Revolution was manifested as the development of computers, microelectronic technology, fiber optics, laser, chip technology, development of atomic energy, information and telecommunication technologies, nuclear science, and biogenetics. This development paved the way for the emergence of the information society and created the opportunity to obtain information much cheaper and easier. With the dominance of machinery and technology, the need for muscle strength gradually decreased and the direction and size of production changed (Taş 2018: 1822; Bulut and Akçacı 2017: 52). The automation of production reduced unit costs, where the amount and diversity of production increased, and consequently this led to a decrease in unit sales prices. While meeting the needs of the increasing population, these developments also expanded the markets for products (MÜSİAD 2017; 34–35).

In this period, where technical revolutions and inventions were experienced almost on a daily basis, R&D (Research and Development) activities came to the fore. In addition, supply chain management entered the economic life, and developments in communication, transportation, and logistics technology accelerated economic globalization (MÜSİAD 2017: 34–35).

However, important problems arose during the Third Industrial Revolution in addition to its positive outcomes. Rapid depletion of world resources ranks first among these problems. Accordingly, the sustainability concept gained importance. Although it is stated that the energy source of the Third Industrial Revolution was nuclear energy; due to environmental concerns, recently, increased importance was attached to renewable energy sources such as sun and wind (Taş 2018: 1822; MÜSİAD 2017: 34). Another problem that arose during that period was urbanization and population growth, along with production increases. This situation led to poor working conditions and low wages. Different social developments such as migration and suburbanization started (MÜSİAD 2017: 39–40).

15.3 The Fourth Industrial Revolution (Industry 4.0)

By the twenty-first century, a novel period started; thanks to the combination of advanced communication, computer, and internet technologies, in which smart machines started to manage both themselves and production processes, and where there was almost no need for manpower. Having been accepted as the Fourth Industrial Revolution, this process is formed upon the increase in the interdependence of products and systems, and aims to connect the physical and virtual worlds (Leyh and Martin 2017: 990).

There is no universal definition of the concept of the Fourth Industrial Revolution, also known as Industry 4.0. However, it is defined as follows by SIMMI 4.0, a Maturity Model: "*Industry 4.0defines the transition from centralized production to a very flexible and self-controlled production. In this production, products and all affected systems and all steps of the engineering process are digitized and interconnected to share and transfer information and distribute it across vertical and horizontal valuechains and even across extensive value networks."(*Leyh et al. 2016*: 1297).* Industry 4.0 concerns today's industrial production as a whole. There are barcodes and chips containing the necessary information on the surfaces of Industry 4.0 products. The data are transmitted online by the scanner and computers and operate the machines properly. Thus, objects can become intelligent/smart and communicate with each other (Sommer 2015: 1513). Figure 15.2. provides the vision on which the Fourth Industrial Revolution has been based.

Basic elements that make up the Fourth Industrial Revolution; Cyber-Physical Systems (CPS), Internet of Things (IoT), Internet of Services (IoS), 3-D Printing, Robotic Systems, Smart Factories, Big Data, Cloud Manufacturing, and Augmented

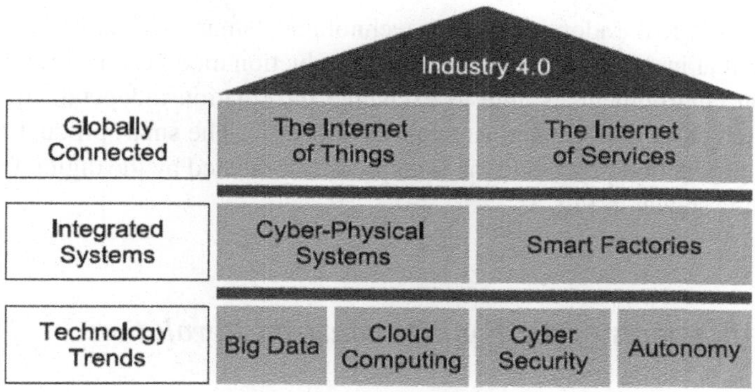

Fig. 15.2 Industry 4.0 vision. *Source* Flynn et al. (2017: 240)

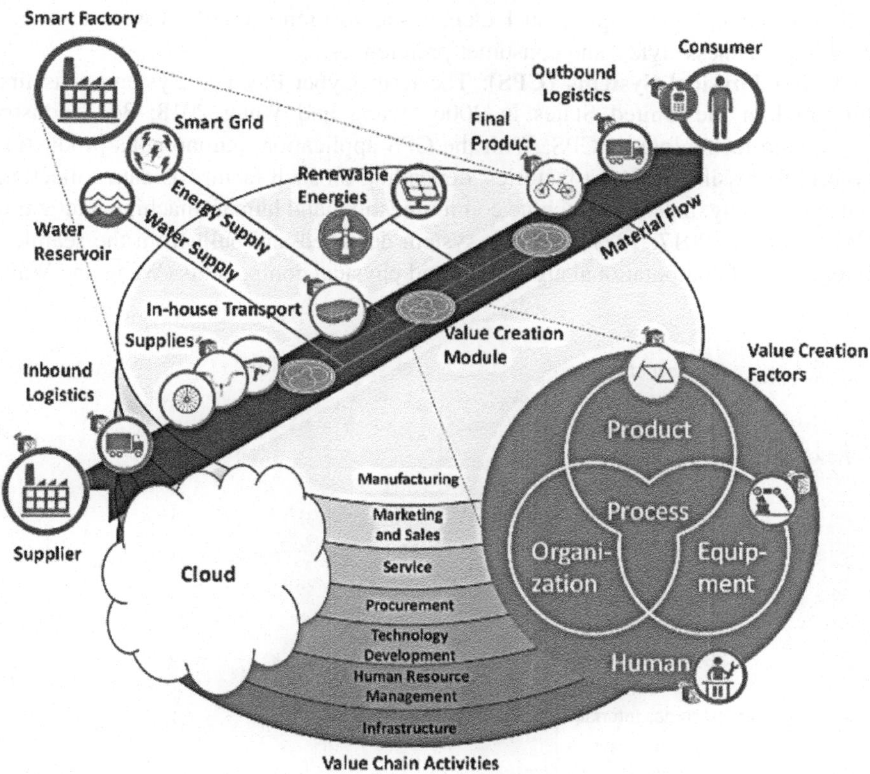

Fig. 15.3 Micro perspective of industry 4.0. *Source* Stock and Seliger (2016: 538)

Reality. Thanks to the adoption of these technologies, smart production processes will be created that include smart machines and production modules that provide a smart production environment, is able to exchange information independently, and can control each other (Pereira and Romero 2017: 1207). The smart production system -from the production process to the final product—created by the digital technology revolution is shown in Fig. 3.

15.3.1 *Elements of the Fourth Industrial Revolution*

The most emphasized technological elements in the Industry 4.0 literature in studies conducted by such organizations as OECD and the World Bank are shown in Fig. 15.4. All these technologies support such elements as cost innovation, software advances, evolving business styles, and consumer preferences.

Cyber-Physical Systems (CPS): The term Cyber-Physical Systems was first proposed in the United States in 2006 (Wang and Wang 2018: 34). Industry 4.0's main technology is CPS. With the CPS application, autonomous production control is possible in smart factories, depending on such factors as data collection, data processing, machine-machine communication, and human-machine interaction (Wagner et al. 2017: 127). CPS is a system designed and built from the seamless integration of computational algorithms and physical components (Wang and Wang

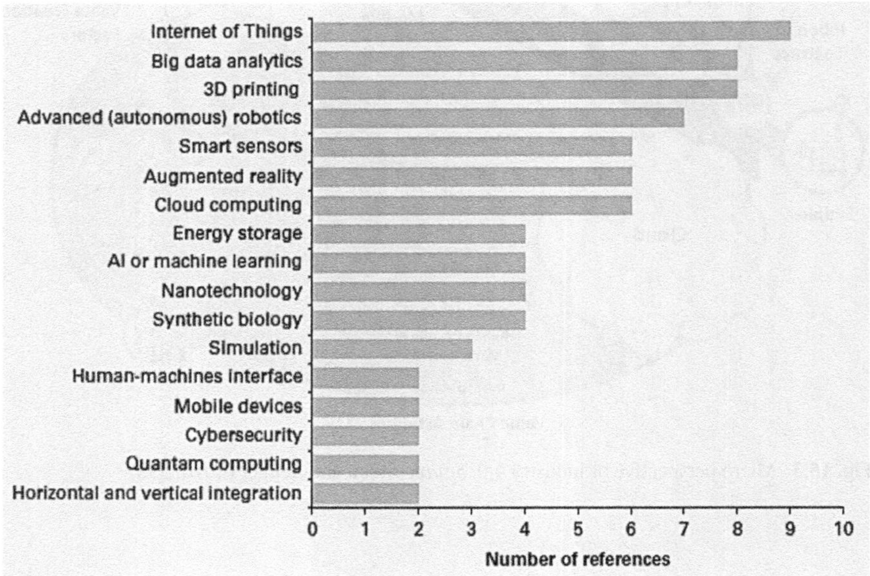

Fig. 15.4 Industry 4.0 Related Technologies. *Source* Hallward-Driemeier & Nayyar, Trouble in the Making? The Future of Manufacturing-Led Development (2018: 95)

2018: 34). In other words, CPS is a system in which the main elements of production such as monitoring, coordination and control are managed by the unified technology formed by computation and communication. This composite technology combines physical systems with cyber technology (Bulut and Akçacı 2017: 57).

Internet of Things (IoT): IoT emerged with the idea of influencing and tracking objects with low-cost sensor technologies, and developed with low-cost computing and internet-based communication technologies with a widespread broadband network. Technically speaking, IoT is defined as "*a collectionof physical artefacts that contain embedded systems of electrical, mechanical, information processing, and communication mechanisms that provide internet-based communication and dataexchange*" (Lane and Schaefer 2017: 3). In other words, IoT is a technological system in which, having been integrated with information processing and communication technologies, the living things and objects around us can transfer data without the need for human-computer interaction (Bulut and Akçacı 2017: 55).

IoT represents the next evolution of the development of the Internet. It also leads many projects such as bridging the gap between the poor and the rich, improving income distribution, using world resources efficiently, and ensuring to be more proactive and less reactive (Evans 2011: 2).

Internet of Services (IoS): IoS is the part of the Internet that provides services and functionality in the form of detailed and web-based software components (Bartodziej 2017: 55). In other words, IoS uses the Internet as a tool to provide and sell services. Therefore, services have been transformed into a commodity subject to trade. IoS creates advanced business models consisting of business networks for service providers and service consumers to provide and consume services. Institutions work together in business networks. The general value of the service is created from the relationship between the company, customers, agents, collectors, and suppliers (Cardosa et al. 2008: 15–16).

3D Printing: 3D Printing is the process of making three-dimensional solid objects using a digital file. 3D printing allows production of complex shapes using less material compared to the traditional production methods ("What is 3D", n.d.). The decision to manufacture with the 3D method is influenced by both the potential to reduce costs and changes in consumer demand and the ability to customize outputs and significantly shorten the time to market (Hallward-Driemeier and Nayyar, Trouble in the Making? The Future of Manufacturing-Led Development 2018 99) 99.

Robotic Systems: Modern robots are characterized as systems that offer flexibility, autonomy, and cooperation. It is estimated that robots will interact with each other and work in smart factories with cost advantage and learn from humans (Kamble et al. 2018: 418). Now, smart robots can identify the materials and tools used in the production process with the help of sensors, know which process they are going to be subjected to, control the quality of the products, find errors in the production process, and transmit them to the relevant units. Therefore, the products in the production process can be processed and followed up with the least error (Taş and Küçükoğlu 2019: 208).

Smart Factories: Smart factories are one of the main factors of Industry 4.0, and they have emerged as a result of the use of smart technologies. In this concept, it is assumed that the entire production will be equipped with sensor-equipped tools and instruments, actors, and autonomous systems. In this way, the use of smart factories in the production process will make production systems more efficient and increase company performance (Lalic et al. 2017: 299).

Big Data: Big data analysis and technologies collect real-time data from many different sources, analyze them comprehensively, and support real-time decision-making. Big data provides advanced manufacturing flexibility, product quality, energy efficiency, and advanced equipment service, as well as supporting new features such as fault finding and predictive analytics. However, high data quality and analytical expertise are required for all these capabilities (Kamble et al. 2018: 417). Data management and distribution are essential to achieving self-aware and self-learning machines. Many research organizations and companies focused on data mining in the so-called Web 2.0 era from late 2004. These researches are not "machine-sourced or industrial data" but on "human-sourced or human-related data". Intelligent analytics and cyber-physical systems, appropriate sensor installations, collection of past and new data, and communication protocols come together in Industry 4.0. Collecting and combining all data is called Big Data (Lee et al. 2014: 4–5).

Cloud Manufacturing (CM): Industry 4.0 and cloud production enable objects (things), services, data, and people in the production area to connect to the production network over the Internet (Kamble et al. 2018: 417). Cloud-based software creates an effect of reducing entry barriers for Industry 4.0 (Erasmus 2019: 235). Cloud (CC) is a service provider as a third party to store data or databases on the Internet. Having a remote location from where production activities take place, the cloud provides improved service speed and easy accessibility for businesses (Kamble et al. 2019: 1320). By applying cloud computing to production, resources are transformed into services and made accessible by participants. In addition, it will be possible to increase the efficiency and respond to changes in customer demands by connecting smart resources to the cloud (Erasmus 2019: 40).

Augmented Reality (AR): AR enriches objects in the real world by using perceptual information generated by the computer. Augmented reality-based applications increase the efficiency of production cells operating with industrial robots. AR application has an important role in performing complex tasks such as finding and maintaining devices and spare parts in the warehouse, and giving critical instructions via mobile phones. Furthermore, AR can also be applied in such areas as quality control, remote assistance, logistics security management, and employee training (Kamble et al. 2019: 1321). The "*overlap of virtual information on the real world view*" is called augmented reality. AR makes it possible to improve a person's perception of reality (Syberfeldt et al. 2017: 9118). In other words, AR means the integration of additional information generated by the computer into the real world. AR applications enable users to access and experience information that has direct spatial relation to their immediate surroundings. This system consists of sensors, AR software, and a suitable screen (Paelke 2014: 1, 3).

15.3.2 Principles and Advantages of Industry 4.0

With the development of information and communication technologies, the Fourth Industrial Revolution has developed a set of unique principles. The basic principles of Industry 4.0 are listed as follows (Taş and Küçükoğlu 2019: 207).

- Interoperability: Cyber-Physical Systems involve people and smart factories connecting and communicating with each other.
- Virtualization: Virtual copy of smart factories is created by combining service data as artificial modeling and simulation.
- Real-Time Capability: It is the ability to analyze data and immediately present the results achieved.
- Modularity: It is the ability of smart factories, which have modules that can meet different and changeable needs, to allow flexible adaptation to these needs.
- Autonomous Management: It is the ability of Cyber-Physical Systems to make their own decisions in smart factories ("Endüstri tarihine kısa", n.d.).
- Service Orientation: It is the provision of cyber-physical systems, people, and smart factory services over the Internet of Services ("Endüstri tarihine kısa", n.d.).

It is possible to list the advantages introduced by the Fourth Industrial Revolution to the economy and business life as follows (Moran 2018).

- Efficient use of resources will significantly reduce production costs by fast and flexible production, less downtime, less quality problems, less resource usage, less material and product waste, and lower operating costs. Efficient use of resources will create an increase in production.
- Productivity and increased production will lead to higher revenues and increased profitability.
- It will improve the ability to share information and collaborate in production and business processes.
- It will facilitate compliance in such fields as quality control, serialization, and data recording.
- It will improve service and experience capability for customers.
- It will create an opportunity for innovation.
- Provide an increased return on investment.

Evidently, the advanced information technologies that emerged in the Fourth Industrial Revolution will enable the implementation of smart factories and further the process to advanced levels. All smart machines, conveyors, and products will be in communication with each other by reconfiguring themselves to ensure product diversity and flexible production. In system operation, the industrial network collects large amounts of data from smart objects and transfers them to the cloud. Thus, system performance is optimized by providing feedback and coordination throughout the system based on big data analysis (Wang et al. 2016: 167). It has been suggested that the positive developments in cost, production, and productivity will lead to a transformation in the labor market. Will smart machines become a substitute for human beings in the business world?

15.3.3 The Problem of Growth and Employment in the Fourth Industrial Revolution

Technological developments from the industrial revolution to the present have led to a significant increase in industrial efficiency. The steam engine strengthening factories in the nineteenth century, the electrification leading to mass production in the twentieth century, the introduction of information technologies in the following years, and the introduction of mobile communication brought many breakthroughs (Rüßmann et al. 2015: 1). Considering the formation processes of Industry 4.0, it is clear that there are some differences between developed countries and developing countries. Table 15.1 shows these differences comparatively.

As seen in Table 15.1, while external targets such as global marketing are dominant during the formation of Industry 4.0 in developed countries, internal targets such as the growth and development of the economy are dominant in developing countries. Additional objectives accompanying the formation of Industry 4.0 in developed countries are associated with “opening up the human potential”. The additional objectives accompanying the formation of Industry 4.0 in developing countries were suggested as “modernization of entrepreneurship”. During the formation of Industry 4.0 in developed countries, the areas of social interest for the expansion of individual production were important, whereas in developing countries, economic areas that focused on the beginning of mass production were important. The level of application of the Industry 4.0 concept in developed countries is national and the state requires the adoption of development strategies. In developing countries, the application area of this concept is institutional and envisages the adoption of the strategies of separate companies (Bogoviz et al. 2018: 161).

Regardless of the degree of development, the digital technological transformation has begun to occur in the vast majority of countries. Therefore, the Fourth Industrial Revolution has a multifaceted and wide impact on the global economy and especially on macro variables such as growth and employment.

15.3.4 Economic Growth in the Fourth Industrial Revolution

As a result of the digital development with the Industry 4.0 revolution, radical changes have occurred in the understanding of production and business models. A new era has begun in the supply chain, production, marketing, and sales processes. Traditional business models such as Taylorism and Fordism were revised, the production process was made flexible, personalized products were targeted, and it was aimed

Table 15.1 Comparative analysis of industry 4.0 formation in developed and developing countries

Criteria of comparison	Developed countries	Developing countries
Dominating main goals of industry 4.0 formation	External goals (global marketing)	Internal goals (growth and development of economy)
Additional (Accompanying) goals of industry 4.0 formation	Opening human potential	Modernization of entrepreneurship
Dominating sphere of interests during industry 4.0 formation	Social: expansion of individual production	Economic: starting massive production
Level of implementation of the concept of industry 4.0	National and state strategies of development	Corporate, strategies of development of separate companies
Influence of industry 4.0 on knowledge economy	Development of knowledge economy	Formation of knowledge economy
Readiness of socio-economic platform of industry 4.0 formation	Formed digital society and digital economy	Digital society and digital economy in the process of formation
Financial barriers on the path of industry 4.0 formation	Absent or low	High
Expected results	Near ten years	Near fifteen years

Source Bogoviz et al. (2018: 161).

to ensure efficiency and optimization, to prevent waste and loss in energy and input (Özsoylu 2017: 58–61). The aim of Industry 4.0 is to increase and expand the long-term competitiveness of the company by increasing the flexibility and efficiency of production through communication, information, and intelligence (Gabriel and Pessl 2016: 133). In this context, factors such as the use of bio-energy resources, automation in production, less waste, less resource use, optimum production and efficiency are becoming increasingly important and businesses investing toward the foregoing direction are encouraged. However, although the rate of return expected by enterprises for their smart factory investments is not attractive, it is obvious that they will not stay away from digital transformation in the future (Toker 2018: 58–61).

As regards economic growth, the economic views are divided into two consisting of economists with technological pessimism and economists with technological optimism. Technological development has a potential deflation effect. Further, capital is preferred over labor, causing wage declines and therefore consumption decreases. However, Industry 4.0 also reduces costs and prices, allowing consumption at low price levels. The global economy, which grew by around 3–3.5% subsequent to the Great Recession, grew by 5% before the 2008 crisis. Today, global economic growth has slowed. This is attributable to many reasons including borrowing, misdistribution of capital, demographic changes, and productivity. In particular, the aging of the population has an effect of slowing down economic growth. Aging societies need to adapt themselves to technological innovations, ensure that older individuals contribute to the workforce, and trigger major increases in productivity through "smart work". Otherwise, slow growth will continue (Schwab 2018).

In Industry 4.0, which we can be considered the rise of the digital industrial revolution, integrating data from process control systems with other data such as cost data will provide optimization in the efficiency, energy, production, and business volume of the businesses. It is expected to create a 10–30% decrease in production costs in enterprises, 10–30% in logistics costs, and 10–20% in quality management costs. Industry 4.0 has made it possible to produce low-cost and high-quality products thanks to the more flexible and efficient manufacturing process it provides. In this case, production is expected to accelerate by 30% with a 20% increase in efficiency. It has been suggested that the aforementioned increase in productivity in production might stimulate industrial growth (McKinsey & Company 2016: 15; Toker 2018: 58–59; Rüßmann et al. 2015: 1; Özsoylu 2017: 60–61).

Industry 4.0's innovative developments in economic systems will create a sectoral balance in national economies by providing simultaneous development in both the service and industrial sectors (Sukhodolov 2019:4). The digital investments that businesses will make will serve as a catalyst for economic growth. The digital and technological development provided during the manufacturing process will be the main element of sustainable growth for both businesses and the country (Toker 2018: 60–61). With the digitization of the global economy, access to products throughout the world will become easier. This will lead to increased demand and economic growth (Feshina et al. 2019: 116).

However, the recent industrial revolution creates many new opportunities for companies, while at the same time creating various challenges due to automation and digitalization processes. One of these challenges is the "dealing with the challenges of knowledge and competence" as a result of technological development. Due to increased automation, the need for highly trained and equipped personnel will also increase. Businesses will have difficulty shifting their employees to more complex workspaces and ensuring that they hold on to changing work environments. Therefore, they will need new strategic approaches in terms of Human Resources Management as well. In addition, virtual and flexible working require new methods of lifelong learning. The complex manufacturing processes have led to an increase in jobs that require higher qualifications. Therefore, businesses should train their employees for more creative, strategic, and coordinated tasks with high responsibilities and help them with acquiring higher qualifications (Hecklau et al. 2016: 1–3). Along with training, the businesses need to strive to develop the right competencies with qualification management by paying attention to complementary skills for the future (Hecklau et al. 2017).

Another economic challenge that will be encountered by the businesses is that technological development, along with globalization, has made markets more "volatile and heterogeneous". With globalization, businesses have to deal with the need to reduce costs in order to compete and continue their product life cycles. To do this, businesses must adapt to many new formations, such as modernizing innovation processes, producing higher services, meeting customer expectations, shifting toward personalized and more flexible manufacturing processes, needing more cooperation, and entering into strategic alliances with suppliers and competitors (Hecklau et al. 2016: 3).

The Industry 4.0 application, which aims to increase competitiveness in the manufacturing and high-tech sector, must make progress in eight key areas including R&D activities and standardization in order to achieve this. These are clearly stated standards for reference architecture, management of complex systems, ensuring a comprehensive broadband infrastructure for the industry, security and safety issues, business organization and business design in the age of the digital industry, special education and continuous professional development, a convenient ordering framework, and resource efficiency (Prause 2015: 160). If all these challenges are overcome and the process is well managed, businesses will acquire an advantage of competition, efficiency, and effectiveness, resulting in economic growth in both sectoral, national, and global terms.

15.3.5 Employment Phenomenon in the Fourth Industrial Revolution

All industry revolutions negatively affected some of the professional groups that existed in the previous manufacturing system and created unemployment during the required adaptation process. However, this has become a temporary situation and the new production system required by the technology revolution has created its own professional groups and its own employment. A similar situation will continue in the Industry 4.0 process, and new employment areas and professions will emerge.

It will be possible to overcome the potential negative effects of Industry 4.0 on the labor market by "*investing in education, promoting entrepreneurship, implementing active labor marketpolicies, financing researchthat promotes employment-creating innovations, introducing laws for automation practices, providing citizens with basic living income, designing policies to promote consumer demand, reducing taxes from low earners and increasing taxes on high earners, reducing taxes from labor and increasing taxes on capital*". According to economists, who held an affirmative approach, technological developments will change the current structure of the labor market in the short term and may cause partial unemployment, but it may increase the quality and quantity of the workforce in the long term. The need for mental power in Industry 4.0 will also increase the general education level and quality of the society, and individuals with a high level and qualified education will be employed in the long term, and the employment structure will become more qualified (Özsoy 2018: 264).

The Fourth Industrial Revolution has led to significant changes in the labor market and will continue to do so. People will still have important roles, the positions they undertake in production will change and the machines will take over some of the muscle work and mental work. The classical process of working, which requires such skills as intelligence, creativity, empathy, and flexibility, will continue, and people will become coordinators in this process. In this process of human and machine network interaction, business content, business processes, work environment, and necessary skills will change and new professions will emerge alongside the professions that become obsolete. In these new professional groups, physical tasks or muscle-based tasks will be less of a center of attention (Gabriel and Pessl 2016: 133).

The social impact of IoT is closely related to the effects of other socioeconomic, political, and technological developments. In addition, formations such as rapid urbanization, new energy sources, climate change, division economy, artificial intelligence, aging societies, geopolitical volatility have an important role to play within this effect. The World Economic Forum predicted that about 7.1 million jobs would be lost in 15 economic areas between 2015 and 2020 during the Fourth Industrial Revolution. Most of the jobs at risk of loss are related to white-collar office and administration jobs. It is anticipated that the process would lead to a demand growth for jobs related to business and financial operations, management, computer, and mathematical roles demand growth and 2.1 million additional jobs would be created. Current estimates show that 45% of operating activities can be automated using already proven technology (Dervojeda 2017).

In the Fourth Industrial Revolution, where robots will replace blue-collar workers and artificial intelligence will replace white-collar, the basic business lines, where many sub-branches based on high competencies will be formed, can be sorted as follows (Toker 2018: 61; Sener and Elevli 2017: 30–33):

- Industrial Software Programmers,
- Information Systems and Internet of Things Solution Developers,
- Industrial Data Analyzer,
- Robot Coordinator, Programmer, Mechanic,
- Manufacturing Technologies Specialist,
- Smart Cities Planner, and
- Product Designers and Manufacturers.

New business lines like this will need more qualified, higher education and high-income workforce potentials. Therefore, education policies that will increase the qualifications of the workforce and take it to the next level as a qualification will gain weight in this process. While the demand for employees who learn complex jobs and are equipped with information communication technologies increases, professions linked to information technologies will become important (Taş 2018: 1826; Özsoylu 2017: 57). Expertise in new communication technologies such as planning, execution, decision-making, control, programming, and error correction will become increasingly important (Gabriel and Pessl 2016: 133).

However, in another respect, with Industry 4.0, more production and wealth will be realized with fewer workers compared to the past. Technological development and digitalization allow "digital companies" to produce at marginal costs of close to zero. This suggests a reverse relationship between technological development and employment (Bulut and Akçacı 2017: 59). The Industry 4.0 process, a period in which there will be no large part of the current tasks and where routine tasks and jobs will be largely automated, indicates that employees'business prospects would depend on their own responsibilities. As a matter of such responsibility, the employee or worker must have "the determination to learn lifelong and evaluate opportunities". In addition, prominent "personal qualifications" will include autonomous movement, cognitive abilities, the ability to improve attitudes, and ethical value systems (Erol et al. 2016: 14).

The Fourth Industrial Revolution set different roles for humanity in both economic and social life in the future. Now people will be employees who develop, design, follow, and control production strategies in intelligent manufacturing processes, instead of working in closed offices under severe physical conditions. The main building block of the Fourth Industrial Revolution is human capital and the management of this capital (Gürün 2019: 86). Therefore, in this process, it is important to ensure equal opportunities in education, to develop technology literacy and thus to create a workforce integrated with the world (Toker 2018: 61).

15.3.6 The Impact of the Fourth Industrial Revolution on Logistics Sector

It is almost impossible to think of production and logistics activities separately. The digital revolution that the Fourth Industrial Revolution will create in the production process will have an impact on all phases associated with the seven rights of the logistics sector (the right product, in the right amount, in the right way, at the right time, from the right source, by the right medium, at the right price).

Logistics, which was considered as planning military supply and troop movements in the nineteenth century and later considered only as storage, packaging, and shipping functions, has a broader meaning today. Logistics faces social responsibilities on such issues as climate change, demographic change, resource efficiency, and sustainability. Today's logistics is defined as "*(…) holistic planning, control, coordination, execution and monitoring of all internal and in-house information and goods flows*". Therefore, the basic principle of logistics is flow orientation through network networks. The second principle of logistics is a holistic perspective on activities, systems, and networks (Bartodziej 2017: 20).

With innovations and applications added to logistics services by CPS, the concept of Logistics 4.0 or Smart Logistics has emerged. Logistics 4.0 covers the combination of CPS, software, and human support and is a result of increased internet use and the development of digitalization, which provides real-time interaction between machine and human. Logistics 4.0 refers to a system that can change market regulations and bring the company closer to customer needs. Therefore, this will increase the level of customer service and make production optimization possible, and reduce the storage and production prices. Successful Resource Planning, Warehouse Management Systems, Transportation Management Systems, Intelligent Transportation Systems, and Information Safety applications are required to operate logistics 4.0 in an effective and strong fashion (Barreto et al. 2017: 1248).

According to another description, Logistics 4.0 is "*the use of Cyber Physical Systems that monitor and control physical processes with feedback loops that often affect computationsof physical processes*" (Karunarathna et al. 2019: 1026).

According to Galindo (2016: 32), logistics 4.0's definition is "*the advancement of standardization through labor savings and IoT evolution*".

15.4 Logistics Sector in Industrial Revolutions

Digital technological development has affected all sectors and stages of the economy as well as the logistics sector. The phases taking place in the logistics sector with industrial revolutions are given in Fig. 15.5. As for Logistics 1.0, the use of sea and land transportation vehicles increased thanks to the steam technology. In Logistics 2.0, cargo handling automation was provided and logistics equipment capable of automatic storage and classification were introduced. Information technologies have been introduced in Logistics 3.0, considered as the systemization of logistics management. The Logistics 4.0 revolution is developing around the digitization of logistics processes (Yilmaz and Duman 2019: 193). With the Fourth Industrial Revolution, the face of the Logistics sector began to change. Yilmaz and Duman (2019: 191) suggested that the purpose of Logistics 4.0 was "*improving effectiveness (flexibility, individualized services, processes and products) and productivity(transparency, process speed, error reduction and merging) and supply chainpartners (wholesalers, retailers, logistics service providers and customers) and enhancing the information communication technologieswith decentralized decision-making structures*".

CPS, which is used to improve the flow of two-way information between executive and decision systems of manufacturing engineering, has also been taken into account in the logistics sector (Timm and Lorig 2015: 3118). The ability to implement the four basic principles of Industry 4.0, namely CPS's interoperability through IoT, information transparency, technical assistance, and autonomous decision-making on Storage, Transport, Packaging, Distribution, Loading–Unloading and Information Services will increase productivity in the logistics sector (Horenberg 2017). With the implementation of the Industry 4.0 concept to the logistics sector, a transition was enabled from "hardware-oriented logistics" to "software-oriented logistics" (Timm and Lorig 2015: 3118). This transformation, which is a result of integration with information technology, involves the exchange of the business's information about production and product both within the business itself and with its customers and suppliers. While, on the one hand, real-time communication of human-machine-parts and products occurs in the production process, on the other hand, all networks will be standardized and information stored in the cloud will become accessible, reducing the error (Shekeli and Minister 2018: 26–27).

Data logistics plays an important role in the realization of information flow and data security is also essential. In this system, which is built on the Internet of Things, all elements of transportation logistics must be compatible with each other and the 4.0 Industry. Flexible, automated, and optimum logistics solutions will be produced

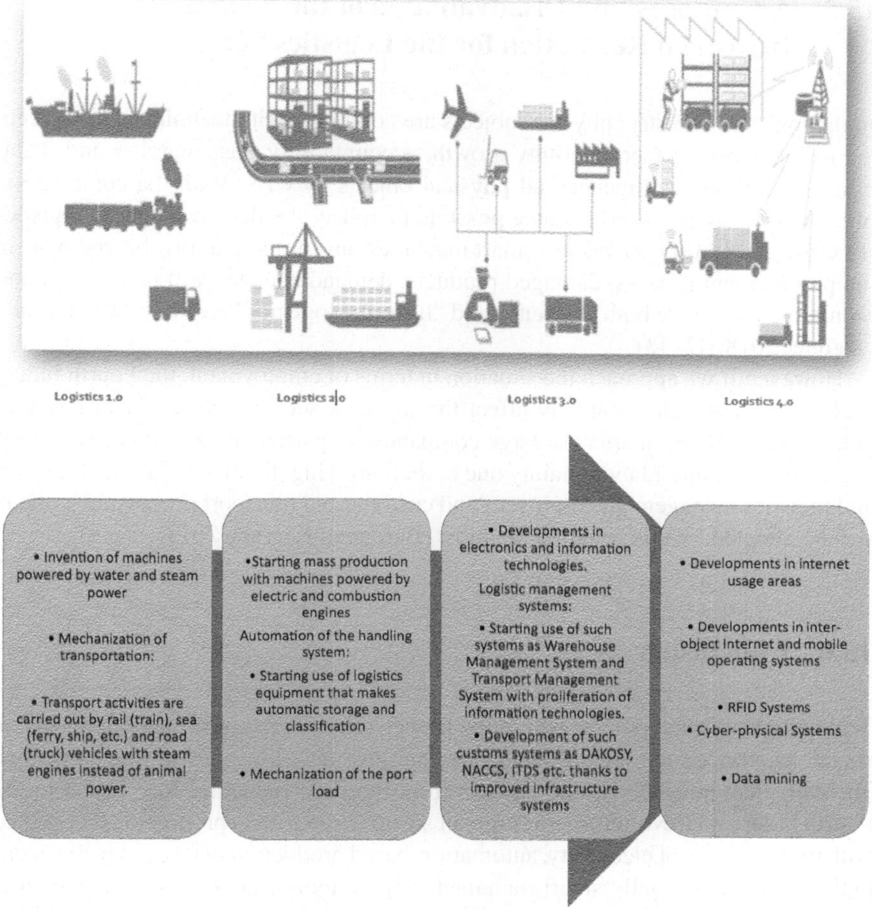

Fig. 15.5 Phases of logistics evolution. *Source* Yilmaz and Duman (2019: 192)

and productivity increases can be experienced as all elements are connected. In addition, the implementation of fully autonomous decision processes in the fields of production and transportation is also very important in the process of advanced automation and integration ("Industry 4.0",n.d., par.9–12).

There are many innovations provided by Logistics 4.0 in the sector. One of them is "Cellular Transportation Systems", which can autonomously move with laser scanners, infrared sensors, and RFID chips in the warehouses. Another development is "Autonomous Transport Robots" that do not need any central control systems ("Industry 4.0", n.d., par.9–12).

15.5 Advantages and Disadvantages of the Fourth Industrial Revolution for the Logistics Sector

In the logistics industry, physical objects are combined with techniques that accept the new potential for productivity growth, accountability, sustainability and scalability, and cyber intelligence and physical objects by CPS. With the combination of existing systems, it has become possible to collect the data available in logistics (Frazzon 2015: 331). Thanks to smart machines, inventory quantity, breakdowns in the procurement process, damaged products, demand changes will be checked and monitored to provide both efficiency and "full-time logistics" services (Özdemir and Özgüner 2018: 42–43).

However, if we approach the situation in terms of employment, the Fourth Industrial Revolution will negatively affect the logistics sector particularly in terms of employment. Three-quarters of large companies, in particular, are expected to go to reduce the amount of labor, mainly due to their growing digital trends. The transport and logistics management employees and customs and transport processing workers will be affected by this situation the most (European Kearney 2015).

15.6 Conclusion

Technological developments within the framework of needs affect human beings on social, economic, and legal dimensions, and give new lifestyles to human beings. Technological advances in the process from the First Industrial Revolution to the Fourth Industrial Revolution (invention of steam engine, mass production that started with the discovery of electricity, automation-based production that began in the twentieth century, and finally smart manufacturing systems equipped with information technology in the twenty-first century) allowed mankind to meet their needs in a more comfortable way.

But will the recent technological age make robots and people competitive in the labor market? Or is this going to be a human-focused information era? Will this technology be able to achieve economic growth? There are both positive and negative views on these issues throughout the world.

In addition to the many economic advantages, such as productivity growth and cost reduction, unfortunately, the downsides of digital technology also arise. The most important disadvantage manifests itself in the labor market. This is because of the fact that high-tech requires some adaptation process. Therefore, in the short term, it is expected that it will have a negative impact on the labor markets, such as job losses. Each new industrial revolution partially or completely neutralizes the sectors provided by its predecessor and creates new sectors. As the computing power continues to grow, sooner or later many professional groups will be the target of automation and the workforce profile will change. The low-skilled workforce will be replaced by a highly qualified workforce. At this stage, while the labor force's performance, creativity, and consciousness regarding the ability to assume responsibility are important, the management of human capital will come to the fore. High technological education, lifelong learning, and continuous development should be supported during the smart manufacturing process. However, creativity, support of new entrepreneurship activities, the implementation of harmonizing training policies will ensure the necessary qualifications in a short time. In addition, it is the responsibility of the employee to guarantee one's employment life in the future. The workforce's ability to adapt to the new business areas and professions as required by technology depends on self-development.

Digital transformation will take over the business world sooner or later. Factories, firms, and companies will have a competitive advantage to the extent that they are able to keep up with digital technological developments and achieve transformation. Thanks to smart production, low-cost advantage, productivity growth, flexible production system, instant response to customer demands, minimizing the margin of error and incompleteness will enable growth in the sector, and therefore all economies will enter into an "economic growth" trend.

The logistics sector, which cannot be considered separately from the production process, has also been one of the sectors most affected by the digital technological development process. Unmanned aircraft, unmanned vehicles, drones, robots, and artificial intelligence are rapidly increasing in the logistics industry. This creates advantages on the one hand and disadvantages on the other. The smart logistics' ability to reduce setbacks and errors to almost zero levels will decrease the costs and increase productivity in the sector. However, the workforce will be disadvantaged in this context. The fact that robots and artificial intelligence control the process in areas such as data entry, shipment, and transport will cause both blue and white collars to lose their jobs in the industry. However, new professions such as robot coordinator and machine advocacy will also apply to the logistics sector and will be able to create employment.

Given the danger posed by Industry 4.0 in the form of robotizing people by depriving them of souls, raising collective awareness towards "human values" and the need for constructive attitude and common sense also come to the fore. Although the digital revolution may seem like a rival of mankind in the business world, it should be noted that human beings themselves made it available and designed such a revolution.

References

Barreto L, Amaral A, Pereira T (2017) Industry 4.0 imlications in logistics: an overview. Manuf Eng Soc Int Conf 2017, MESIC 2017 13:1245–1252. https://doi.org/10.1016/j.promfg.2017.09.045

Bartodziej CJ (2017) The concept industry 4.0 an empirical analysis of technologies and applications in production logistics. Berlin, Germany. https://link.springer.com/gp/book/9783658165017#aboutBook

Bogoviz AV, Osipov VS, Chistyakova MK, Borisov MY (2018) Comparative analysis of formation of ındustry 4.0 in developed and developing countries. EG Popkova

Dervojeda K (ed) (2017) *How IoT will impact our jobs*. https://ec.europa.eu/digital-single-market/en/blog/how-iot-will-impact-our-jobs
Bulut E, Akçacı T (2017) Endüstri 4.0 ve inovasyon göstergeleri kapsamında Türkiye analizi. ASSAM-UHAD 7:50–72
Cardosa J, Voigt K, Winkler M (2008) Service engineering for the internet of services. Filipe J, Cordeiro J (ed) ICEIS 2008 15–27 https://www.researchgate.net/publication/220711282
Dombrowski U, Wagner T (2014) Mental strain as field of action in the 4th industrial revolution. Variety Management in Manufacturing. In: Proceedings of the 47th CIRP Conference on manufacturing systems, Procedia CIRP. 17, pp 100–105. https://www.sciencedirect.com
Endüstri 4.0 ile lojistiğin değişen yüzü. (n.d.). https://www.endustri40.com/endustri-4-0-ile-logistigin-degisen-yuzu/
Endüstri tarihine kısa bir yolculuk. (n.d.). https://www.endustri40.com/endustri-tarihine-kisa-bir-yolculuk/
Erasmus J (2019) The application of business process management in smart manufacturing: towards dynamic allocation of manufacturing resources. Technische Universiteit, Eindhoven, Eindhoven
Erol S, Jäger A, Hold P, Ott K, Sihn W (2016) Tangible industry 4.0: a scenario-based approach to learning for the future of production. In: 6th CIRP Conference on learning factories, 54, 13–18.https://doi.org/10.1016/j.procir.2016.03.162
European Kearney (2015) Digital supply chains: increasingly critical for competitive edge. WHU Logistics Study 2015. https://www.kearney.com
Evans D (2011) The internet of things, how the next evolution of the internet is changing everything. Cisco Internet Business Solutions Group (IBSG). https://www.cisco.com/go/ibsg
Feshina SS, Konovalova OV, Sinyavsky NG (2019) Industry 4.0-transition to new economic reality. In: Popkova EG, Ragulina YV, Bogoviz AV (eds) Industry 4.0: Industrial revolution of the 21st Century, studies in systems, decision and control, vol 169, pp 111–120. https://doi.org/10.1007/978-3-319-94310-7_11
Flynn J, Dance S, Schaefer D (2017) Industry 4.0 and ts potential ımpact on employment demographics in the UK. In: Gao J et al (eds) Advances in manufacturing technology XXXI, pp 239–244. https://doi.org/10.3233/978-1-61499-792-4-239
Frazzon EM (2015) Big data applied to cyber-physical logistic systems: conceptual model and perspectives. Braz J Oper Prod Manage 12:330–337. https://doi.org/10.14488/BJOPM.2015.v12.n2.a12
Gabriel M, Pessl E (2016) Industry 4.0 and sustainability aspects with a special focus on future of work and ecological consequences. ANN Fac Eng Hunedoara Int J Eng 14(2), 131–136
Galindo LD (2016) The challenges of logistics 4.0 for the supply chain management and the information technology. https://semanticscholar.org/paper/
Gürün F (2019) Endüstri 4.0 ve beşeri sermayenin geleceği. *Sosyal Siyaset Konferansları Dergisi*, 76, 67–88. https://doi.org/10.26650/jspc.2019.76.0004
Hallward-Driemeier M, Nayyar G (2018) Trouble in the Making? the future of manufacturing-led development. The World Bank Group. www.worldbank.org
Hecklau F, Galeitzke M, Flachs S, Kohl H (2016) Holistic approach for human resource management in industry 4.0. In: 6th CLF-6th CIRP Conference on learning factories, CIRP 54, 1–6. https://doi.org/10.1016/j.procir.2016.05.102
Hecklau F, Orth R, Kidschun F, Kohl H (2017). Human resources management: meta-study-analysis of future competences in industry 4.0. https://www.researchgate.net/publication/327262309
Horenberg D (2017) Application within logistics 4.0. 9th IBA Bachelor thesis conference. University of Twente, The Faculty of Behavioural, Management and Social Sciences. https://essay.utwente.nl/72668/1/Horenberg_BA_BMS.pdf
Kamble SS, Gunasekaran A, Dhone NC (2019) Industry 4.0 and lean manufacturing practices for sustainable organisational performance in indian manufacturing campanies. Int J Prod Res 9(5):1319–1337. https://doi.org/10.1080/00207543.2019.1630772

Kamble SS, Gunasekaran A, Gawankar SA (2018) Sustainable industry 4.0 framework: a systematic literature review identifying the current trends and future perspectives. Process Saf Environ Prot 117:408–425. https://doi.org/10.1016/j.psep.2018.05.009

Karunarathna N, Wickramarachchi R, Vidanagamachchi K (2019) A study of the implications of logistics 4.0 in future warehousing: a Sri Lankan Perspective. Int Conf Ind Eng Oper Manage 1024–1035. www.ieomsociety.org/ieom2019/papers/274.pdf

Lalic B, Majstorovic V, Marjanovic U, Delić M, Tasic N (2017) The effect of industry 4.0 concepts and e-learning on manufacturing firm performance: evidence from transitional economy. In: IFIP International Federation for Information Processing, Published by Springer International Publishing AG 2017. APMS 2017, pp 298–305. https://doi.org/10.1007/978-3-319-66923-6_35

Lee J, Hung-An K, Sahanhu Y (2014) Service innovation and smart analytics for industry 4.0 and big data environment. Product Services Systems and Value Creation. In: Proceedings of the 6th CIRP Conference on industrial product service systems, 16. pp 3–8. https://doi.org/10.1016/j.procir.2014.02.001

Leyh C, Bley K, Schäffer T, Forstenhäusler S (2016) SIMMI 4.0—A maturity model for classifying the enterprise—wide IT and software landscape focusing on industry 4.0. In: Proceedings of the Federated Conference on computer science and information systems, ACSIS 8:1297–1302. https://doi.org/10.15439/2016F478

Leyh C, Martin S (2017) Industry 4.0 and lean production—A matching relationship? An analysis of selected industry 4.0 models. In: Proceedings of the Federated Conference on computer science and information systems 11:989–993. https://doi.org/10.15439/2017F365

McKinsey & Company (2016) Industry 4.0 after the initial hype, where manufacturers are finding value and how they can best capture it. 15–32. https://www.mckinsey.com

Moran K (2018) Benefits of industry 4.0. SI controls thinking ahead. https://slcontrols.com/benefits-of-industry-4-0/

MÜSİAD (2017) *2017 Lojistik sektör raporu: Endüstri 4.0 ve geleceğin lojistiği*. https://www.musiad.org.tr/uploads/yayinlar/arastirma-raporlari/pdf/lojistik-raporu.pdf

Özdemir A, Özgüner M (2018) Endüstri 4.0 ve lojistik sektörüne etkileri: Lojistik 4.0. *İşletme ve İktisat Çalışmaları Dergisi*, 6(4):39–47. https://doi.org/10.32479/iicd.147

Özsoy CE (2018) Industry 4.0 and its potential impact on employment. J Curr Res Bus Econ 8(2), 249–270. https://doi.org/10.26579/jocrebe-8.2.16

Özsoylu AF (2017) Endüstri 4.0. *Çukurova Üniversitesi İİBF Dergisi*, 21(1):41–64

Paelke V (2014) Augmented reality in the smart factory: Sporting workers in an industry 4.0 environment. In: Proceeding of the (2014) IEEE Emerging Technology and Factory Automation (ETFA). Barcelona 1–4. https://doi.org/10.1109/ETFA.2014.7005252

Pereira AC, Romero F (2017) A review of the meanings and the implications of the industry 4.0 concept. Procedia Manufact 13:1206–1214

Prause G (2015) Sustainable business models and structures for industry 4.0. J Secur Sustain Issues 5(2):159–169. https://doi.org/10.9770/jssi.2015.5.2(3)

Ragulina V, Bogoviz AV (ed) (2019) Industry 4.0: ındustrial revolution of the 21st century, studies in systems, decision and control, 169, 154–169. https://doi.org/10.1007/978-3-319-94310-7

Rüßmann M, Lorenz M, Gerbert P, Waldner M, Justus J, Engel P, Harnisch M (2015) Industry 4.0: the future of productivity and growth in manufacturing industries. BCG The Boston Consulting Group. https://www.innovasyon.org

Schwab K (2018) *Dördüncü Sanayi Devrimi* (2nd edn) (Trans. Z. Dicleli). Optimist Yayınları

Sommer L (2015) Industrial revolution-industry 4.0: Are German manufacturing SMEs the first victims of this revolution? J Ind Eng Manage 8(5):1512–1532. https://doi.org/10.3926/jiem.1470

Sözen M, Mescioğlu T (2019) Endüstri 4.0'ın itici güçlerinin Türkiye ve Çin üzerindeki etkileri. Int J Soc Inq (12)1:287–315. https://dergipark.org.tr/tr/pub/ijsi/issue/46304/582087

Stock T, Seliger G (2016) Opportunities of sustainable manufacturing in industry 4.0. Procedia CIRP 40:536–541

Sukhodolov YA (2019) The notion, essence and peculiarities of industry 4.0 as a sphere of industry. In: Popkova EG, Ragulina YV, Bogoviz AV (eds) Industry 4.0: Industrial revolution of the 21st

Century, Studies in Systems, Decision and Control, vol 169, pp 3–10. https://doi.org/10.1007/978-3-319-94310-7_1
Syberfeldt A, Danielsson O, Gustavsson P (2017) Augmented reality smart glasses in the smart factory: product evaluation guidelines and review of available products. IEEE Access 5:9118–9130. https://doi.org/10.1109/ACCESS.2017.2703952
Şekkeli ZH, Bakan İ (2018) Endüstri 4.0'ın etkisiyle lojistik 4.0. J Life Econ 5(2):17–36. https://doi.org/10.15637/jlecon.247
Şener S, Elevli B (2017) Endüstri 4.0'da yeni iş kolları ve yüksek öğrenim. *Mühendis Beyinler Dergisi*, 1(2):25–37
Taş HY (2018) Dördüncü sanayi devriminin (endüstri 4.0) çalışma hayatına ve istihdama muhtemel etkileri. *Opus Uluslararası Toplum Araştırmaları Dergisi*, 9(16):1817–1836. http://opusjournal.net. https://doi.org/10.26466/opus.479123
Taş HY, Küçükoğlu M (2019) Endüstri 4.0 (dördüncü sanayi devrimi) sürecinde yeni nesil girişimler hakkında bir inceleme. *4. Uluslararası Gap Sosyal Bilimler Kongresi*, 205–212. https://www.researchgate.net/publication/338127162
Thames L, Schaefer D (2017) Industry 4.0: an overview of key benefits, technologies and challenges. In: Thames L, Schaefer D (eds) Cybersecurity for Industry 4.0, Analysis For Design and Manufacturing, Springer Series in Manufacturing, Birmingham, UK, pp 1–35
Timm IJ, Lorig F (2015) Logistics 4.0-A challenge for simulattion. In: Yilmaz L, Chan WKV, Moon I, Roeder TMK, Macal C, Rossetti MD (eds) Winter simulation conference. https://doi.org/10.1109/wsc.2015.7408428
Toker K (2018) Endüstri 4.0 ve sürdürülebilirliğe etkileri. Istanbul Manage J 29(84):51–61
Türk Dil Kurumu (TDK). https://sozluk.gov.tr
Wagner, T., Herrmann, C. and Thiede. (2017). Industry 4.0 impacts on lean production systems. In: The 50th CIRP Conference on manufacturing systems, Procedia CIRP 63:125–131.
Wang L, Wang XV (2018) Cloud-based cyber—physical systems in manufacturing. https://doi.org/10.1007/978-3-319-67693-7
Wang S, Wang J, Zhang D, Li D, Zhang C (2016). Towards smart factory for industry 4.0: A self organized multiagent system with big data based feedback and coordination. Comput Netw https://www.elsevier.com/locate/comnet
What is 3D Printing? (n.d.), https://3dprinting.com/what-is-3d-printing/
Yılmaz Ü, Duman B (2019) Lojistik 4.0 kavramına genel bir bakış: Geçmişten bugüne gelişim ve değişimi. *Bilecik Şeyh Edebali Üniversitesi Sosyal Bilimler Enstitü Dergisi*, 4(1):186–200. https://doi.org/10.33905/bseusbed.465962
Zambon I, Cecchini M, Egidi G, Saporito MG, Colantoni A (2019) Revolution 4.0: ındustry vs. agriculture in a future development for SMEs. Processes 7(36):2–16. https://doi.org/10.3390/pr7010036

Duygu Yücel is Assistant Professor of Economy at the Trakya University Edirne Social Sciences Vocational School. Before she worked as a Research Assistant at Istanbul University Social Sciences Institute between 1999–2002. Between 1995–2003, she worked as a Research Assistant at the Department of Economics Theory, Faculty of Economics and Administrative Sciences, Trakya University. She received her PhD in Economics Program at the Institute of Social Sciences, Istanbul University in 2011. Her research focuses on macroeconomic theories, economic growth, employment, income distribution and poverty. Assistant Professor Yücel has authored and contributed to book chapters, articles on employment, poverty, exclusion, income distribution. In addition, she has experience in many economic conferences and has conducted research in the field of women's studies and holds certificates in entrepreneurship.

Index

İ. İyigün and Ö. F. Görçün (eds.), *Logistics 4.0 and Future of Supply Chains*, Accounting, Finance, Sustainability, Governance & Fraud: Theory and Application, https://doi.org/10.1007/978-981-16-5644-6

The manufacturer's authorised representative in the EU is Springer Nature Customer Service Centre GmbH, Europaplatz 3, 69115 Heidelberg, Germany. If you have any concerns regarding our products, please contact ProductSafety@springernature.com

Printed and bound by CPI Group (UK) Ltd, Croydon, CR0 4YY
30/09/2026
02225246-0010